Some Thoughts about the Evolution of Human Behavior

Some Thoughts about the Evolution of Human Behavior

a Literature Survey

Arthur J. Boucot

With edits by
John M. Saul and John B. Southard
(students of the author from the first day of his teaching career)

ARCHAEOPRESS ARCHAEOLOGY

Archaeopress Publishing Ltd
Summertown Pavilion
18-24 Middle Way
Summertown
Oxford OX2 7LG
www.archaeopress.com

ISBN 978-1-78969-903-6
ISBN 978-1-78969-904-3 (e-Pdf)

© Arthur J. Boucot and Archaeopress 2021

John M. Saul and John B. Southard thank Janice A. Glaholm for editorial assistance.

All rights reserved. No part of this book may be reproduced, or transmitted, in any form or by any
means, electronic, mechanical, photocopying or otherwise, without the prior written permission of the
copyright owners.

This book is available direct from Archaeopress or from our website www.archaeopress.com

Contents

Preface

I am a geologist and a paleontologist. Why would I, with no prior experience in archaeology or anthropology, or with the benefit of any academic training in those disciplines, enter into a consideration of the morphological and cultural changes associated with the remains of *Homo*? I have had extensive experience with the morphology and taxonomy of marine invertebrates, earlier Paleozoic through Triassic, with the biogeography of such organisms, with their paleoecology and with trying to understand their evolutionary relations. This background permits me to consider the corpus of material developed over two centuries by archaeologists and anthropologists from a somewhat different perspective.

I am particularly concerned to consider whether the overall behavioral conservatism displayed by marine invertebrate taxa, as well as many terrestrial taxa, is also characteristic of *Homo* from its first recognition in the Late Pliocene to the present (Boucot, 1990; Boucot and Poinar, 2010). Are the many major cultural changes documented so thoroughly by generations of archaeologists and anthropologists best viewed as changes in technology rather than as truly organic, evolutionary changes? In other words, are the overall basics of human behavior conservative and unchanging?

I have tried to extract data from the rich archaeological and anthropological literature, Late Pliocene and younger, useful in addressing this question. Some scholars may protest that an outsider with no experience or training in archaeology and anthropology has no business in muddying the waters. But I suggest that an outsider with a fresh point of view might be able to contribute to additional understanding of what the archaeological and anthropological record provides.

Acknowledgements

I am greatly indebted to Deborah Carroll, Jessica Layford, Kathy Varbel, and the other members of the Oregon State University Interlibrary Loan Office for their very friendly, capable work on my behalf; without their work this treatment would not have been possible.

Kathryn M. Nichols provided much encouragement and moral support.

Dr John M. Saul of Paris provided a very useful review of an earlier manuscript draft.

Introduction

The following is a sample of what information the anthropological and archaeological record provides on the antiquity of human behaviors; these are merely examples, and are not exhaustive by any means. Conard (2005) concluded that truly 'modern' behaviors first appear in the Aurignacian and its equivalents; not everyone will agree with his conclusion although the compilation he made is impressive, but the sampling problem is always with us (think here of such rarities as musical instruments!). D'Errico et al. (2003; see also d'Errico, 2003) represents the opposing view, and also provides a comprehensive compilation. Henshilwood and Marean (2003; see also Bar-Yosef and Vandermeersch, 1993, for an earlier discussion of the problem) contribute to the discussion, maintaining that we are far from having all the data necessary to properly answer the basic question. Much of the argument might boil down to the first appearance of fully modern graphic art in the European Aurignacian and graphic art of equivalent age in Indonesia, and whether or not this might be a preservational artifact.

John A. Moore, a distinguished biologist, addressed many aspects of the same problems we consider here in his 1985 'Science as a way of knowing – human ecology'. His concerns were humankind's destructive effects on the natural environment that makes life possible. He was especially concerned with the destruction of soils and water supplies, to which one must add today the damage to the world's oceans. As a background to his ecological concerns, he briefly reviewed what was known in 1985 about human history and cultural development since the first appearance several million years ago of *Homo*. His review is very perceptive. I have covered the same ground with the addition of information provided by many workers in the past thirty years, with the results closely paralleling Moore's. As a paleontologist and geologist, I have added a number of items he did not cover, but again, our conclusions are similar.

The review by McBrearty and Brooks (2000; see their Figure 13; see also Brooks, 1996) of behavioral evidence available to 2000 provides entrée to the literature and a critical evaluation of that literature. Potts and Sloan (2010; see also Schick and Toth, 1993) provided a well-illustrated running account of many of the factors and types of evidence involved with human behavior and evolution.

The term 'cultural evolution' might just as well be termed 'technological evolution', since the developments in stone tools are just as much technology as they are culture; technology is not a synonym for the developments brought on only during the industrial revolution.

Thesis

After reviewing the voluminous anthropological and archaeological literature available to me, I have arrived at a conclusion concerning their evolutionary-behavioral significance. I conclude that the quantum evolutionary gap between the great apes and the hominins led to the initial and continuing presence of the basic human behaviors. These are the basic behaviors that sharply separate us from the great apes and their direct ancestors.

In the following treatment I will consider those basic behaviors that have left a significant ancient record. The antiquity of the record of these basic behaviors is a consequence of many sampling factors. I will try to summarize these sampling factors while considering each basic behavior. For some of these basic behaviors the evidence is of great antiquity and appears to be very reliable. For others the record is very limited, largely a consequence of various sampling factors. It would be a serious mistake to consider the lower time limit of each basic behavior as the 'true' lower limit without considering the sampling factors involved.

The sampling situation is seriously impacted by the fact that the bulk of late nineteenth- and twentieth-century anthropological and archaeological work has been centered on Western Europe, with Eastern Europe and Asia, with the exception of the Middle East, attracting far, far less attention.

First Appearance of 'New' Behaviors or New Technologies?

While trying to better understand the evolutionary-behavioral implications of the anthropological–archaeological record I have been struck by the apparently 'sudden' appearance of specific human behaviors. The question arises whether these 'sudden' appearances are real, or alternatively, represent the incompleteness and other defects of that record. For example, the first evidences of human use of fire occur far back in the Paleolithic with the finding of hearths beneath which there is fire-baked clay. Are these 'first' hearths evidence of the earliest human use of fire, or merely artifacts of a defective record compounded by limited investigation of potential hearth sites? Was the use of fire by humans a transitional rather than a sudden event? Berna et al. (2012) demonstrated the use of fire in an Early Acheulian South African cave site (Wonderwerk Cave) reliably dated 1.0 million years; this is currently the oldest reliable demonstration of the use of fire. Present evidence about the first appearance of 'fire' is inconclusive, undoubtedly a sampling artifact, although Wrangham's (2009) conclusions regarding the significance of *Homo* as small-mouthed, and with other cranial features that are consistent with a diet of cooked food, necessarily place possession of fire at the very first appearance of *Homo*. When one adds to this the need of the developing human infant for 'baby food' (see the section on Infant Nutrition) that involves cooking, the use of fire coinciding with the appearance of the *H. erectus–H. sapiens* lineage is very positive.

Tryon and McBrearty (2002) suggested that the transition from the Acheulian to the Middle Stone Age in Africa is not abrupt, and that the abandonment of handaxes and cleavers of Acheulian type was followed by their replacement with various points. However, this 'transition' may have involved abrupt origination at a single site, a very limited area, followed by 'transitional' spread to distant areas; the record is inadequate to solve this question.

The major changes in stone-tool sequences, and the evidence of geographically diachronous appearance for some of them, Oldowan to Late Paleolithic, fit this model. Barham and Mitchell (2008) suggest that the Oldowan appearance coincides with significant climate change that might have been causal rather than coincidental. Lordkipanidze et al. (2013) described a *Homo erectus*-type skull from Dmanisi where Oldowan-type tools have been found. However, Panger et al. (2002) raise the possibility that pre-Oldowan hominids might also have been capable of generating stone tools, since hand anatomy and phylogenetic relations are in agreement with this possibility despite the absence of positive evidence.

The artistic capabilities present in the Aurignacian, far earlier than the Neolithic, as discussed below, may be a sampling artifact owing to the absence or non-recognition of Aurignacian-comparable cave art in Africa and Asia except for Sulawesi (Aubert et al., 2014) and Borneo (Aubert et al., 2018).

D.E. Lieberman (2013, p. 203) emphasized that it was the advent of the Neolithic, with agriculture and village life, rather than the hunter-gatherer style of life, that made civilization possible. Does his suggestion conflict with the more gradualistic view of human cultural development provided by McBrearty and Brooks (2000)? Not necessarily. This might be a sampling problem. Moore (1985, p. 532) quoted Sir Leonard Wooley concerning the advent of the Neolithic 'instead of having to live where food abounded [man] made it abound where he lived; this says it all!'

The initial Neolithic 'appearance' of evidence for farming is an excellent example. The evidence suggests that farming first appeared, following the Natufian, in the Early Neolithic pre-pottery interval of the Near East, in a geographically limited area. From this region it spread outwards both east and west, reaching northern Europe only very late in what could be considered the Iron Age. The point is that 'new' behavioral–technological innovations may well originate in one restricted area from which they spread out (diffused) later in time, i.e., in a geologically very short time span.

When one comes to the 'sudden' appearance of the short-lived Copper Age, followed quickly by the Bronze Age, we are on more solid ground. Both Craddock (1995) and Černych (1992) indicated a Sixth-Millennium BC initiation of the Copper Age. Craddock (1995) made it clear that the initial use of native copper was quickly followed by the development of technology to smelt copper ores, particularly copper carbonates like malachite and sulfides like chalcopyrite. The question arises: was this 'sudden' appearance triggered in a single locality, possibly by a single person or group, or alternatively over a very short interval by different human groups? There is no good answer to this question, but its global Old-World suddenness is unquestionable; it is a real phenomenon, not a sampling artifact.

The initial appearance of the Bronze Age and its subsequent spread is another good example (Černych, 1992). Černych (1992, Figures 2, 3, p. 2-3) made the obvious point that new technologies do not simultaneously affect all human populations. For example, until very recently natives of the northern fringes of Eurasia and Arctic North America were considered to be still in the Late Paleolithic!

Turning to the Industrial Revolution and the torrent of inventions it involved leading to the present day, it is clear that in many prominent cases a single individual or a small group was responsible for the innovative techniques, followed by their widespread adoption far from the original site.

Taxonomy

Stringer (2002) reviewed the evolutionary relationships between various forms of *Homo*; his summary was reasonable as of 2002, but Dennell (2009) produced an even more comprehensive account of what is currently 'known' and what major areas of ignorance remain concerning the evolution of *Homo*. The hominids are not as closely related to chimpanzees and gorillas as has been suggested earlier. White et al. (2009) described *Ardipithecus ramidus* from the earlier Pliocene of Ethiopia as not bipedal and more primitive than *Australopithecus*, to which it is possibly related (see Lovejoy et al., 2009b, and D.E. Lieberman, 2013). Spoor et al. (2015) and Villmoare et al. (2015) provided convincing evidence that early *Homo, H. habilis*, and a related form dated at 2.8 million years are more advanced than *Australopithecus* and might be related to the *H. erectus-H. sapiens* lineage although distinct from it.

Berger et al.'s (2015) account of the new species *Homo naledi* may well make necessary some changes in our concepts of human evolution, but until the material is reliably dated and subjected to further study this is uncertain.

Sawyer et al. (2015) discussed the relatively recent Russian hominid discoveries at Denisova Cave in the Altai Mountains of southern Siberia. The skeletal material consists of a finger bone, a toe bone, and several molars; the molars differ significantly in their morphology from those of Neandertals and *H. sapiens*. They described the DNA evidence suggesting that the Denisovans are distinct from Neandertals and *H. sapiens* and shared a common ancestor with them. A certain amount of DNA derived from Denisovans is present in modern humans from Southeast Asia and Oceania (Reich et al., 2011), with the inference that archaic Denisovans lived over a broad geographic range from Siberia to tropical Asia.

D.E. Lieberman (2013) provided a very insightful up-to-date account of current thinking about hominin evolution and the relations of the various species currently thought to belong to *Homo*. Boaz and Ciochon (2004) review the possible relations of *H. erectus* in time from other homininid taxa.

Lieberman and Bar-Yosef (2005) provided a useful account of the evolutionary relations of the 'specific' taxa belonging to *Homo*.

Tomasello (2009) pointed out, using studies with young infants, that cooperation may be one of the most critical human traits separating us from other taxa, although others have similarly appealed to language capability.

Harmand et al. (2015) described stone tools 3.3 million years old from Lomekwi 3, West Turkana, Kenya. The only 'contemporaneous' homininid of this age is *Kenyanthropus*, which is not well known. *Kenyanthropus* has small molars, a more *Homo* trait, but until information about its rib cage, pelvis, feet, etc. are known it is too soon to be sure about its affinities. If remains of this taxon are eventually found at the same horizon as the Lomekwi 3 stone tools, then a case might be made about the tool's maker. In any event these stone tools from Lomekwi 3 significantly predate the oldest Oldowan tools.

Quantum Evolution

George Gaylord Simpson (1953), in many ways the leading paleontologist of the Twentieth Century, provided the concept of 'quantum evolution' to explain one of the leading dilemmas facing the concept of organic evolution. Darwinian evolution suggests that one species gives rise to another, that there are no 'gaps' morphologically or otherwise between one species and its descendant species. Therefore, in principle, there should be a continuous chain of species leading from an ancestral family and/or higher taxon to its descendant family or higher taxon. But neither the fossil record nor the present biota provides such species-to-species chains between supposedly ancestral families and higher taxa to their descendant family and higher taxa. For example, there are no species intermediate between the members of the dog and cat families or from their putative, older ancestral family. This is very troubling. Simpson, an experienced taxonomist, was well aware of this dilemma. It demanded an explanation consistent with modern biology. He suggested that the 'missing' species chains between ancestral and descendant families could be explained by appealing to the following: 1) very small populations; 2) highly endemic location; or 3) very rapid evolution. *If* these three qualifications be granted, it is reasonable that both the fossil record and the evidence of the present should provide no positive evidence for the 'missing' species. He pointed out that this concept is in agreement with the concepts of genetic theory with rates of evolution being inversely related to overall population size (see his Figure 47, upper figure, for a graphic depiction).

Hominin Evolution

The evolution of the hominins can be explained by the concept of quantum evolution. There is no 'chain' of intermediate species leading from a putatively ape-like ancestral family to the hominins. The hominins are classed as a subfamily within the Hominidae, but this is an arbitrary artifact of classification that does not truly indicate the skeletal and morphological features, digestive-tract morphology and physiology, life history, cognitive changes, and other items that separate the lineage of *Homo erectus, H. sapiens,* and *H. neanderthalensis* from the great apes, i.e., from gorillas, chimpanzees, bonobos, *Australopithecus* plus *Ardipithecus,* and *Homo habilis.* Again, there are no known intermediate species involved in this revolutionary development.

Gagneux and Varki (2001; Varki, 2001) pointed out that there must be a great variety of still largely unknown genetic differences between humans and the great apes. They did produce one distinct difference in a common mammalian cell-surface sugar where humans and the great apes are very distinct.

The selective factors responsible for the episode of quantum evolution that produced the hominins several million years ago are very uncertain.

One can provide a list of the significant items involved in this quantum-evolution 'leap' from the great-ape category to the hominin category. Such a list includes: making and use of fire for cooking, supported by skeletal evidence of *Homo*'s small mouth; reduced size of post-canine teeth; reduced chewing capability indicated by lack of muscle-attachment areas for the appropriate muscles; indications of a longer small intestine and shorter colon suggested by the absence of an expanded chimpanzee–ape-type pelvis for support of a paunch; and the need for soft, mushy food for infants beginning at about six months; relatively parallel-sided rib cage as contrasted with the funnel-shaped chimpanzee–ape type; shoulder anatomy designed for accurate throwing and clubbing; skeletal modifications supportive of bipedalism; hands with a shortened thumb modified for use of fingers in tool manufacture and use; foramen magnum positioned beneath the skull rather than behind it; a female pelvis requiring a 90° rotation of the foetus during delivery; and 90% right-handedness as contrasted with 50:50 for the great apes. The level of sexual dimorphism in *Homo* is much lower than in the great apes and most other primates. Wrangham and Carmody (2010) pointed out that humans are just about the only primates that sleep on the ground, their safety ensured by using fire to deter predators. Human females are sexually receptive at any time, in contrast to the great apes, whose sexual relations are restricted to the brief estrus interval. Humans need to clean the anal region after defecation owing to the close proximity of their limbs due to their bipedal posture. A long menopause among humans contrasts with the situation among the great apes. Absence of penile spines and facial vibrissae in *Homo* also contrasts with the great apes and other primates.

McBrearty (2007) made it clear that the Upper Paleolithic European replacement of the Neandertals by the Cro-Magnons is a biogeographic event, and is not of an evolutionary character.

Data

Summary

Any consideration of the antiquity of human behavior begins with the skeletal evidence because it is in many ways the most basic, comprehensive, and best documented. The information in the following sections, 'Hand' to 'Mortality', reviews skeletal and other evidence item by item, with most items first appearing with *H. erectus*.

There are some 'gradual' aspects to evolution within *Homo,* including a progressively lower level of prognathism, an increase in brain size, an increase in height, and a lower level of sexual dimorphism.

Skeletal Features

Skeletons of *H. sapiens* are characterized by:

- a foramen magnum situated 'beneath' the skull rather than at its posterior as in the great apes and other primates;
- relatively long legs and short arms, in contrast to the long arms and short legs of the great apes;
- a parallel-sided rib cage among hominins as contrasted with the funnel-shaped rib cage in the great apes;
- a narrow pelvis of the hominins as contrasted to the relatively broad pelvis in the great apes, useful in supporting the large paunch;
- a hominin female pelvis morphology requiring the foetal head to rotate 90° during birth, as contrasted with the great apes for which the broad female pelvic opening does not require such rotation;
- non-prehensile toes in hominins, as contrasted with prehensile toes in the great apes;
- a relatively small mouth in hominins with small post-canine teeth, as contrasted with a far larger mouth in the great apes, with evidence of far more massive jaw muscles and a well-developed sagittal crest associated with their largely herbivorous diet that required extensive chewing of tough materials (with the weak jaw muscles of hominins correlating with a softer diet that probably featured foods softened by cooking);
- relatively large brains in hominins, as contrasted with those of the great apes;
- hominin shoulder morphology modified for accurate throwing and clubbing, as contrasted with its lack in the great apes;
- a large, 'globular braincase, much of which expands behind a centrally placed foramen magnum' in hominins (D.E. Lieberman, 2011);
- brow ridges absent or subdued as in *H. neanderthalensis*, as contrasted with the great apes;
- a short face with a pronounced chin in *H. sapiens*, as contrasted with the great apes; and
- a well-developed, protruding nose with downward-opening nostrils as contrasted with the great apes, with certain of these distinctions less evident in the case of *H. erectus*.

Much of this involves a progressive 'loss' of the prognathous condition in favor of the 'small' face of *Homo*.

D.E. Lieberman (2011) reviewed the many skeletal differences between the human and chimpanzee skulls and discussed the basic differences between the human pharynx and that of all other primates.

Spoor et al. (1994) showed that the early homininid labyrinthine morphology, beginning with *Homo erectus*, is unlike that present in *Australopithecus* or the great apes, with the implication that *Homo* was more specialized for bipedalism, rather than climbing, while *Australopithecus* and the great apes retained climbing capabilities.

D.E. Lieberman (2011) pointed out that the presence in *Homo* of a nuchal line indicates that there is a nuchal ligament for the first time, absent in *Australopithecus* and the great apes, important for balance in long-distance 'endurance running', which in turn makes such running useful as a hunting strategy in running down large herbivorous prey.

D.E. Lieberman (2011) makes the point that sexual dimorphism decreased significantly from the great ape and *Australopithecus* situation to the human condition.

Hand

Barham and Mitchell (2008, their Figure 3.5) made the important point that the hominin hand features relatively short fingers combined with an elongated thumb, making possible the use of the hand for tool making and tool use in a manner impossible for the higher anthropoids. (Their figure features a chimpanzee hand for comparison.) They also pointed out that it is important to recognize that there is a difference between tool use and tool making, with the hominin hand capable of both in a 'modern' manner. Marzke and Shackley (1987; see also Marzke, 1997) add additional material to the questions regarding human hand capabilities.

Susman (1994) analyzed the implications of the functional morphology of the human thumb as contrasted with that of other higher anthropoids. He concluded that only in members of our genus is there reliable functional morphological evidence for the capability of using tools in a complex manner and that available evidence from the last 2.5 to 2.6 million years suggests that *Australopithecus afarensis* lacked this capability, as does the chimpanzee.

Ward et al. (2014) described evidence from Kenya for a modern human-like hand of probable Acheulian age that might have belonged to *H. erectus*.

Skinner et al. (2015) described trabecular-bone form in *Homo* and *Australopithecus africanus* that suggest that the latter may have had human-like hand-use far earlier than *Homo*.

Handedness

The study by Toth (1985; see also Frost, 1980) of the antiquity of handedness in hominids has direct bearing on the questions being dealt with here. He summarized data indicating that about 10% of a human population is left-handed. He mentioned that this phenomenon is involved with hemisphere lateralization in humans, where special functions are assigned to the different hemispheres – involving such things as motor ability and linguistic abilities, among others. He then went on to show how he had been able to discriminate between stone flakes made by left-handed workers from those made by right-handed individuals both now and in the earlier Pleistocene. He made a good case that sites, as ancient as 1.4 to 1.9 million years, yield flake samples showing that the percentage of left-handed versus right-handed makers remained approximately constant from then until now. He pointed out that behavioral data from non-hominid populations, including other higher primates, except for hominids, show a 50:50 ratio of handedness, i.e., no evidence for hemisphere lateralization except in hominids. This is certainly evidence favoring behavioral constancy, incidentally of taxonomic value at the subfamily level. Legge (1981, p. 101) provided evidence from deer-antler use from the Middle Bronze Age Grimes Grave locality indicating that left antlers of red deer were preferred over right antlers, which correlates with the use of the left antlers by right-handed flint miners, i.e., further support for the dominance of right-handedness.

Bermudez de Castro et al.'s (1988) description of inclined scratches on the middle and earlier Upper Pleistocene upper incisors indicates a great predominance of right-handedness among humans. The behavior consists of grabbing a piece of meat between the front teeth with one hand, while using a sharp flint flake to slice the meat off into a bite-sized bit just in front of the lips – with an occasional 'slip' accounting for the inclined scratches. Having observed during a 1947 visit to North Greenland the dexterous manner in which Polar Eskimos are able to grasp a chunk of blubber in one hand and front teeth, and then slice off a bit with a razor-sharp ulu (woman's knife; plural uluit) right next to the lips, I can personally testify to the feasibility of the process.

Bradshaw (1988) provided an excellent account of lateral asymmetry in humans and its implications for both speech and behavior of various types. Coren and Porac (1977) reviewed evidence dating back some five thousand years indicating a predominance of over 90% of right-handedness, based on data derived from artistic efforts. Dart (1949) suggested that australopithecines were dominantly right-handed based on damage to baboon skull interpreted as prey (but see Brain, 1958, for an alternative explanation that ascribed the data to the activities of leopards).

Stoddart (1990, p. 66) mentioned that far more axillary androsterone is produced in the right axilla of right-handed people whereas in the left-handed far more is produced in the left axilla. No connection between this axillary production and right-handedness and left-handedness is suggested.

Shoulder

There is an important literature on the anatomical characteristics of humans regarding their abilities to accurately throw objects and to use the hands for clubbing (see Young, 2003, for an example; also Atwater, 1979). Darlington (1975) inferred that human anatomy and the resultant ability to accurately throw stones may well have been important early on in their evolution. There is a parallel literature on the anatomical characteristics of humans for throwing spears (see Churchill and Rhodes, 2009, for an example).

Roach et al. (2013; see also Roach et al., 2012, and Larson, 2007, 2007b) described the functioning of the shoulder in *Homo* that makes high-speed throwing possible, i.e., in the use of projectile weapons. Ruben (oral communication, 2014) pointed out that the same shoulder anatomy is also well adapted to bludgeoning motion, which certainly agrees with the evidence from the past of badly damaged skulls!

Brain Size and Morphology

Snodgrass et al. (2009) provided a useful account of what was involved in the remarkable degree of encephalization characteristic of hominids, including anatomical changes, digestive physiology, and overall changes in life histories.

Park et al. (2007, Table 1; see also Schoenemann, 2006, Figure 2) compared the cranial capacities of hominids, with *Australopithecus afarensis* at 342–350 cubic centimeters; robust australopithecines at 494–537 cc; *Homo habilis* at 509–775 cc; *H. erectus* at 727–1225 cc; archaic *H. sapiens* at 1100–1586 cc; Neandertals at 1200–1750 cc; and modern *H. sapiens* at 1349 cc.

D.E. Lieberman (2013, Table 2) provided the following brain-size estimates: *H. erectus* 600–1200 cc; *H. heidelbergensis* 900–1400 cc; *H. neanderthalensis* 1170–1740 cc; and *H. sapiens* 1100–1900 cc. Table 1 in D.E. Lieberman (2013) has brain-size estimates for *Ardipithecus* 280–350 cc, and gracile *A. afarensis* 400–550 cc.

McHenry (1994, Table 3) provided the following brain-size estimates: *Australopithecus afarensis* 384 cc; African *H. erectus* 820 cc; *H. sapiens* 1250 cc.

McHenry (1994b, Table 1) provided the following brain-size estimates: *A. afarensis* 384 cc; early *H. erectus* 804 cc; late *H. erectus* 980 cc; and *H. sapiens* 1350 cc.

Harvey and Clutton-Brock (1985, Table 1) tabulated information on adult brain weights showing that *Homo sapiens* brain weight is about three times that of the chimpanzee and more than twice that of the gorilla.

Holloway (1972) described an australopithecine endocast from Swartkrans, comparing it with that of other australopithecine endocasts and with pongids. He concluded that it is of homininid rather than pongid form.

Schoenemann (2006) reviewed many aspects of brain evolution and morphology as they apply to human behavior in all of its complexity, making some penetrating suggestions but without arriving at firm conclusions. This is an area where future work should eventually provide more insight.

Leonard et al. (2003) made the important point that *Homo* has a much smaller skeletal muscle mass relative to the other higher primates and that it is associated with their much larger brain (their Figure 4), relating this to a higher diet quality in *Homo* than in comparative higher primates.

Leonard et al. (2007b, Table 18.3) pointed out the importance in brain development in the infant *Homo sapiens,* emphasizing the massive growth during the first 18 months, followed by a marked slowing down into adulthood. There is also an infant increase in body fat followed by a falloff later.

Pellett (1990, Table 1) reviewed human protein requirements with the tallying of the very great requirements of infants as contrasted with adults.

Infant Nutrition

Wrangham (2009, see above) pointed out the obvious critical fact that human skull morphology, including the teeth, made a diet of cooked food (requiring fire!) necessary. Cooked food requires far less chewing than the tough plant diet of the great apes. Milton (1989, Table 3.2) pointed out the supporting evidence provided by the relatively lengthy small intestine and short colon of humans as contrasted with the relatively short small intestine and lengthy colon of the great apes. Together with the broad pelvis of the great apes that supports their large paunch – in contrast to the short pelvis and lack of a paunch in humans – this further supports Wrangham's position.

All of the above correlates with the infant requiring a very high-protein diet after weaning while the brain is still growing rapidly, and a need for suitable 'baby food' from the beginning. The implication is that suitable food for both *Homo erectus* and *H. sapiens* infants needed to be 'mushy', also implicating the use of stone-boiling to 'soften' protein-rich meat and large-seeded legumes to the point where they could be handled by the infant. Stone-boiling would provide the 'mush' needed by the developing infant whereas the 'steaming' of foods would not be regularly capable of providing soft mush. D.E. Lieberman (2011, Table 4.1) pointed out that the eruption of the first molar can be relatively late, with the eruption of the second molar not occurring before at least six years, i.e., chewing can occur late in infant development well after weaning.

Another 'solution' to the problem of providing the infant with adequate protein and glucose very early on for brain development is provided by 'premastication' (Pelto et al., 2010). The protein can be provided by meat or other animal protein and large-seeded legumes, with underground carbohydrate-rich storage-organs as sources of glucose, softened or not by cooking, and made into a mush in the human mouth, where it is mixed with saliva. Maternal milk becomes inadequate as an exclusive source of nutrition for the developing infant by approximately six months (Sellen, 2007; Dewey, 2003), but teeth suitable for chewing erupt only at about 18–24 months, or later. Pokutta and Howcroft (2013) described isotopic evidence for the use of supplementary foods being used at about six months in infant remains from the Únětice Culture of southwestern Poland, dated between 2100 and 1600 BC. Premastication is the practice by an adult, most commonly the mother, of prechewing foods suitable for the infant to swallow, and providing them directly mouth-to-mouth or otherwise. This also commonly implies that the premasticated foods will have been 'softened' before chewing to make them more easily digestible with the use of fire, as in stone-boiling. Dewey (2003) emphasizes that at six months puréed, mashed, and semi-solid food may be introduced; by eight months 'finger foods'; and by twelve months 'family foods', with food that might cause choking avoided, such as nuts, grapes, and raw carrots. Human saliva will also help the infant by providing various substances essential for infant development.

Hardy et al. (2015) discussed the potential role of carbohydrate-rich underground plant storage-organs in providing a rich, reliable source of carbohydrate starch easily convertible into the glucose needed for infant brain-growth once they had been cooked and then chewed by the caregiver, where salivary amylases aided in the starch conversion to glucose. Available data suggest that the various salivary amylases necessary for the conversion of starch to glucose appear very early and might have been present from the first appearance of *Homo*. (Samuelson et al., 1990, proposed a 'recent origin' for the human amylase genes.) They also point out that sources of meat protein from hunting would have been far less reliable for a steady infant diet than would carbohydrate-rich underground plant storage-organs. Meisler and Ting (1993) provided more information about the history of the human amylase genes. This information further supports Wrangham's proposal concerning the use of fire for cooking from the very beginning.

Cummings and Englyst (1995) discussed the various types of food carbohydrates and their utility as food, and commented about the digestibility of various types and also of their value when warm due to cooking, as opposed to raw or too cool after having been cooked.

Stevenson and Allaire (1991, and their Table 3) summarized the development and learning of normal feeding and swallowing in the human infant from the initial transition from breastfeeding to the use of puréed food, 'baby food', at about six months, and through the many transitions to an 'adult' diet at about 18 to 24 months. They emphasize that the transition from one food class to another during the infant's growth involves a high degree of learning with the help of the mother or other close relative. Anyone who has observed the changes in infant diet during these early growth stages will have been impressed by the high level of patience needed by the mother in 'teaching' the infant to take one food class after another through time until the adult diet stage is reached.

Kennedy (2005) discussed the need for higher protein intake, emphasizing animal products, for the developing foetus, the mother during lactation, and for the weanling, concluding that early weaning time in humans and lengthy 'childhood' are critical adaptations. Leonard et al. (2007) emphasized the body fatness of *Homo sapiens* infants until about 18 months postnatal with corresponding dietary needs. Bard (1995) indicated that weaning in chimpanzees occurs at about five years. Dettwyler (1995, Figure 2.2) indicated that the 'normal' age at weaning for humans in traditional societies is nearly three years. Harvey and Clutton-Brock (1985, Table 1), cited weaning age for chimpanzee at 1460 days, for gorilla 1583 days, and for *Homo sapiens* 720 days. Nishida and Turner (1996, p. 960) reported that chimpanzees begin to solicit solid food from their mothers at about age six months.

Lonsdorf et al (2014) reported that chimpanzee infants do not use solid food until six months after birth, similar to human infants, in relying exclusively on breastfeeding. Lonsdorf (written communication, 2015) reported that after six months chimpanzees begin to eat 'soft plant foods and fruit' and that mothers 'do not exclusively premasticate food for them, but do share bites of fruit, etc., based on the begging behavior of the offspring'.

Bard (1995) indicated that in contrast with other primates, neither human nor chimpanzee infants are able to support their own weight for the first few months, and need maternal support. Bard (1995) also finds that chimpanzee maternal parenting is learned from co-occurring conspecifics.

Legs and Bipedal Walking

Sellers et al. (2005) concluded that *Australopithecus afarensis* was capable of sustained efficient bipedal walking at gaits within the range of modern humans and greater than that of quadrupedal knuckle-walking speed in chimpanzees.

Lovejoy (1988) described the great similarity of the femur and hip of *Australopithecus afarensis* to those of *Homo* and major differences with the chimpanzee, indicating that *A. afarensis* was strictly bipedal and did not have a prehensile foot as in chimpanzee. In support of this conclusion, he also cited differences with the knees and ankles. Lovejoy et al. (2009) pointed out that the foot of *Ardipithecus ramidus* makes it a more reasonable ancestral type to *Australopithecus afarensis,* which eliminates gorilla and chimpanzee types from that role.

Ward et al. (2011) provided skeletal evidence that the fourth metatarsal and arches of *Australopithecus afarensis* are strictly of human rather than pre-human type, i.e., strictly modified for bipedal locomotion.

In an innovative paper, Spoor et al. (1994) found that the labyrinthine system of the inner-ear bones indicates that *H. erectus* is similar to *H. sapiens* but that *Australopithecus* is intermediate between gorilla and chimpanzee with the suggestion that it combined bipedal locomotion with climbing ability, i.e., is intermediate between the great apes and humans.

Feet (Prehensile Toes)

Homo and *Australopithecus* are unique in lacking the opposable big toe present in gorillas and chimpanzees and *Ardipithecus.*

Bennett et al. (2009; see also Crompton and Pataky, 2009) described undoubted footprints of human hominin from the earlier Quaternary (1.5 million years old) of Kenya. These footprints may have been made by *Homo ergaster* or *H. erectus.* They are the oldest known hominid footprints. Leakey (1981, 1987; see also White, 1980; Robbins, 1987) described hominid footprints, from Tanzania, presumably made by *Australopithecus afarensis.*

Rib Cage

Ruff (1993, Figure 2) makes the point that the overall bony morphology of *A. afarensis* is much broader in both the rib cage and the pelvis than is the case with *H. erectus,* i.e., far more ape-like, which would correlate with a largely vegetarian diet and an ape-like digestive tract with a large paunch needed for support both 'above' and 'below'. Overall, *H. erectus* is far more slender than *A. afarensis.* Wheeler (1993) pointed out the 'funnel-shaped rib cages' and 'wider pelves' of the australopithecines, which are also shared with *Pan, Gorillo,* and *Pongo,* as contrasted with the more parallel-shaped rib cage and narrower pelvis of *Homo.* Presumably these basic skeletal differences are needed to support the far larger paunch, with its much longer large intestine needed for the far more vegetarian diets of the australopithecines and great apes. These are major skeletal distinctions.

Legs and Arms

The legs of *Australopithecus* are relatively long, as contrasted with those of earlier hominids. They are combined with relatively long arms that are more ape-like.

McHenry and Coffing (2000) concluded that the limbs of *Australopithecus* were intermediate between the great apes and *Homo*.

Pelvis and Spine

Lovejoy et al. (2009b) pointed out that the pelvis of *Ardipithecus ramidus* is a mosaic of features that is unlike that of the chimpanzee but more primitive than that of *Australopithecus*. The pelvis of *A. ramidus* is a mosaic useful for climbing and upright walking (and no more knuckle-walking) with a prehensile foot useful for climbing, while further changes in *Australopithecus* made for more effective walking but eliminated the specializations useful for climbing.

The pelvis of *Australopithecus* is relatively human-like, as contrasted with that of the chimpanzee and other great apes (Marchal, 2000), but is distinct from that in *Homo*, and there are some differences between *H. erectus* and Neandertals as compared with modern humans.

Berge (1984) concluded that the pelvis of *A. africanus*, a gracile form, is more like that of *Homo* than is the case with the robust forms such as *A. robustus*.

Lovejoy (2005) made the case for the presence of spinal lordosis in *Australopithecus* as in *Homo*, although absent in apes. He also made the case that the changes associated with spinal lordosis were not accompanied by similar pelvic changes that would enable the birth of a large-brained foetus rather than a small-brained ape-like form.

Neck and Position of the Foramen Magnum

There is a progressive shift in the position of the foramen magnum from a posterior skull position in the great apes to a position in the mid-anterior of *Homo* that facilitates bipedal walking and running.

Tooth Morphology

Leonard et al. (2007, Table 2) indicated that post-canine teeth in *H. erectus* and *H. sapiens* decrease markedly in size from that in australopithecines, presumably due to significantly different diets. McHenry and Coffing (2000) concluded that post-canine teeth became reduced in size in the change from *Australopithecus* to *Homo*. D.E. Lieberman (2011) discussed the relatively thin enamel on post-canine teeth in *Homo*, as contrasted with the great apes and *Australopithecus*, which correlates with an implied major change to soft foods from fibrous and 'hard' foods that require extensive chewing.

Prognathism

D.E. Lieberman et al. (2004) provided some experimental evidence from hyrax raised on soft and harder foods, showing that soft, i.e., cooked, foods do have an effect on the form of the face. This implies that jaw morphology and chin in *H. sapiens,* with its far less prognathic form than in ancestral types, does correlate with a diet of cooked food. (Changes in dentition among modern humans may also be implied.) Rightmire (1990) made it clear that *H. erectus* lacks the sagittal crest so characteristic of the great apes, whose diet of tough vegetable foods require massive jaw muscles. He noted that *H. erectus* has a reasonable level of prognathism.

Sexual Dimorphism

Lockwood et al. (1996) concluded that *Australopithecus afarensis* has a degree of sexual dimorphism intermediate between that of gorillas and orangutans on the one hand, and humans and chimpanzees on the other. Harvey and Clutton-Brock (1985, Table 1) tabulated female weights versus male weights, which showed a very high level of dimorphism in the chimpanzee and gorilla compared to *Homo sapiens.* McHenry (1994, Table 1) compared estimated weights for male and female, a measure of dimorphism, for *Australopithecus* and *H. erectus,* showing the marked decrease in sexual dimorphism with the latter, and suggested that sexual dimorphism in early hominids is more than in modern humans but much less than in *Gorilla* and *Pongo.* McHenry (1994, Table 1) shows *A. afarensis* with a level of body-size dimorphism only slightly above that of *Pan,* with the level of sexual dimorphism markedly reduced in *H. erectus.*

McHenry and Coffing (2000, Table 1) concluded that sexual dimorphism in size was strong in *Australopithecus* compared to *Homo: A. afarensis* male 151 cm (5 feet); *A. afarensis* female 105 cm (3.5 feet); *H. erectus* male 180 cm (6 feet); *H. erectus* female 160 cm (5.3 feet); *H. sapiens* male 175 cm (5.8 feet); *H. sapiens* female 161 cm (5.4 feet).

Leutenegger (1974) described the strong degree of sexual dimorphism in the male and female hominid pelvis, controlled by the need to cope with the delivery of the infant.

Size and Robusticity

Ruff et al. (1993b) and Ruff (1993) provided information about the height of the few known Australopithecine and *H. erectus* specimens known from complete skeletons. For *A. afarensis,* the height is 107 cm (3 feet, 5 inches) and for *H. erectus,* 185 cm (6 feet, 1 inch) and 170 cm (5 feet, 6 inches) for a second specimen. The extreme rarity of relatively complete pre–*H. sapiens* skeletons precludes the use of complete skeletons in attempts to understand significant changes in hominid heights through time. Instead, the literature provides numerous estimates of hominid weights through time as judged from various correlations between different, isolated, skeletal elements of various sorts, with weight as a proxy for 'size'.

Ruff (1993) made the point that cold-climate, high-latitude *H. sapiens* have relatively compact, short bodies as contrasted with low-latitude, warm-climate humans, with Neandertals being more like the cold-climate modern types in this regard, which correlates well with their geographic distribution and its inferred climate over time.

Feldesman et al. (1990) used a technique for estimating human height based on femur lengths; from data in their Table 6, they estimated the height of *H. erectus* at 152 to 187 cm (5 feet to 6.2 feet), *H. neanderthalensis* at 143.8 to 180.5 cm (4.8 feet to 6 feet), Cro Magnon at 152.8 to 199.3 cm (5.1 feet to 6.6 feet), and *A. afarensis* (Lucy) at 105.6 cm (3.5 feet).

Ruff et al. (1993b) measured postcranial robusticity in *Homo* by using data from changing bone thickness, and showed that it has decreased markedly from *H. erectus* to modern *H. sapiens*. During the 'early' part of *Homo* evolution there is a great expansion of brain volume while robusticity declines more gradually. 'Causes' are elusive, although changing technologies that required less physical effort might be involved.

Chirchir et al. (2014) discussed a marked decrease in limb-joint trabecular density in 'recent modern humans' as contrasted with older *Homo sapiens* and homininid precursors, which they ascribe to the significant change in 'technological and cultural innovations' beginning with the advent of farming. The exact cause is uncertain, but one wonders whether dietary changes might have been involved.

McHenry (1994) also concluded that there was a significant size increase from *A. afarensis* to *H. erectus*. McHenry and Coffing (2000) concluded that *Australopithecus* is relatively small compared to *Homo*, as did Ruff (1993).

Chang (1986, p. 6–8) discussed the Chinese occurrences of *H. erectus* dating from 200,000 to more than 1,000,000 years BP. Zhu et al. (2008) described *H. erectus* remains from Southwest China of approximately 1.7 million years age.

Baculum

Humans are the only primates lacking a baculum. There is no fossil record of a baculum in any higher primate, but the small size to be expected in view of that in modern great apes such as the gorilla (Davis, 1951) and chimpanzee makes it possible that they have been overlooked during preparation. Dixson (1987) cited 12.5 mm in *Gorilla gorilla*, 8.5 mm in *Pan paniscus*, and 6 mm and 7.8 mm in *Pan troglodytes*, suggesting that size-reduction in baculum characterizes the great apes and *Homo* as an ancestral trend that reaches its ultimate in *Homo*.

Hearing

P. Lieberman (1968) made it clear that the vocal tracts of chimpanzees and gorillas are distinctly different from those of humans, lacking the capability to make the vocalizations necessary for human speech; vocal tracts develop in human infants at about six weeks. D.E. Lieberman (2011) referred to a paper by Martinez et al. (2004) describing inner-ear morphological attributes consistent with modern human hearing by at least the Middle Pleistocene. D.E. Lieberman (2011) indicates that the presence of the language capability from the beginning of the *Homo* lineage is a real possibility. Quam et al. (2015) concluded that *Australopithecus* and *Paranthropus* had hearing capabilities intermediate between those of modern chimpanzees and humans; unfortunately, there are not yet any data for *Homo erectus*.

Moggi-Cecchi and Collard (2002) described a Sterkfontein stapes, suggesting that early humans had hearing more like those of the great apes and *Australopithecus* than of modern humans or Neandertals.

Mortality

Trinkhaus (1995) provided evidence that Neandertal mortality was high in prime-age adults and low in older adults. When combined with the analysis of Berger and Trinkaus (1995) of trauma affecting Neandertals, this indicates the probability that the trauma resulted from hunting large, dangerous ungulates, and also helps explain the mortality pattern.

Soft-Tissue Evidence with Some Skeletal Basis

Summary

The evidence for this item is based entirely on the correlation between skeletal morphology, *H. erectus* to *H. sapiens,* and modern digestive anatomy.

Digestive-Tract Morphology and Physiology

Milton (1989, Table 3.2) emphasized the relatively small-volume colons contrasted with the large-volume small intestine in humans versus the reverse situation in the other higher primates. This correlates with major dietary differences, namely, plant-food dominance in the other higher primates versus omnivory, combined with cooking, which makes nutrients more easily available in humans.

Milton (1999) provided a very comprehensive account of the digestive anatomy and food requirements of humans, chimpanzees, and gorillas. Her account includes the fact that the small intestine of humans has a much larger volume than the colon, with the reverse being true of the gorilla and the chimpanzee. She also emphasized that as size increases in the gorilla, the quality of the vegetation consumed decreases. Further, that the chimpanzee uses higher-quality vegetation with a corresponding use of some animal food, including ants and termites. She also emphasized the need for developing human infants after weaning for more protein-rich diets. She suggested that the modern situation in humans may have included *H. erectus.* Also see Milton's later papers. [The author left a note to himself to review this section.]

Soft-Tissue Evidence Supported by Molecular Information

Penile spines

The evidence for this item is purely molecular.

Reno et al. (2013) described the molecular evidence for the absence of the penile spine/vibrissa enhancer in humans, Neandertals, and Denisovans that is present in the great apes and other primates. Muchlinski (2010) provided a comprehensive summary (her Table 1) of vibrissa presence and absence in mammals. Hill (1946), while describing penile spines in the chimpanzee, commented, 'A further point needing emphasis is that the discovery of spines on the glans penis of the Chimpanzee still further widens the gap between it and Man with respect to that part of the anatomy'.

Smell

The evidence is largely molecular.

Using molecular data, Gilad et al. (2003) showed that the human sense of smell is significantly lower than that of the great apes. In general, it is clear that vision is the primary primate and human sense organ, with hearing second, followed at some distance in humans by the sense of smell.

Stoddart (1990, 2015) considered the human sense of smell at length, with its long-term biological basis, and also considered the nature of the axillary and other apocrine glands with the functional significance of their odoriferous products. He also considered the human sense of smell as contrasted with that of other primates. Stoddart (1990, p. 6–7) commented on the difference between chimpanzees and humans: 'Genetically we are closely related but the gulf of behavioral cultural and intellectual differences that separates us is of fathomless depth'. Stoddart (1990, p. 142–206; 1998; 2015) considered many aspects of both perfumes and incense with evidence from Classical Antiquity to the present about sources and their natures and the fact that much of their odor is similar to human underarm axillary odor.

Bartosiewicz (2003) pointed out that the relatively poor human sense of smell can quickly become accustomed to the noxious smells associated with various trades, contrasting this with the far more sensitive character of most animals.

The smell of sewer gas from toilets was finally eliminated by Alexander Cummings' 1775 patent for the sewer trap!

Sexual Receptivity

The skeletal evidence about relative sizes is very positive, but inferences about causation are speculative.

Wrangham et al. (1999) made the important point that the relatively lower level of sexual dimorphism in Homo contrasts with that of the great apes and also for Australopithecus. This can be explained most readily by suggesting that the uniquely human trait of continuous sexual receptivity among females was made possible by the use of fire and underground plant storage-organs, a contrast to the condition in the great apes and other primates where receptivity is restricted to those times where ovulation, etc. occurs with dietary changes. This would then result in more continuous male attention to the female's needs for reproduction. Possibly critical is the need for combined female-male pair-bonded couples to protect the young until they are sufficiently mature, which increases the fitness of both the males and females, in contrast to situations in which the male roves about to copulate with any available female.

Soft Tissue Variables with No Skeletal Evidence

Summary

There are a number of important items occurring in modern higher anthropoids, gorillas, chimpanzees, and humans in particular that are present only as soft-tissue evidence. IF one is willing to assume that these variables were present early in *Homo,* and some of them possibly in *Australopithecus* as well, they then provide even more evidence for the wide gulf separating the great apes from *Homo* and possibly in part from *Australopithecus.* These are considered below. Exceptional here is the very positive skeletal evidence for the obstetric implications of the human female pelvis.

Vision

Surridge et al. (2003) compiled the evidence bearing on the unique trichromatic vision of gorilla, chimpanzee and human, and Old World primates in general. Very important is the forward placement of the human eyes, which provides for binocular vision, an essential for so many human activities, as is also the case for many other primates.

'Hairlessness' and Skin

Man is sometimes referred to as the 'hairless ape'! Wheeler (1981) provided a cogent set of arguments indicating that bipedal posture, lack of a carotid rete, and poorly developed nasal turbinates are opposed by hairlessness (except on the head, where the hair protects the brain from overheating). Combined with very functional sweat glands and upright posture that presents a much smaller area to direct sunlight, these modifications reduce the danger of overheating the brain. Since present methods throw no light on the presence or absence of body hair beginning with *H. erectus,* we are left with the supposition that hair may have been absent since the 'beginning' as part of the quantum-evolution origins of *Homo.* Montagna (1985) pointed out that *Homo* is not truly hairless, since the hair has been mostly 'miniaturized'. Montagna (1985) also pointed out that human skin, as well as that of chimpanzees and gorillas, is relatively 'thick', with an abundance of 'elastic fibers' and an axillary organ for the production of odor, and that human females are unique among mammals for having enlarged breasts when not lactating, a possible sexual attractant.

Sweat Glands

Folk and Semken (1991) considered the evolution of sweat glands in mammals. They conclude that human 'hairlessness' and bipedalism are involved with the use of eccrine sweat glands for cooling, and that the apocrine glands in the axilla are associated with sebaceous glands that secrete oil (see Stoddart, 1990, Figure 3.1, for a diagram of the human hair-follicle and associated glandular structures). These axillary and sebaceous glands together secrete various fatty acids that when used by bacteria produce intense, smelly products that in humans are also attractive to mosquitoes. There are a few apocrine glands in the pubic and anal regions in humans but most in the axilla. Leidal et al. (1982) discussed the nature of the human axillary sweat glands. Zeng et al. (1991, 1996) described the characteristic odors produced by the human axillae in males and in females, finding them to be similar. Stoddart (1990) reviewed the areas where human odors are present, including the axillae, the male and female genital areas, and the anal region, which begin to function with puberty in humans. Rennie et al. (1990) discussed some aspects of the bacterially generated compounds present in human axillary sweat. Leyden et al. (1981) considered the microbiology of the human axilla and some of the associated odoriferous compounds, and Miyazaki et al. (2003) considered the chemistry of some of the odoriferous compounds generated by microbial metabolism of the axillary region. Apes and chimpanzees have sweat glands in their axillae but fewer than in humans, while female humans are better endowed than males (Stoddart, 1990, p. 7). Stoddart (1990, p. 70) summarized some of the evidence for human axillary odors having in a role in male–female sexual attraction.

Estrus

Estrus is considered to be absent in humans, as contrasted with chimpanzees and gorillas (Graham, ed., 1981), but Miller et al. (2007) presented some evidence concerning the sizes of lap-dancer's 'tips' suggesting that it may be largely but not completely hidden in humans. Lap dancers notwithstanding, the presence of estrus in gorillas and chimpanzees with prominent lordosis in the female is the real thing.

Obstetric Implications (see also Pelvis and Spine, above)

Rosenberg (1998) pointed out that the human female pelvis, as contrasted with that of the great apes, leads to difficult births, with face-down presentation that needs help from other females as contrasted with 'easy' solitary births of the great apes. This situation certainly implies a high level of communal cooperation in the birthing activity. Beausang (2000) discussed birthing evidence of various types, including items that should be looked for. Rosenberg and Trevathan (1996) discussed rotation of the foetal skull during delivery and the accompanying need for help from others in *Homo sapiens* while pointing out that there is inadequate evidence for earlier *Homo* species, and concluded that birth in australopithecines was probably simpler than in *Homo*. Their paper is well illustrated to make these points clear; they suggested that the female human pelvis, with its unique morphology that makes delivery complicated, was probably selected as one aspect of bipedalism, in contrast to the knuckle-walking of the great apes. Thus, it appears that in *Homo* the foetus is born before it has completed its gestational state, unlike the conditions for chimpanzee and gorilla, in which the newborn is able to grasp its mother's hair. This situation is presumably caused in an evolutionary sense by the need to birth before the foetal skull is too large to permit passage through the birth canal. In some ways this is reminiscent of marsupial births, which occur before gestation is complete, but where the infant continues to develop in the safety of the mother's pouch. In humans, birth produces altricial young at a somewhat foetal stage of development, as contrasted with the precocial stage in the great apes. The rotation of the foetus during this complex delivery results in a 'backwards-facing' delivery, which necessitates assistance from familiars, the 'midwife' situation, since the mother is unable to assist in making sure that the umbilical cord does not interfere with the infant's ability to breathe and also to clear any obstructing mucus that might interfere with breathing.

Leutenegger (1974) discussed the difficult birth in *Homo* based on the small size of the birth canal compared to the size of the foetal skull and the need to rotate the foetal skull during delivery, with all of this also involving the trade-off in the pelvic structure needed for bipedal locomotion as contrasted with pelvic structure needed for delivery.

The unique morphology of the human female pelvis contrasts with the simpler structure in chimpanzees and gorillas.

Life Span

Harvey and Clutton-Brock (1985, Table 1) tabulated data indicating that life span in chimpanzees is about 44 years, about in 39 in gorillas, and 70 in *Homo sapiens*.

Menopause

Barham and Mitchell (2008, p. 155; see also Hawkes et al., 1998) pointed out that long menopause is a uniquely human characteristic, not shared with the higher anthropoids; this may reflect the support role of grandmothers as an adaptive advantage in feeding and caring for children. They also point out the human pattern of early weaning, high fertility, and late maturation (adolescence), plus support from relatives. They also suggested that the existence of language was probably vital in maintaining these relationships with kin and grandmothers, as well as the ability of males to conduct hunts and the like.

Hawkes et al. (1997; see also O'Connell et al., 1999) treated various aspects of the evolutionary explanation for long postmenopausal life span. They pointed out that early weaning, of course, requires the continual feeding of the infant, in contrast to the situation in the higher anthropoids, and that grandmothers play a very significant role in feeding the infant. O'Connell et al. (1999) also emphasized the potential role of underground plant storage-organs and cooking in making higher-quality food readily available to infants (see also Nelson, 2010, and Thoms, 2009, for the soup-making and stew-making capability from the very beginning).

Behavioral Features

Birth Interval

Galdikas and Wood (1990) concluded that birth interval in human is shorter than that in other higher primates. Wrangham (2007, p. 311) noted that gorillas have a birth interval of about 3.9 years, chimpanzees about 5.0 to 6.2 years, and humans 9 months, indicating that human reproductive rate is much greater. Gestation in humans is about 266 days, about 237 days in chimpanzees, and about 257 days in gorillas; in general, gestation period is a function of adult size, i.e., very fast in mice, very slow in elephants.

Pair Bonding

Stoddard (1998, 2015) discussed human pair bonding at some length. Humans live in small, gregarious groups of pair-bonded members. The question is: Why do pair-bonded members of the opposite sexes stay together, and why don't the males commonly seek copulation with other females in the group? Stoddard suggested that pair-bonding substantially increases the chances of infant survival, thus increasing the fitness of both partners. Also involved is that ovulation is concealed in humans with no odoriferous cues about estrus, which decreases the chances for 'outside' copulations, all of this being 'a uniquely human feature'. Stoddart (2015) pointed out that the absence of a functional vomeronasal organ in the great apes and humans ensures that human ovulation time is hidden, thus reducing the chances of 'outside' copulations.

Division of Labor

Barber (1994) prefaced her consideration of 'women's work' with a telling quotation from Brown (1970) that says it all. Brown pointed out the obvious that women's child-bearing and child-rearing role constrained them to activities that did not interfere with breast feeding, caring for, and guarding children. This resulted in an emphasis for women on those activities, in contrast to those emphasized by men, which were compatible with this role, including preparing food and searching for vegetable foodstuffs, wild and later cultivated, and later still the preparation of textiles and clothing.

Endicott (1999) reviewed gender relations present in a large number of hunter-gatherer societies. She made it clear that there is large variation in the nature of gender roles from one society to another. One cannot assume that women always behave in a minority role or are subject to male dominance.

Other Behavioral Features

Included here are the following: circumstantial evidence favoring about a 10:1 ratio of right-handedness in hominins as opposed to a near 50:50 ratio in the great apes; and the relationship of the hominin thumb plus its musculature and tendons that make tool-making and allied delicate uses of the hand possible, which contrasts with the absence of this capability in the great apes; and the rapid growth of the newborn's skull, indicative of rapid increase in brain size initially, which requires a high-protein diet provided by 'mushy' food that has been softened by cooking. Also significant is the possible presence of language in *Homo* and possibly in *Australopithecus,* as contrasted with the great apes. Kay et al. (1998) found that the size of the hypoglossal nerve (cranial nerve XII) is significantly larger in *Homo* than in chimpanzees and gorillas, with the dimensions of *H. sapiens*, including Neandertals and various species of *Homo* of the past being in the human range, with the implication that human linguistic ability was significantly greater than that of the great apes or *Australopithecus*; this nerve is involved with the tongue, suggesting its influence on linguistic ability.

Physiological Features

Among the major physiological features are the following: the presence of a lengthy large intestine in the great apes, where much of the digestion of their predominantly herbivorous diet takes place, as contrasted with the relatively short large intestine of the more omnivorous homininds, where the bulk of the digestive process takes place in the small intestine; relative hairlessness in the hominins, as contrasted with the great apes, with hairlessness combined with abundant sweat glands and a bipedal posture that exposes far less of the body to direct sunlight, helping to maintain a lower body temperature, with hair on the head helping reduce cranial temperatures; presence of menopause combined with hominin birth at a far less mature stage than in the great apes with the result that grandmothers can assist in child rearing while the mother becomes pregnant more frequently, with an extended interval of childhood in hominins, as contrasted with the great apes.

Potts (1998) surveyed the varied evidence provided by paleoclimatology in the broadest sense, concluding that it is probably too early to subscribe to any one explanation of concerning 'key' environmental factors governing selection among the ancestors of humans. However, it needs to be emphasized that on a short time scale terrestrial animals and plants have no difficulty in tracking their favored environment(s), as is evident from the pollen record of the later Quaternary as well as from the associated animals, whereas natural selection operating on a long time-scale is certainly capable of modifying organisms to more favorably suit their changed environment(s).

Diet

Summary

The information on human diet is badly biased by preservational phenomena. The Oldowan data is almost exclusively from bones with cut marks and long bones broken to secure marrow, with little or no positive information about plant foods, small animals, insects, fruit, or nuts, since the preservational environment is dominated by the deposition of bones to the exclusion of other materials. Things improve with the Acheulian but do not really become comprehensive in regard to plant, fruit, and nut materials until the Bronze Age.

Trying to figure out the diet(s) of the earliest *Homo* is largely an inferential business. It makes sense, however, to strongly suspect or conclude that the earliest humans living in a warm low-latitude environment would certainly have taken advantage of any and all the available vegetable foods, leafy, fruit, and nut. The emphasis in the archaeological–anthropological literature is largely devoted to accounts of cut marks on prey bones (Bunn, 2007), with the inference that the earliest humans were mostly dependent on meat. But this view does not make much sense in view of the availability of various vegetable foods plus the small teeth and enamel-type characteristics of *Homo* and the dietary requirements for important plant components, including carbohydrates that yield starch. *Homo* is clearly an omnivore who has depended on relatively soft foods from the beginning, with diverse cooking methods implied. There is also the obvious fact that successful hunts are not as reliable on a daily basis as is the availability of edible plant products.

Wrangham (2009; see also 2007, and Wrangham et al., 1999 plus Wrangham and Conklin-Brittain, 2003) provided a highly attractive hypothesis based on skeletal data, including smaller lower jaws with smaller post-canine teeth (see D.E. Lieberman et al., 2004, for comment about the effects of soft foods on jaw size). Wrangham emphasizes significantly larger brains beginning with *Homo erectus*, the effects of cooking on foodstuffs, the time necessary to chew raw versus cooked foods, the physiological needs of brains versus other systems (see Aiello and Wheeler, 1995, and Aiello and Wells, 2002 for discussion), and other correlates such as the time needed to digest cooked versus raw foodstuffs, and the use of fire for cooking beginning in the Paleolithic. He pointed out that roots, corms, and tubers could be baked, roasted, or steamed, with meat roasted, or smoked, etc. Nelson (2010) and Thomas (2009) discussed the utility of stone-boiling in food-processing. Organ et al. (2011) provided data showing that food processing time, chewing, etc. increases with body size in primates, including chimpanzees and apes, except in *Homo,* where the small size of the molars correlates with much lower food processing time in the mouth. Wollstonecroft (2011) reviewed food processing possibilities used by *Homo* to make plant-food nutrients more readily available, including cooking, chopping, grating, soaking, etc.

It occurs to me that 'pure' meat eating as a way of life would require that the raw meat be cut into pieces small enough to be swallowed safely without choking, but that cooking meat would make it much easier to process into pieces small enough for safe swallowing.

Supplementary to Wrangham's (2009) conclusions is the basic human digestive anatomy (Milton, 1989, Table 3.2) featuring a relatively small-volume colon with a large-volume small intestine in contrast to the situation in the other higher primates, which correlates with the profound differences in diet – largely plant-based in the higher primates versus far more omnivorous in humans.

Aiello (1997) summarized her thinking concerning the enlargement of the human brain, done in part by reducing the relative size of the glucose-expensive digestive tract, increasing the input of meat protein, and a high-quality diet of easily digested food (cooking implied), with large group-size implied. To this one might add the consumption of protein-rich legumes from the wild, including peas and beans of various types. One might add that in addition to protein is the need for glucose, readily derived from carbohydrate-provided starch present in certain plant parts.

The relatively lengthy small intestine and short colon characteristic of *Homo,* as contrasted with the reverse in the great apes, is further evidence for omnivory in humans, as contrasted with herbivory in great apes.

Bar-Yosef (2004) emphasized the more ready availability of skeletal remains with evidence for carnivory. See Foley (2001) for maximization of carnivory as opposed to herbivory in the *Homo* diet. Vegetable foods must have played an important role in *Homo*'s diet; Bar-Yosef (2004) reviews some of the limited evidence of vegetable foods in early diets.

Hardy et al. (2015) analyzed dental calculus from Qesem Cave, Israel, associated with post-Acheulian, pre-Mousterian Lower Paleolithic *Homo* that includes soot, probably derived from smoky fires, and starch granules plus organic compounds derived from plants. All of this indicates the use of fire in a smoky environment plus important food components of plant rather than meat origin.

Women and children were certainly occupied with gathering fruits, seeds, nuts, and tubers for food while the males were hunting mammals for meat. O'Connell et al. (1999) emphasized the potentially important role that cooked underground plant storage-organs (corms, roots, tubers) may have played beginning with *Homo erectus.*

Australopithecines and Prehominins

A number of authors have made attempts to better understand the diets and associated behaviors of pre-hominin hominoids and hominins (see the following for examples: Peters, 1979; Covert and Kay, 1981; P.W. Lucas et al., 1985; Scott et al., 2005; Strait et al., 2009; Ungar et al., 2010; Constantino et al., 2010; Ungar and Sponheimer, 2011; Cerling et al., 2011; P.W. Lucas et al., 2013), but it is unclear how closely their conclusions fit the diet(s) of *Homo*, although they do seem to suggest that pre-hominin hominoids were vegetarians.

Henry et al. (2012) used plant phytoliths from *Australopithecus sediba* calculus to indicate the use of various plant foods.

Brain (1988) interpreted a number of long-bone and antler artifacts showing evidence of wear and scratches at their tips as digging sticks for obtaining vegetable foods from the Early Stone Age of Swartkrans Cave, with the inference that a robust australopithecene was involved.

Laden and Wrangham (2005) reviewed evidence that indicates the strong possibility that australopithecines were using underground plant storage-organs as fallback foods and possibly as basic foods. The evidence consists of their thicker tooth enamel than present in the great apes, of some co-occurrences with mole-rat fossils, the latter being a group that uses underground plant storage-organs as a basic food type beginning in the Miocene, and possible evidence of digging-sticks and weights.

Oldowan

Plummer (2004) reviewed the evidence of Oldowan artifacts, diets (with emphasis on meat, including marrow and brains, etc.), including plant materials, and commented on climates of the time, tool making, and tool materials.

Bunn (2007) reviewed the evidence provided by cut marks on long bones and percussion marks on broken bones to support the use of meat and marrow in the Oldowan interval.

Barham and Mitchell (2008) reviewed the evidence provided by hominins contemporary with Oldowan tools that shows from both tooth morphology and isotopic evidence from tooth enamel that these taxa were omnivorous, i.e., not chiefly herbivorous, as is characteristic of the higher anthropoids.

Lemorini et al. (2014) used use-wear on stone implements from Kanjera South, Kenya, to indicate their use for both butchery and on plant tissues, including wood, grit-covered plant tissues that might represent underground plant storage-organs, and the stems of grasses or sedges, implying a variety of plant uses.

Acheulian

Backwell and d'Errico (2001) provided evidence that bone tools were used by early hominids at the Swartkrans Cave site to forage for termites.

Goren-Inbar et al. (2002b) discussed pit-marked stone from the Acheulian of Gesher Benot Ya'aqov, Israel, associated with nut debris, concluding that they were used to break open nuts and pitted stones.

Lepre et al. (2011) dated the appearance of the Acheulian at 1.76 million years ago. Yamei (2000) described Acheulian Mode 2 technologies from South China, far south of where they had been previously thought to occur.

Middle Paleolithic

In a highly innovative paper, Sistiaga et al. (2014) demonstrate, using faecal biochemical markers preserved in 50,000-year-old Neandertal feces, that both plant and animal food was utilized at a Spanish site. Henry et al. (2011) cited evidence preserved in dental calculus of varied food items in Neandertals from Iraq and Belgium of diets that included dates, legumes, and grass seeds, and Lev et al. (2005) cited acorns, pistachio nuts, almonds, legumes, and others at Kebara Cave, Mount Carmel. Hardy et al. (2012) discussed evidence preserved in dental calculus from Neandertals in a Spanish cave with evidence of starch grains from starchy food and also from camomile and yarrow, medicinal plants.

Barton et al. (1999) cited wild olives, stone pine nuts (*Pinus pinea*), molluscs, and mammals from Neandertal sites in Gibraltar.

Richards and Trinkhaus (2009) summarized much of the literature concerning nitrogen-isotope information, indicating that Neandertals subsisted in largest part on a protein-rich meat diet, while also agreeing that the isotopic data do not indicate the role of non-meat protein in diets. Hardy (2010; see also Hather et al., 1994) pointed out that total reliance on meat by Neandertals would have led to serious dietary problems and that plant foods were certainly available to them.

Buckley et al. (2014) used biochemical evidence from starch grains on Mesolithic, pre-Mesolithic, and Neolithic teeth from Central Sudan to indicate the use of purple nutgrass (*Cyperus rotundus*) tubers as food, which in turn reduced the incidence of dental caries.

Zvelebil (1996) summarized a mass of circumstantial evidence suggesting the widespread use of plant foods in the later Mesolithic of Europe before the Neolithic.

Upper Paleolithic

Karkanas et al. (2004) described some Aurignacian clay hearths from southern Greece that were probably used in relatively low-temperature cooking of both animal and plant foods.

Revedin et al. (2010) described Upper Paleolithic evidence of starch preserved on grinding tools, indicating the possible use of varied plant sources.

Aranguren et al. (2007) described starch grains on grindstones probably derived from *Typha* and various wild grass seeds from a Gravettian site in Italy dated approximately 25,000 years BP.

Close and Wendorf (1990) reviewed the various foods known from 18,000 years BP in the Mahgreb and the Nile Valley, including terrestrial snails in the Mahgreb and catfish plus various wild plant types in the Nile Valley as well as *Unio*.

Piperno et al. (2004) described an Upper Paleolithic Israeli site with a grinding stone from which starch grains of wild barley (*Hordeum*) and other grasses were recovered, with the possible presence of a primitive oven used for baking, with an age of about 20,000 years BP.

Natufian

Bar-Yosef and Valla (eds., 1991) discussed plant and animal foods derived from hunting and gathering from the Natufian.

Weiss et al. (2008) described an Epipaleolithic Israeli brush-hut site with a concentration of plant food seeds and a grinding stone at one side of the hut.

Mesolithic

Buckley et al. (2014) used biochemical evidence from starch grains on Mesolithic, pre-Mesolithic, and Neolithic teeth from Central Sudan to indicate the use of purple nutgrass (*Cyperus rotundus*) tubers as food, which in turn reduced the incidence of dental caries.

Zvelebil (1996) summarized a mass of circumstantial evidence suggesting the widespread use of plant-foods in the later Mesolithic of Europe prior to the Neolithic.

Hansen (1991) described Late Mesolithic and Neolithic botanical remains from Franchthi Cave in the Argolid, emphasizing the presence of abundant wild fruits, nuts, and grain seeds in the Late Mesolithic, but no evidence of domestication during the interval from 25,000 to 13,000 years BP, while in the interval from 13,000 to 11,000 years BP there is evidence for use of wild cereals, legumes, and nuts. By 8,000 years BP domesticated emmer wheat, barley, and lentils appear, with grain seeds abundant together with the remains of domesticated sheep and goats; the inference is that domesticated crops in the Near East by 9,000 years BP was followed by later spread to southeastern Europe.

Neolithic

Miller (1991) provided a useful site-by-site summary of many of the Near Eastern localities with evidences of plant cultivation, taxon by taxon, for the Neolithic and later intervals, and Kroll (1991) did the same for southeastern Europe, Küster (1991) for central Europe south of the Danube, Knörzer (1991) for Germany north of the Danube, Wasylikowa et al. (1991) for East-Central Europe, Hopf (1991) for South and Southwestern Europe, Bekels (1991) for Western Europe, Greig (1991) for Britain, and Jensen (1991) for the Nordic countries; these papers provide a very comprehensive view of Neolithic and younger crop presences, even including *Cannabis* from a few regions.

Janouchevitch and Markevitch (1971) provided a brief summary of evidence dealing with grain, legumes, and fruit from the Neolithic of the region from the Bug River to the Prut River.

Costantini (1989) discussed the change from the Sicilian Mesolithic to the Neolithic in terms of the incoming of cultivated grains and other foods.

Close (1995) discussed the widespread use of wild grass seeds across the Sahara in the very early Neolithic.

Darby et al. (1976) provided an extensive discussion dealing with Egyptian wheats and barley during historical time.

Tegel et al. (2012) listed the dietary items identified from an Early Neolithic German water well, including einkorn and emmer wheat, peas, lentils, linseed, poppy, strawberry, sloe (blackthorn), apples, raspberries, hazelnuts, bladder cherry, and black henbane.

Food Processing

Querns, Grinders, and Grindstones

Kraybill (1977) reviewed much of what is known concerning grinding and pounding with stones ('mortar and pestle, etc.'), while pointing out that they undoubtedly were used by at least Early Paleolithic time, with the implication that their ubiquity at many sites suggests the overall importance of plant foods in pre-agricultural diets.

Kraybill (2011; see also Watts, 2014) reviewed much of what is known concerning grinding and pounding with stones ('mortar and pestle, saddle querns, rotary querns, millstones [including single millstones against which the object to be ground is rubbed, abraded; see Runnels, 1985, for Late Neolithic and Early Bronze Age Aegean examples], etc.'), while pointing out that they undoubtedly were used very early, by at least Early Paleolithic time, with the implication that their ubiquity at many sites suggests the overall importance of plant foods in pre-agricultural diets. Curtis (2001) dealt with Egyptian grinding, milling, and sieving of grain and Near Eastern grain milling. Barker (2006, p. 74–75) emphasized the fact that both hunter-gatherers and agriculturalists use querns for either wild seeds or domesticated ones. Watts (2014b) discussed saddle querns and rotary querns in some detail, followed by an extensive account of British querns through time. Watts (2014) pointed out that saddle querns appear in the earlier Neolithic, followed by rotary querns beginning in the Iron Age. Martinez, 2002, documented a Spanish rotary quern from the Fifth Century BC. Dubreuil (2004) described experiments she conducted using Natufian grinding slabs with various plant and mineral substances, comparing her results with actual Natufian grinding slabs that indicated significant differences between the various substances being ground. It needs to be kept in mind that the mortar and pestle crush the material whereas querns grind it! De Beaune (2002) illustrated and discussed Mousterian, Upper Paleolithic, and Natufian querns and mortar and pestle.

Molleson (2000, 1989) discussed the skeletal changes in females caused by continual use of the saddle quern for grinding grain.

Frankel (2003) described the Olynthus Mill of Classical Greece, beginning in about the Fifth Century BC, which spread to Greek colonies elsewhere in more coastal Mediterranean regions and into the Black Sea region; this mill was more effective than the saddle quern but less effective than the rotary quern that supplanted it. (Storck and Teague, 1952, Figure 37, provided an excellent account of this mill type, which they call a 'lever mill', and also discussed querns in some detail; see also Curtis, 2001.) Williams-Thorpe and Thorpe (1993) refer to these mills as 'hopper-rubbers'; they reviewed the record of saddle querns and younger types from the Mediterranean region Neolithic to Roman periods and commented on trade in these millstones. Curtis (2001) described the rotary hand quern of the time just before the Common Era and shortly after, followed by the animal-driven rotary mills, before the advent of more modern types. Wikander (2000, p. 371–400) discussed water mills from Roman times and later. [Paragraph reconstructed from the author's notes – Eds.]

Henshilwood et al. (2011) described a unique abalone shell with associated red ocher, a probable grindstone, and other materials, suggesting that they represent 100,000-year-old South African Middle Stone Age remains of 'paint' that also included 'oily' material derived from marrow and bone fat. Van Peer et al. (2003) identified ocher remains on possible grinding stones from the Early to Middle Stone Age of Sai Island, northern Sudan.

Cut Marks Made by Humans on Bone

McPherron et al. (2010) discussed evidence for stone-tool cut marks on bone dated at 3.39 million years ago at Dikika, Ethiopia, suggesting that *Australopithecus* was most likely responsible, since *H. erectus* does not occur this early in time; no stone tools are associated with this find, although they were concluded to have been used here. However, in the absence of any hominid remains one can suggest that an austropithecine was involved here, relying on the basis of known biostratigraphy OR that this represents an earlier occurrence of *H. erectus* than known elsewhere.

Potts and Shipman (1981) described various cut-mark types from Olduvai Gorge, Tanzania, indicating that Oldowan-type tools were used to process carcasses for their meat. Domínguez-Rodrigo and Barba (2006) showed that hominid meat-processing of Oldowan age at one Tanzanian locality, at least, did not involve scavenging.

There is extensive literature, evidence, and comment on mammal bones with cut marks made by stone tools. This information almost leads to the conclusion that early *Homo* depended largely on hunting for the protein part of the diet. Carmody et al. (2011) made the point that cooking of food increases its energy gain as well as palatability and edibility (see first paragraph of the Introductory paragraph on human behavior for comments on Wrangham, 2009).

Domínguez-Rodrigo (2002; see also Pickering and Domínguez-Rodrigo, 2006) reviewed the evidence, pro and con, about early human meat scavenging versus hunting.

Dominguez-Rodrigo et al. (2005) described Late Pliocene cut marks on ungulate bone that were undoubtedly made by humans of some kind, using sharp tools to remove flesh. Milo (1998) described very convincing cut marks on bovid bones from the Middle Stone Age Klasies River Mouth, South Africa. Hoffecker et al. (2010; see also Holliday, 2010) described an Upper Paleolithic Russian site, Kostenki, with abundant evidence of cut marks butchery and breaking of long bones for marrow.

Sources of Food

There is extensive information about the use of large mammals for food, but much less on the use of lagomorphs and other small mammals, birds, tortoises, fish and shellfish, and insects. Stiner et al. (2000) provided an account of the use of small game in the Paleolithic of the Mediterranean region that helps to partly fill this gap in our understanding.

Cauvin (2000) reviewed the origins of agriculture in the Levant and its subsequent spread to adjacent regions, including Cyprus, in some detail, including the initial presence of rectangular dwellings in contrast to the earlier Natufian and Khiamenian circular types.

Evidence for Cultivation

The basic reference for the presence of cultivated plants used for food and other purposes from West Asia, Europe, and the Nile Valley is the scholarly compilation by Zohary and Hopf (1993) of their modern, wild, and pre-domestication ranges, and archaeological records. For 'soft-bodied' plants such as those of the cabbage family or even garlic, the archaeological record is very limited, chiefly using such things as Egyptian and Sumerian records, whereas for more 'hard' types such as the pits of date palms, the record is more rewarding. However, Evershed et al. (1991) showed that some leafy vegetables, such as members of the cabbage family, can be identified by using epicuticular wax components preserved in potsherds. Many records, particularly for various grains and legumes, go back into the Neogene (23,000 years BP) but none earlier.

Chang (1986, p. 80, Table 5) listed the principal Chinese cultivated plants.

Hancock (2012) provided a view of crop evolution from the molecular perspective that importantly supplements and extends the work of Zohary and Hopf (1993) and includes taxa with which they did not deal.

Evans (1964, Plate IV, p. 265) briefly discussed Minoan saffron harvesting and gardens on Crete.

Zohary and Hopf's (1993) examples include the following:

wheat (*Triticum*), 8000 BC, Near East
barley (*Hordeum*), 8000 BC, Near East
rye (*Secale*), Early Neolithic, Near East
oat (*Avena*), 2000 or 1000 BC, Europe
broomcorn millet (*Panicum*), 5000 BC, Georgia
foxtail millet (*Setaria*), 6000 BC, North China
sorghum (*Sorghum*), 2000 BC, Oman
rice (*Oryza*), 3000 BC, India
pulses; pea (*Pisum*), 7000 BC, Near East
lentil (*Lens*), 7000 BC, Near East
chick pea (*Cicer*), 7000 BC(?), Near East
fava bean (*Vicia faba*), 6800–6500 BC, Near East
bitter vetch (*Vicia ervilia*), Early Neolithic, Near East

grass pea (*Lathyrus*), probably Early Neolithic, Near East or Southeastern Europe
oil & fiber; flax (*Linum*), 7000 BC, Near East
hemp (*Cannabis*), 2500 BC, China; 1000 BC Anatolia, southern Russia
narcotic properties, 1000 BC, India
Old World cottons (*Gossypium*), 2000 BC, Pakistan
poppy (*Papaver*), Middle and Late Neolithic, Europe (oil seeds)
false flax (*Camelina*), 3000–2000 BC, Europe
other cruciferous oil crops (Cruciferae), 3000 BC, Iraq
Sesame (*Sesamum*), 1000 BC or earlier, India, Near East
fruit trees and nuts; olive (*Olea*), 3700–3500 BC, Near East
grape vine (*Vitis vinifera*), 4000 BC, Levant
fig (*Ficus carica*), 4000 BC, Levant
Sycamore fig (*Ficus sycomorus*), 3000 BC, Egypt
date palm (*Phoenix dactylifera*), 4000 BC, Mesopotamia
pomegranate (*Punica*), 3000 BC, Near East
apple (*Malus*), 1000 BC, Levant
pear (*Pyrus*), 500 BC, Europe
plum (*Prunus domesticus*), BC–A.D., Europe
cherries (*Prunus avium* and *cerasus*), 100 BC, Turkey
apricot (*Armeniaca vulgaris*), 100 BC, Iran or Armenia
peach (*Persica vulgaris*), 2000 BC, China
quince (*Cydonia vulgaris*), BC–A.D., Mediterrean Basin
citron (*Citron medica*), 1200 BC, Cyprus
almond (*Amgdalus communis*), 3000 BC, Levant
walnut (*Juglans regia*), 3000 BC, NE Turkey, Caucasus, Iran
chestnut (*Castanea sativa*), uncertain records
hazels (*Corylus avellana* and *maxima*), BC–A.D., widely distributed
pistachio (*Pistacia vera*), BC–A.D., Near East
vegetables and tubers; watermelon (*Citrullus lanatus*), 2000 BC, Egypt
melon (*Cucumis melo*), 3000 BC, Egypt
leek (*Allium porrum*), 2000 BC Mesopotamia
garlic (*Allium sativum*), 2000 BC, Levant
onion (*Allium cepa*), 2000 BC, Egypt
lettuce (*Lactuca sativa*), 2000 BC, Egypt
chufa or rush nut (*Cyperus esculentus*), 3000 BC, Egypt
cabbage (*Brassica oleracea*), BC–A.D., Europe
beet (*Beta vulgaris*), 1000 BC, Levant
condiments; coriander (*Coriandrum sativum*), 2000 BC, Levant
cumin (*Cuminum cyminum*) and dill (*Anethum graveolens*), 2000 BC, Syria
black cumin (*Nigella sativa*), 1000 BC Levant, Egypt
saffron (*Crocus sativua*), BC–A.D., Levant
dye crops; woad (*Isatis tinctoria*), uncertain
Dyer's rocket (*Reseda luteola*), 1000 BC, Mediterranean Basin and SW Asia
madder (*Rubia tinctorum*), 1000 BC, Egypt
true indigo (*Indigofera tinctoria*), uncertain, India, near East
safflower (*Carthamus tinctorius*), 2000 BC, Egypt

Kislev et al. (2006) described what they conclude are the remains of domesticated figs of the Early Neolithic, approximately 11,000 years BP, from the Jordan Valley, a very early occurrence.

Stöllner et al. (2003) cited caraway and aniseed in fecal remains from the Iron Age Dürrnberg-bei-Hallein salt mines, as well as poppy seed, 'gold-of-pleasure' (*Camelina sativa*), and flax.

Grain and Legumes

Rice culture in Southern and Southeast Asia (Glover and Higham, 1996) dates from the end of the Third Millennium BC in India, while in Southeast Asia it dates back to 8000 years BP. Chang (1989) tabulated records for cultivation in southern Asia as far back as 7000 years BP, and inferred dates between 10,000 and 15,000 years BP. Fuller and Ling Qin (2010) considered the question of rice domestication and the disappearance of oaks for acorns at about 6000 BC in the Lower Yangtze region. Fuller et al. (2007) concluded that wild rice and other foods were collected by foragers, following which pre-domestication cultivation began about 5000 BC and rice domestication about 4000 BC. Sharma (1985) cited cultivated rice in India during the 'earliest Neolithic', but see Glover and Higham (1996, pp. 416, 419) for critical review.

Postgate and Powell (1984) provided an introduction to cereal cultivation in Sumer that is of interest for the Bronze Age.

Barton et al. (2009) suggested several centers of agriculture beginnings in China, with a southern one centered on rice and a northern one centered on millets, with a beginning at about 7000 years BP. Zhang and Kuen (2005) cited short-grain, japonica rice from Central China in the Early Neolithic. Bellwood et al. (1992) concluded that rice culture in the Yangtze Valley began by 6000 BC and much later elsewhere in Southeast Asia and India.

However, Fuller et al. (2007) suggested that the Lower Yangtze rice culture really began only about 4000 BC but that pre-domestication began about 5000 BC, with the latter associated with other 'subsistence foods' such as nuts, acorns, and water chestnuts, with hunter-gatherers using wild rice even earlier.

An Zhimin (1989) briefly summarized the presence of millet in North China. Lu (1999) emphasized that North China millet-growing was well established by 6000 BC, and probably originated during the preceding two millennia.

Glover and Higham (1996) reviewed evidence for rice occurrence from South, Southeast, and East Asia, with the oldest dated 17,000–19,000 years BP (their p. 421), i.e., Late Paleolithic; from Central China from 5000–3900 BC (their pp. 426, 435 and Table 23.2) for cultivated rice in general; and 8000 years BP for the Yangtzi River region. [Reconstructed text – Eds.]

Yan (1992) summarized the information about the domestication of millets in North China and of rice in South China during the Neolithic, and also commented on the probable presence of domesticated pigs.

Darby et al. (1976) detailed the evidence for the presence in ancient Egypt of various vegetables and fruits.

Vigne (2008) cited grain production at 10,000 to 9300 BC. Cucchi and Vigne (2006) discussed the commensal relation of the house mouse with grain storage beginning in the pre-agricultural Natufian and extending from the pre-ceramic Early Neolithic to the present, and also discussed the introduction of the house mouse to Mediterraean islands by humans. Peters et al. (2005, Figure 5) reviewed the evidence for Early Neolithic (Pre-Pottery Neolithic A, 'PPNA') plant domestication of grain and pulses in the upper Euphrates–Tigris Basin. The presence of numerous mortars and pestles, as at some Natufian sites (Edwards, 1991; see also Bar-Yosef and Vall (eds., 1991), is consistent with the presence of grain. Van Zeist and Casparie (1968) described wild einkorn wheat and barley seeds from an Early Neolithic northern Syrian site associated with rectangular dwellings. Haaland (1992) discussed the very late appearance of domesticated sorghum in the Nile Valley, emphasizing that the appearance of the Neolithic way of life varied from place to place.

Kislev and Bar-Yosef (1988) surveyed occurrences of legumes in the Near Eastern Early Neolithic and discussed the possibilities for domestication.

Henry et al. (2009) showed that starch grains subjected to cooking could still be identified to species in many cases, particularly if the cooking period was minimal; this approach could be used to identify species from dental calculus, stone tools, and pottery.

Bread and Noodles

Währen (1989) dealt with the history of bread and baked goods from the Neolithic to Roman times in Europe, while Robinson (1999) reported carbon-14-dated 5500-year-old barley bread from Oxfordshire, with the earliest breads probably being unleavened flatbreads baked on stones. Geller (1993) described evidence of Fourth Millennium BC bread, and beer made from it, in Egypt, and Egyptian bread making in general, including ovens. Darby et al. (1976; see also Curtis, 2001, for an extended description of bread making) provided an extended discussion of various ancient Egyptian breads, Near Eastern breads, including Sumerian. Behre (1991) briefly cited First Millennium BC bread from Ipweger Moor, northwestern Germany.

Haaland (2007) made the important point that ovens used for bread-making appear in the Near East in the later Pre-Pottery Neolithic (Pre-Pottery Neolithic B, 'PPNB'). He also pointed out that in the Near East there were grains being used containing the gluten needed for bread making, while in North Africa at this time the local millets and sorghums lack gluten, hence no possibility for bread making (flatbreads excluded). Curtis (2001) dealt with Near Eastern ovens from the Third Millennium BC.

Lu Houyuan et al. (2005) described a unique occurrence of Late Neolithic millet noodles from Northwestern China, clearly a minimum age.

'Roots, Tubers, Leaves' [The author's text for this section was not found – Eds.]

Cooking

Nelson (2010) and Thomas (2009) made a very positive case that 'stone boiling', done by dropping fire-heated stones into bark, basketry, skin, and paunch containers, including those positioned in shallow pits in the ground, are entirely adequate for the boiling needed to make soups and stews! Nakazawa et al. (2009) proposed that abundant fire-cracked rocks at an Early Magdalenian Cantabrian site were cracked after being dropped into the water used for stone-boiling; this possibility is interesting but needs further support. Ragir (2000) pointed out that fire, i.e., baking, tends to detoxify underground plant organs and make them suitable for human food.

Hodder (2006, p. 120–121) indicated that in the absence of suitable stones near his Early Neolithic Turkish site, clay-ball aggregates, some of which were cracked, may have been used as 'stones' heated in an oven then raked out, covered with leaves on top of which food for cooking was placed, followed in turn by another layer of leaves with the food being steamed in this manner. Atalay (2006) discussed the various possible uses of the Çatalhöyük clay balls in detail, and considered their use as heated 'stones' placed in skins or baskets filled with water and raw foods for cooking, as well as for parching of grains, etc., and Hodder (ed., 2007) discussed many of the hemispherical ovens and also clay-ball occurrences.

Molleson et al. (1993) made the point, using data from Abu Hureyra in northern Syria, that there was much less tooth wear by consumers of cooked food, as contrasted with earlier consumers of raw food, seeds in particular.

Gathering and Processing Fruits, Seeds, and other Edible Plant Parts

Acheulian

Goren-Inbar et al. (2000, 2002, 2002b) listed and described edible fruits and nuts, with evidence of nut cracking from pitted hammers and anvils from the Acheulian of Gesher Benot Ya'aqov, Israel.

Middle Pleistocene

Goren-Inbar et al. (2002) described reliable early Middle Pleistocene evidence for the use of hammer and anvil stones for cracking the shells of 'nuts' of various kinds from an Israeli site, and reviewed similar evidence from elsewhere.

Lev et al. (2005) described seeds and fruits used by late Mousterians from the Kebara Cave, Mt. Carmel.

McLaren (1992) discussed wild plum pits from a Mousterian Neandertal site in Syria, and Akazawa (1987) discussed hackberry (*Celtis*) seeds from the same site, a hearth.

Opperman and Heydenrych (1990) cited corm remains from a Middle Stone Age cave in the Northeastern Cape, South Africa. Van Peer et al. (2003) identified starch remains on possible grinding stones from the Early to Middle Stone Age of Sai Island, in the Nile, in northern Sudan.

Late Paleolithic

Hillman et al. (1989, 1989b) surveyed remains of edible plants from the Late Paleolithic of Wadi Kubbaniya, Egypt, while providing a provocative account that could suggest extensive earlier human use of plant foods.

Revedin et al. (2010) described starch grains from European (Italian, Czech, Russian) Gravettian pestles and mortars indicating the processing for flour of various wild plant materials.

Close and Wendorf (1990) cited the various plants used in the Nile Valley at 18,000 BP.

Fagan and Van Noten (1971) cited the various edible plants present at Gwisho, Zambia, during the Late Stone Age.

Mesolithic

Kajale (1989) described the occurrence in Sri Lanka of the remains of wild breadfruit, wild banana, and *Canarium* nuts in the Late Mesolithic.

Natufian

Kislev et al. (1992; see also Kislev et al., 2002) commented on the abundant remains of wild fruit and cereal at an Epipaleolithic site on the Sea of Galilee. Hillman et al. (1989a) recovered fruits and seeds from a Syrian Epipaleolithic site indicating pre-agricultural occupation at 8000 to 7000 BC, all suggesting a pre-domestication origin.

Neolithic and Bronze Age

Galili et al. (1997; see also Breton, 2006, for 6000 BP examples; Curtis, 2001, for the Near East) discussed the evidence for the production of olive oil, probably from wild olives, in the Late Neolithic–Early Chalcolithic Seventh Millennium BC of Israel; this is the oldest certain evidence for the production of olive oil. Curtis (2001) reviewed the production of olive oil in Classical Antiquity. Zohary and Spiegel-Roy (1975) reviewed Near Eastern Chalcolithic presence of olive stones, Late Neolithic figs and dates, Bronze Age pomegranates, and Bronze Age cultivated grapes in the Near East and the Aegean region. Kislev et al. (2006) found cultivated figs in the Jordan Valley from the Early Neolithic (11,400–11,200 years BP). Darby et al. (1976) discussed various spices used by the ancient Egyptians.

Mbida et al. (2006; see also Mbida et al., 2004; de Langhe et al., 2006) reviewed the evidence for cultivated bananas and plantains going back 4000 years BP from Cameroon, West Africa, based on phytoliths preserved at cultural sites. Lejju et al. (2006) discussed banana phytoliths from cultivated sites in Uganda from the Sixth Millennium BC. Neumann and Hildebrand (2009) critically reviewed the evidence and indicate that the Uganda occurrence is very questionable, and that the Cameroon occurrence is possible but needs more work. Bananas were introduced from tropical southern Asia in one way or another. Denham et al. (2004) discussed evidence of cultivated bananas from the New Guinea Highlands at about 6000 years BP.

Barham and Mitchell (2008, Chapter 9) reviewed the evidence of various Mid-Holocene and Neolithic agricultural practices in Africa, noting their patchy nature and common association with hunter-gathering activities as well as the different crops being raised – a very complex story in detail.

Hunting and Fishing

Bettinger (2009) provided a series of models dealing with how hunter-gatherers make decisions about which prey and resource to pursue in order to maximize resource success, such as when to switch while hunting from one prey species to another.

D'Errico et al. (2012; see also Fagan and Van Noten, 1971, for the Late Stone Age at Gwisho, Zambia) provided South African evidence for the use of poison-tipped weapons approximately 24,000 years ago.

Marshall (1986) stressed the importance, when trying to decide between evidence of scavenging versus hunting, of comparing the bone modification patterns on small bovids versus large bovids, while concluding that cut marks from Olduvai are not distinct from Neolithic cut marks.

Bramble and Lieberman (2004; see also Lieberman et al., 2007) made the important point that *Homo* is capable of long-distance 'endurance running' in 'persistence hunting' that enables the hunting down by tiring ungulate prey to the point of hypothermia when they collapse and can then easily be killed (if not already tired to the point of death) and dismembered; they suggested that this character might have been present in *Homo erectus*. Steudel-Numbers and Wall-Scheffler (2009) raised some questions about this hunting possibility, used by southern African Bushman until recently, and possibly still used. Pickering and Bunn (2007) emphasized that persistence hunting requires a high level of tracking capability.

Animal

Leonard and Robertson (1997) considered the relationship between diets and activity for *Homo* and *Australopithecus* as well as for the chimpanzee, and concluded that the probable heightened activity level for *Homo* correlated with a diet including a lot of meat as contrasted with the more vegetarian diet of the others.

Mammal

Oldowan

Pickering and Domínguez-Rodrigo (2006) reviewed the evidence for Oldowan-age hunting and use of large-mammal carcasses as contrasted with mere scavenging, using the evidence of cut marks on long bones.

Acheulian

Rabinovich et al. (2008; see also Bar-Oz and Munro, 2007, and Munro and Bar-Oz, 2005, for the preference for marrow from the long bones) provided a detailed account of Acheulian, early middle Pleistocene butchery, involving a number of steps, with the implication that it represents hominid technology extending much farther back in time, i.e., very conservative. Goren-Inbar et al. (1994) discussed a butchered elephant skull from the Israeli Acheulian that might indicate the use of the brain as food.

Movius (1950) described a German Acheulian sharpened yew hunting spear found between the ribs of a straight-tusked elephant. Jakob-Friesen (1956) described this Acheulian mammoth-find from northwestern Germany with accompanying flint artifacts that may have been involved in the hunting and butchering of the animal. Thieme and Veil (1985) provided details about the locality, the hunting spear, and the flint artifacts associated with the mammoth, of Eemian age, during the last interglacial. Dennell (1997) and Thieme (1997) discussed the technological–cultural significance of a wooden throwing spear from a 400,000-year-old German Paleolithic locality and its implications for the presence of an advanced hunting technology far earlier than previously assumed. Thieme (2005) described the occurrence at Schöningen, Lower Saxony, of a number of wooden hunting spears with the skeletons of numerous horses and a few hearths plus one stick that might have been a spit over which horse meat was smoked.

Middle Paleolithic

Klein and Cruz-Uribe (1996) discussed the evidence of coastal seals and of bovids from Middle and later Stone Age South African sites.

Boëda et al. (1999) described the presence of a Levallois point embedded in the cervical vertebrae of a wild ass from the Mousterian of Syria. Milo (1998) described Middle Stone Age South African bovid remains, including even that of a large and dangerous extinct buffalo with an embedded stone-weapon point, and widespread evidence of butchering that involved a group of individuals.

Berger and Trinkaus (1995) reviewed evidence for a high incidence of Neandertal injuries that might have resulted from the hunt.

Speth and Tchernov (1998) made a good case for Neandertals from Kebara Cave, Israel, having hunted large, 'dangerous' prey.

Mania et al. (1990) reviewed the evidence of hunting for large Middle Paleolithic mammals in Eastern Germany, providing numerous details.

Scott (1980) documented evidence from a headland cliff site on Jersey replete with Middle Paleolithic skeletons of mammoth and wooly rhinoceros and an age-structure indicating that a large group of female mammoths and their young had been driven over a cliff, where they died and were then butchered at a time Jersey was connected to the European continent.

Upper Paleolithic

Soffer (1985) discussed various Russian Plain Upper Paleolithic ungulate remains, including abundant mammoth materials.

Pringle (1997) discussed the presence of various net-weaving types from the Gravettian of Dolni Vestonice that were probably parts of nets used to trap small mammals, whose bones are abundant at this site.

Straus (1993) made the point that Upper Paleolithic Europeans took advantage of topographic 'traps' to successfully hunt groups of herd animals.

Soffer (1985) discussed the widespread animal bones preserved at Central Russian Plain Paleolithic hunting(?) sites.

Natufian

Lieberman (1993) discusses seasonal mobility and migratory hunting versus more sedentary hunting from a specific site, with the Natufians providing a good example of the latter.

Neolithic and Late Paleolithic

Simmons (1998, 1999; see also Reese, 1996) summarized information from Cyprus suggesting that Neolithic hunters, possibly in the Tenth Millennium BC, exterminated the endemic pygmy hippopotamus and pygmy elephants. Peters et al. (1999) reviewed the distribution through time of the Early Neolithic pre-ceramic interval of animal husbandry in the Northern Levant.

Legge and Rowley-Conway (1987) described the presence of 'walled' traps ('Desert Kites') used in the Late Paleolithic and Early Neolithic hunting of gazelle herds in the Near East.

Invertebrates, Chiefly Shellfish

Middle Stone Age

Marean et al. (2007; see also Thackeray, 1988, for data from several sites) discussed the Middle Stone Age use of shellfish at a South African Middle Pleistocene coastal site, and Klein et al. (2004) did the same for a Middle Stone Age and Late Stone Age site in the Western Cape with information as well about tortoises and some mammals, but little on fish. Singer and Wymer (1982) gave detailed information about the extensive Mesolithic shell middens, featuring intertidal shelled invertebrates plus evidence of a few vertebrates, including probable whale and dolphin carcass material, which had probably washed ashore.

Parkington (2001) emphasized the importance of shellfish in Middle Stone Age South African diets, and speculated that women were possibly the chief 'collectors' and that the nutritive value of shellfish materially aided in pregnancy and subsequent lactation, as contrasted with similar nutritive value from mammalian prey hunted by the males. He also added some information about coeval Mediterranean shellfish consumption.

Brink and Deacon (1982; see also Klein et al., 2004, above) described a Middle Stone Age shell midden with marine shells from South Africa. Thackeray (1988) described shell middens from the Middle Paleolithic of South Africa. Voigt (1982) provided a fairly comprehensive view of molluscan usage from the intertidal zone, based on remains found in nearby caves of various ages in the Klasies River area, with inferences concerning different water temperatures, the overall message being that molluscan resources were heavily utilized during an extended period of time.

Upper Paleolithic

Kuhn et al. (2001) described nearshore mollusks used as food from the earliest Upper Paleolithic of the Levant.

Neolithic

Close and Wendorf (1990) cited shell middens containing *Helix* from the Mahgreb at 18,000 years BP. Pearl oyster middens of the Third Millennium BC were cited from Bahrain by Donkin (1998).

Fish

Lower Paleolithic

Stewart (1994) described the various occurrences of fish remains at Olduvai Gorge, with some showing marks of cutting by humans.

Middle Stone Age

Yellen et al. (1995) and Brooks et al. (1995) documented an abundance of fish remains as well as barbed bone points that were probably used by *Homo sapiens* to catch large catfish from shallow waters at Katanda (Upper Semliki Valley, eastern Zaire) during the Middle Stone Age. Robbins et al. (1994, 2000) described evidence of Middle Stone Age fish remains and barbed bone points from Botswana.

Muñoz and Casadevall (1997) discussed the remains of freshwater fish from a cave in northeastern Spain, with dates extending from the Mousterian through much of the Upper Paleolithic.

Bonsall et al. (1997) emphasized the abundance of freshwater fish at the Iron Gates on the Danube in both the later Mesolithic and the Neolithic, although remains of many mammals, particularly red deer and wild pig, are also present in varying abundances. Cook et al. (2001) used nitrogen-isotope evidence to indicate the importance of fish in their diets, with a tapering off in the Neolithic, probably owing to greater use of cereals.

Late Stone Age

Klein et al. (2004) discussed the abundance of fish at the Late Stone Age level of the Ysterfontein 1 Middle Stone Age site with molluscs and ostrich shell far more abundant from the Middle Stone Age materials.

Close and Wendorf (1990) reviewed the various foods known from 18,000 BP in the Mahgreb and Nile Valley, including terrestrial snails in the Mahgreb and catfish plus various wild plant types in the Nile Valley as well as *Unio*. Greenwood (1968) described some freshwater Nubian fish remains with an age of almost 21,000 years BP.

Robbins (1974) described various bone harpoons and barbless bone 'points' from a 'Late Stone Age' fishing settlement in the Lake Rudolf [Turkana] Basin, Kenya.

Natufian

Kislev et al. (1992) commented on the abundance of fish-bone remains and also animal remains at an Epipaleolithic site on the Sea of Galilee. Nadel (ed., 2002) discussed an Epipaleolithic site, Ohalo II, on the shore of the Sea of Galilee, with brush huts for living quarters and possibly for food storage, with abundant evidence of fish remains, also reporting string possibly used for nets and/or fish traps, and 'sinkers' for possibly anchoring the nets or traps, as well as the remains of terrestrial plants and prey animals of various food types.

Some of the papers in Bonsall (ed., 1985 [1989]) illustrated bone and antler barbed points that were used for fishing.

Neolithic

Powell (2003; see also Sampson, 1998, for fish hooks from here) discussed abundant fish remains at the Cave of the Cyclops in the Cyclades, eastern Aegean, Mesolithic, of Eighth to Ninth Millennia BP in age. Garfinkel et al. (2002) commented extensively about remains of marine fish from a Neolithic Israeli Pottery site. Fish hooks made of gold have been found in the Mekong.

Bronze Age

Marshall (1931) dealt with fish hooks from Mohenjo-Daro.

Birds

Hardy et al. (2001) emphasized the presence of bird remains at two Paleolithic Crimean sites. Cassoli and Tagliacozzo (1997) described numerous bird bones from an Italian Epigravattian site that showed cut marks and evidence of burning.

Honey and Beeswax

It is easy to assume that honey and beeswax from wild bees were obtained by humans from a very early stage in their history, but proof of their existence is difficult to obtain. Presumably wild-bee honey was collected by humans from their first appearance in the Lower Paleolithic, although positive evidence is lacking. Regert et al. (2001) used biochemical methods to demonstrate the presence of beeswax from Neolithic (6899–6700 or 5450–5350 years BC) Macedonian ceramics (obviously, a minimum age), and reviewed the identification of beeswax biochemically associated with various archaeological materials, chiefly ceramics. Heron et al. (1994) provided similar evidence from a Neolithic Bavarian potsherd locality dated 3700–3340 BC. Needham and Evans (1987) provided biochemical evidence of honey on the surfaces of

Neolithic British potsherds. Crane (1983) provided a comprehensive account of various types of beehives from the present, plus a brief account of archaeological evidence, the oldest from the Altamira cave paintings, which would suggest an Upper Paleolithic age for collecting wild-bee honey. Crane and Graham (1985; see also Kardara, 1961, p. 265) discussed the beehive evidence from Classical Antiquity. Mazar et al. (2008) described Tenth Century BC Iron Age beehives from the Jordan Valley, which were identified by means of both morphological and biochemical evidence of beeswax residues. Crane (1999, pp. 2, 171; see also his Table 20.3A, p. 163; see also Curtis, 2001) cited horizontal beehives from Egypt between 5000 and 3000 BC as well as horizontal pottery hives from the Eastern Mediterranean (Table 23.2A) from Classical Antiquity. Crane (1999, p. 530) cited the use of beeswax as early as 3500 and 3000 BC in the lost-wax casting process. Higham (1996) cited the use of wax in the lost-wax casting process for bronze articles from the Bronze Age of Southeast Asia.

Alcoholic Beverages, Including Wine and Beer Plus Various Additives

Summary

It is probable that the production of alcoholic beverages was largely associated with the first appearance(s) of containers, chiefly ceramic. However, it is also reasonable to suspect that earlier humans from the beginning may have appreciated the pleasures of alcohol induced by eating fermented fruits of one sort or another right off the plants or where they had dropped to the ground.

McGovern (2009) provided a very useful, relatively comprehensive account of the distribution and characteristics of alcoholic beverages through time, together with information about added flavoring materials, hallucinogenic additives, etc.

McGovern et al. (2005, 2004) reviewed the evidence for the presence of alcoholic beverages in China from the Early Neolithic and later, using both biochemical analysis of residues from ancient pottery and morphological interpretation of types of pottery and bronze vessels.

Wine

It is important to know that the earliest evidence of wine is based on chemical analysis of residues remaining on or in pottery remains, with the caveat that before the presence of pottery, wine use would probably remain unrecognized. McGovern et al. (1996; see also Berkowitz, 1996, 7000 years BP) provided reliable chemical evidence for the presence of resinated wine in a Neolithic pottery jar from the Zagros; this is the oldest evidence of wine, but it is obviously a minimum age. Badler (1996; see also Michel et al., 1993) discussed biochemical evidence of wine and the morphologies of ceramic jars from Godin Tepe, Iran, dated to the Fourth Millennium BC.

Darby et al. (1976; see also Nicholson and Shaw, 2000; also Curtis, 2001) dealt with ancient Egyptian wine making and consumption, and also discussed date wine, palm wine, pomegranate wine, and fig wine. Curtis (2001) dealt with Near Eastern and Hellenistic wine making during the Third Millennium BC.

Valamoti et al. (2007) described evidence for the use of wild grapes and possible wine cups from Fifth Millennium BC northern Greece. Zhang and Kuen (2005) cited biochemical evidence from Central China for the presence of wine or beer in the Early Neolithic, approximately 9000 years BP.

McGovern, Fleming, and Katz (eds., 1996) surveyed the evidence for wine, wine making, wine drinking, and grapes from the Near East, Egypt, and the Mediterranean region, during Neolithic to younger times – a very comprehensive treatment.

Ahlström (1978) described some Palestinian bedrock wine presses and vats, but dating is lacking, although an Early Bronze Age is considered for some.

Mead

Dickson (1978) made a good case, based on pollen analytic data, for the presence of mead in Bronze Age Fife, Scotland.

Beer, Ale, and Allied Beverages

Geller (1993; see also Nicholson and Shaw, 2000, for more details; also Curtis, 2001) described Fourth Millennium BC evidence of beer and breweries from Egypt. Curtis (2001) dealt with Near Eastern, Sumerian, and other beer and beer-making. Michel et al. (1992; see also Michel et al., 1993) used calcium oxalate preserved in internal grooves of a jar from the Late Uruk Period, about Fourth Millennium BC, as evidence of beer. Dineley (2004) reviewed the barley-malting process both in Britain and elsewhere with a discussion of common additives such as meadowsweet and even henbane, with evidence of it during the Neolithic, much younger in Britain than on the Continent. Nelson (2005) published an extensive account about beer in Europe and the Near East that contains a wealth of information from Classical Antiquity onwards. Dineley and Dineley (2000) discussed the evidence for the presence of ale of various types with flavorings and provided an in-depth account of the entire beer-making process that is applicable to 4000 BC and younger Bronze Age British localities. Samuel (1996) discussed extensive evidence provided by beer residues obtained from Egypian pottery far back in time. Maeir and Garfinkel (1992) described bone and metal Bronze Age implements as beer strainers; beer strainers are essentially long straws (metal, bone, or plant material) that enable the drinker to suck up the beer while avoiding the surface 'scum' of plant materials. Darby et al. (1976) discussed beer making and use in ancient Egypt, with malting involved for the grains. Darby et al. (1976) also dealt with the evidence for 'Booza' in ancient Egypt, which is prepared from crumbled bread that is treated with yeast and then fermented yielding a beer-like beverage reminiscent of Russian kvass, except that the Russians use rye bread rather than wheat or barley bread. Katz and Voigt (1986) discussed beer making and drinking from Ur and also c. 4000 BC at Tepe Gawra in Northern Iraq with 'straws' of some sort used to suck up the beer from below surface 'scum'.

Distilled Beverages

Forbes (1970) reviewed what is known about distillation apparatus, with most evidence suggesting that it has existed only for the past few thousand years, and that distilled spirits probably existed for far less time, with widespread usage beginning only about a thousand years later. Ryšánek and Václavů (1989) described a central Mesopotamian pottery distillation apparatus from about 3500 BC and a Slovakian example of similar type from about 1500 BC, but there is no information regarding just what was being distilled.

Farming

Summary

The advent of farming with the Neolithic is a puzzling event (see Scarre, 2005, for a discussion of the various possibilities applicable to different parts of the world). Why humans waited so long before discovering farming is the subject of considerable speculation. The geographic spread of farming was a relatively slow event, with many areas still not involved, most notably, the hunter-gatherers of the present.

Barker (2006) discussed the advent of farming at some length without arriving at any particular 'cause', although he does conclude that climate change was certainly heavily involved in one way or another.

Haaland (1992; see also 1999), among others, made the important point that cultivation should be kept separate conceptually from domestication, while considering the late presence of domesticated sorghum in the Nile Valley.

Morris (2015, p. 140) commented 'Foragers, I observed, overwhelmingly lived in small, low density groups, and generally saw political and wealth hierarchies as bad things. They were very tolerant of gender hierarchies, and (by modern lights) surprisingly tolerant of violence. Farmers lived in bigger, denser communities, and generally saw steep political, wealth, and gender hierarchies as fine. They had much less patience than foragers, though, for interpersonal violence, and restricted its range of legitimate use more narrowly. Fossil-fuel folks live in bigger, denser communities still. They tend to see political and gender hierarchy as bad things, and violence as particularly evil, but they are generally more tolerant of wealth hierarchies than foragers, although not so tolerant as farmers.' Morris notes that the size of major political units increases greatly from approximately 500 BC (p. 154, Figure 5.6).

Harris (1989, Figure 1.1) provided a diagram useful in understanding the basic transition from hunter-gathering to agriculture.

Bar-Yosef (1998; see also 'Leiberman', 2013, p. 182–183) made a case that the transition in the Levant from hunter-gatherer to agricultural status was locally pushed by the climate changes of the Younger Dryas, which lowered yields of wild cereals and pulses, forcing the adoption of domestication and all that followed in terms of population structure and technologies, that is from the Natufian to the Early Neolithic; evidence includes sickles with microliths that have a surface gloss (see Bar-Yosef and Valla, eds., 1991) caused by the adhering phytolith opaline silica left behind from the stems of the cereal-grasses. Bellwood (2005) summarized the evidence suggesting that the Natufians may have practiced a limited level plant domestication–cultivation, followed by widespread use in the subsequent Pre-Pottery

Neolithic A (9500–8500 BC) of the Early Neolithic. Tchernov (1994) made the point that the Natufians were the first markedly sedentary populations, and even though not agriculturalists they attracted commensals such as mice and barn owls, as well as having domesticated dogs, with the implication that sedentism eventually led in the direction of plant domestication.

Bocquet-Appel and Bar-Yosef (2008) edited a volume dedicated to the Neolithic Demographic Transition, in which from place to place and at different times sedentary populations greatly increased in size, density, and reproductive rate. The cause(s) probably include the major change in food supply, largely featuring cereals, food storage, and/or a sedentary lifestyle.

Snir et al. (2015) described an Israeli occurrence of proto-weeds at a 23,000-year-old hunter-gatherer campsite together with evidence of the collection of wild cereal seeds, i.e., early evidence for the use of pre-domestication cereal seeds.

Bogucki (1988) extensively reviewed the evidence concerning Neolithic settlement and farming in North-Central Europe. Bogucki and Grygiel (1983) described some earlier Neolithic farming sites from Poland.

Constantini and Constantini Biasini (1985) discussed the advent of farming in Baluchistan, citing the evidence from Mehrgarh in western Pakistan near Quetta for Seventh Millennium BC, pre-pottery Neolithic farming. Possehl (1997) listed the grain types found in the earlier Neolithic of Mehrgarh, including barley and wheat species plus jujube and dates. Costantini (1981) described evidence of pre-pottery Neolithic wheat and barley at Mehrgarh.

Zvelebil (ed., 1986) considered in some detail, region by region, local transitions from hunter-gathering to farming in Eurasia.

The relatively late Neolithic transition from hunter-gatherer to agriculture in the circum-Baltic region was complex, including many variables, probably taking place over several thousand years, and much later than the earlier Neolithic initiation of farming in southeastern Europe; it probably involved far more than just the substitution of one life style for another (Zvelebil, 1996).

Sharma et al. (1980) provided an overview of Epi-Paleolithic to Neolithic farming in Northern India and some details about animal husbandry and the Neolithic pottery present.

Sherratt (1986b, Figure 5) provided a diagrammatic view of the geographic origins and geographic expansions of such things as crop cultivation, use of the ox for traction and plowing, and tree crops, horse riding, wheeled vehicles, domestic camels, etc.

However, the far younger initiation of agriculture at various locales in the New World, and possibly in Eastern Asia, is not explained by the same climate changes.

Larsen (1995) made it clear that, with the Neolithic advent of agriculture, human health was negatively affected in many ways, including increasing levels of caries, increase in birthrate and population size, increase in physiological stress, decline in nutrition, and change in activity types and workloads. The popular concept that quality of life increased with the advent of agriculture is incorrect from the physical standpoint.

Van Andel and Runnels (1995) provided evidence that Neolithic farming in Europe began in Southeastern Europe, starting possibly with Thessaly, emphasizing that certain suitable soil types associated with flood plains were selected, and that subsequent colonization tended to occupy more distant but similar soil types to the west and northwest, i.e., this was not a smooth wave but rather a somewhat disjointed jumping forward from spot to suitable spot.

Bar-Yosef and Meadow (1994; see also Vigne, 2008) suggested that agricultural practices in the Levant began at about 10,000–12,000 years BP, and included cereals as well as other plants and also, slightly later, with animal domestication. Rollefson et al. (1992) cited grains and pulses from the Pre-Pottery Neolithic B at Ain Ghazal, Jordan. Broodbank and Strasser (1991) dated the presence on Crete of agriculture by the late Eighth or early Seventh Millennium BC, complete with domestic animals, despite the overwater distance involved. Zohary (1996; 1989, for wheat, barley, pulses, and flax) suggested 8000 years BC for wheat and barley in the Near East. Garrard et al. (1996) suggested domesticated lentils and other pulse from the Eighth Millennium BC in the Near East. Harris (ed., 1996) dealt with the spread of agriculture and pastoralism from the Near East into Europe. Constantini (1989) indicated that the change from Mesolithic foraging to Early Neolithic cultivation of seed crops in the Neolithic occurred between the Seventh and Fifth Millennia BC. Jochim (2000) discussed the Mesolithic–Neolithic transition in south-central Europe from hunter-gatherer to agriculture. Gignoux et al. (2011) suggested, using evidence from mitochondrial DNA, that agriculture began about 12,000 years BP in Africa, Europe, and Asia. Vigne (2008, p. 186) pointed out that the spread of the 'Neolithic package' occurred over a considerable interval in Europe, reaching northeastern Europe only in the Fourth Millennium. Bellwood (2005) emphasized that the selection of non-shattering grains from among wild stocks could have led to the ultimate domestication of grain. Miller (1992) surveyed many Near Eastern Pre-pottery Neolithic sites in regard to both grain and legumes that indicated legumes as an important part of the harvested materials.

Barker (2006) discussed two kinds of foraging: 'routine foraging' for readily available non-seasonal resources, and 'collecting' of more seasonal resources. He provided a diagram (his Figure 11) inferring a transition from foraging to farming and animal domestication with an intermediate stage of wild-food cultivation (causation is elusive and maybe multi-factorial, differing from region to region).

Hays (1975) discussed the presence of pastoralism in the Sahara region in the Neolithic and its disappearance with increasing aridification.

Legge and Rowley-Conway (2000) provided evidence from an Early Neolithic Pre-Pottery site on the Euphrates for domestication of goats and sheep, with marked decrease in hunted gazelles as the source of animal protein.

Shennan (2008; see also Bandy, 2008) considered the demographic possibilities provided by the Neolithic Demographic Transition in conjunction with the larger populations made possible by the domestication of plants and animals. Hershekovitz and Gopher (2008) described some key demographic differences shown by Natufian and Early Neolithic populations relative to mortality rates, ages at death for females and males, and incidence of certain aspects of health.

The later development of agriculture in the New World, separately in South America from that in western North America, which is separate in turn from the developments in Eastern North America, is discussed in Scarre (2005).

Barham and Mitchell (2008, Chapter 9) discussed the reasons for the relative absence of agriculture in Southern Africa owing to the absence of useful cultivars and the barrier posed for importation by the Sahara.

Crane (1999, p. 472) cited Herodotus (c. 485–425 BC) concerning hand pollination in Babylon.

Jones (2012) edited a volume, 'Manure Matters', that discussed in some detail the use of manure, Neolithic and more recent, while considering how to recognize its presence, including such things as proximity to human villages and towns, biochemical 'signatures', bulk chemical composition, abundance of domestic animal bones, and so forth. This is clearly an important area for serious consideration.

Domestic Animals and 'Pets'

Sherratt (1981) made the point that before the use of domestic animals for milk, wool, and other secondary products, they were used chiefly as a source of meat, and also that they were not used in China to any extent in farming.

Tchernov (*in* Bar Yosef and Valla, eds., 1991) pointed out that the presence of many commensal animals (house mouse, barn owl, house sparrow, and others), although not a matter of domestication, is consistent with the sedentary Natufian life style in the Near East. Davis (1987, p. 126–154) considered the variable evidence bearing on animal domestication.

Horses, Donkeys, Asses

Anthony (2007, p. 200; see also Sherratt, 1981, p. 273; Anthony and Brown, 1991) suggested that horse domestication appeared after 4800 BC, and horseback riding by about 4200 BC (p. 221) or 4500 BC (p. 237), whereas Levine (1999) suggested the end of the Third Millennium BC; Levine considered many of the difficulties in trying to find the earliest occurrences of horse riding. Zarin (1976) critically reviewed the problems associated with identifying Third-Millennium-BC Mesopotamian horses, asses, and hemiones (onagers) using the available skeletal and pictorial evidence available to him. Outram et al. (2009) made a case for the use of bits in Kazakhstan at about 3500 BC, which implies the use of bridles. Rudenko (1970) described and illustrated various items including bridles, saddles, saddle cloths, bits, breast straps, and cheek pieces from the Iron Age Pazyryk burials. The Honorable Lady Maria de Andalusia (edited by Mark S. Harris, 2008) credits the invention of the stirrup to the Chinese about 302 A.D.

Oates (2003) cited domesticated horses from the late Third Millennium BC. Vilà et al. (2001) suggested that horse domestication may have occurred at many widespread sites. Olsen (2006) discussed the horse situation with all its complexities, while dealing with a Fourth-Millennium BC northern Kazakhstan site where domesticated horses were present.

Sherratt (1981, p. 274–275; Sherratt, 1997, p. 209) indicated the use of the ass or donkey in the Third Millennium BC, but suggested the Fourth Millennium. Oates (2003) cited domestication of the donkey in the Fourth Millennium BC. Clutton-Brock (1999, p. 115) pointed out that skeletal evidence is not helpful in discriminating between wild asses and early domesticated asses, as is also true with horses, while indicating the presence of domesticated asses as early as 2800 BC.

Camels

Sherratt (1981, p. 274–275) indicated the use of both the dromedary camel and the Bactrian camel in the Third Millennium BC. Sapir-Hen and Ben-Yosef (2013) made a case for domesticated dromedaries first appearing in the southern Levant at about the Tenth Century BC, probably being used to transport copper from deposits in Jordan and nearby Sinai. Clutton-Brock (1999, p. 156; see also Artzy, 1994, for trans-Arabian trade by camel) suggested camel domestication by at least 2600 BC.

Goats, Sheep

Garrard et al. (1996; see also Clutton-Brock, 1981) suggested that domestication of goats and sheep in the Near East from the Seventh Millennium BC. Legge (1996; see Fernández et al., 2002, for DNA evidence) concluded that goat domestication in the Near East dated back to at least the Ninth Millennium BC. Vigne (2008) cited the mid–Ninth Millennium BC. Zeder and Hesse (2000) used morphological evidence to conclude that domesticated goats in the Zagros dated to 10,000 years ago. Bar-Gal et al. (2002) used morphological and genetic data to make the case that in Israel domestic goats appeared in the early Pottery Neolithic (PN), whereas in the preceding Pre-Pottery Neolithic B (PPNB) there are transitional forms from wild to domesticated. Zeder (2006) reviewed the various morphological criteria for recognizing domesticated goats from wild goats, concluding that it is difficult to use them authoritatively. Clutton-Brock (1999, p. 74) used data from Jericho and early Neolithic Greece to estimate 7000 to 8000 and 7200 BC; she also suggested (p. 78) that at about 8000 BC there was a 'switch' from depending more on wild game to goats.

Dogs and Cats

Clutton-Brock (1981, p. 42) noted the presence of an Iraqi dog at about 12,000 years BP. Vigne (2008; see also Tchernov in Bar-Yosef and Valla, eds., 1991, and Tchernov, 1997, for Natufian dogs) cited 18,000 to 12,000 BC in Eurasia for dog domestication. Savolainen et al. (2002) used genetic data to conclude that domesticated dogs probably originated in East Asia approximately 15,000 years BP. Vigne and Guilaine (2004) considered the problem of early domestication and noted that immigrants to Cyprus in the end-Ninth and Eighth Millennia BC brought along dogs, and cats (see Vigne et al., 2004, for domestic cats in the Cyprian earlier Neolithic), as well as foxes, some of which undoubtedly became feral as there were no native precursors on Cyprus. The presence of Early Neolithic domesticated cats on Cyprus (Vigne et al., 2004) suggests their possible presence as pets. Pennisi (2004) discussed a cat buried with a 9500-year-old human, presumably its owner, on Cyprus. Using genetic evidence, Vilà et al. (1997) suggested that dogs originated more than 100,000 years ago, which would have made them potential associates of hunter-gatherers. Were dogs and humans a 'mixed herd' as are zebras and wildebeest? Sablin and Khlopachev (2002) made a case, based on morphological evidence, for the presence of domesticated dogs in the Bryansk region about 13,000–17,000 years ago. Using genetic data, Wayne et al. (2006) estimated that domesticated dogs appeared in the Late Paleolithic, i.e, definitely pre-Neolithic. Bar-Yosef and Valla (eds., 1991; see also Tchernov, 1997) discussed the presence of domesticated dogs in the Natufian. It is clear that the 'time' when dogs were domesticated is a difficult one.

Pigs

Clutton-Brock (1981, p. 72) suggested domestication by 7000 BC. Vigne (2008) cited the mid-Ninth Millennium BC. Giuffra et al. (2000) provided genetic evidence indicating that Asian and European pigs evolved independently from discrete Asian and European wild boars. Albarella et al. (2006) discussed the many problems involved with trying to understand where and when pigs were domesticated; pigs may have been domesticated at many different sites over time. Larson et al. (2005) used molecular data to suggest multiple origins of various pig types from wild ancestors in various parts of Eurasia. A key element may have been the spread of the notion that pigs could be domesticated.

Cattle

Clutton-Brock (1981, p. 50, 66) suggested cattle domestication as early as 7000 years BC. Vigne (2008) cited the mid-Ninth Millennium BC. Götherström et al. (2005) suggested that cattle domestication occurred approximately 10,000 years ago in Europe and Asia and may have also involved some interbreeding with wild aurochs. MacHugh et al. (1997) made a case that African and Eurasian domesticated cattle had separate origins. Bradley and Magee (2006) reviewed the genetic bases involved with cattle domestication and cited some of the PPNB archaeological evidence. Clutton-Brock (1999, p. 87) cited remains from Çatal Hüyük dated 6200 BC.

Milk and Dairy Products

Sherratt (1981, p. 275–282; see also Davis, 1987) discussed lactose tolerance in Europeans and intolerance in other groups, with evidence of milk use dating to the Fourth Millennium BC. Beja-Pereira et al. (2003) found possible evidence of gene–culture coevolution between cattle and humans affecting lactose-tolerant human societies of the European Neolithic. Tishkoff et al. (2007) described the genetic evidence for lactose tolerance in certain African groups that developed independently from the Europeans relatively rapidly. Vigne (2008, p. 194–195), however, pointed out that genetic evidence for the human consumption of milk was absent during the Mesolithic but became increasingly prevalent during the Neolithic, beginning in the Early Neolithic in Europe and the Near East. Simoons (1979) reviewed evidence for lactase malabsorption in Eurasia and discussed the prevalent use of fresh milk in ancient dairying rather than cheese and other products that do not produce lactase-intolerance problems.

Legge (1981, Figure 52) made it clear that evidence from the age distribution in cattle from some Middle Bronze Age sites, including Grimes Grave, showed that milk production was more important locally than meat production. Greenfield (1988) used mortality data derived from skeletal material for domestic animals to conclude that the arrival of important milk production occurred in the post-Neolithic, though Bogucki (1986) had discussed similar mortality data from the Neolithic Linear Pottery Culture.

Craig et al. (2005) described biochemical evidence for the presence of milk products in the Early Neolithic of Central and Eastern Europe (5900–5500 BC). Copley et al. (2003) described the biochemical evidence for milk products from Britain dating from the Early Neolithic to the Iron Age. Dunne et al. (2012) described biochemical evidence of dairy products from the Fifth Millennium BC Saharan rock-art sites suggest the presence of dairy cattle and milking. Evershed et al. (2008) provided biochemical evidence for the presence of dairy products from the Near East and southeastern Europe dating back to the Seventh Millennium BC. Ryan (2005) reviewed the later Bronze Age evidence for milking from Egypt and the Near East with the implication that the practices probably had considerable antiquity. Simoons (1971) reviewed the pictorial evidence from the Near East on seals, etc., and from Egypt and North African rock drawings, indicating that milk and its products date back to at least the Third Millennium BC. Dudd and Evershed (1998) reviewed the biochemical means of recognizing ancient dairy products on pottery. Outram et al. (2009) provided evidence for the use of mare's milk in Kazakhstan at about 3500 BC.

Cheese

Bogucki (1984, 1986, 1988) concluded that widespread ceramic sieves of the European Linear Pottery Culture from France to the Ukraine, dating back to 5400 years BC, were for cheese production and took this as more evidence for the early use of milk products, with the implication that even lactose-intolerant people would have had no problem digesting cheese or yogurt. Bogucki and Grygiel (1983) described a sieve, possibly used in cheese making, from a Polish earlier Neolithic site. Using biochemical evidence from sherds, Salque et al. (2013) made a convincing case that at least some of the sieve-like Linear Pottery Culture pots were used in cheese making. Curtis (2001) dealt with Near Eastern cheese making in the Third Millennium BC.

Yang et al. (2014) provided biochemical evidence that cheese was present in Early Bronze Age Xinjiang, about 3800 years BP; they also commented that this cheese, together with a fermented alcoholic beverage, are the earliest evidences of human use of microorganisms for transforming food properties. Of potential interest is the paper of Geetha et al. (1996) citing various plant parts that have been used to curdle milk in India, Japan, and Nigeria.

Storage Facilities

Kuijt (2008) reviewed the importance of food-storage facilities for the Levantine Neolithic Demographic Transition, pointing out how critical they must have been for both the earlier, pre-Transition Natufians, who used wild foods as well as domesticated foods during the later time interval (see especially his Tables 1 and 2). Hodder (2006, Figure 27, p. 57; also Figure 54, p. 131; see also Hodder, ed., 2007) discussed Neolithic storage bins at Çatal Hüyük as small ones used for individual house storage.

Soffer (1989) reviewed the question of storage in the Eurasian Late Paleolithic, and indicated that good evidence was present on the Dnepr region of the Russian Plain for meat storage, and also commented on later Natufian items but did not find such evidence in the Late Paleolithic of France, Germany, or Central Europe.

Chang (1986, p. 90) cited 5000-year-old Chinese underground grain-storage pits.

Curtis (2001) dealt with Egyptian cereal-storage granaries, Near Eastern silos, and beehive granary complexes, as well as Roman grain-storage facilities.

Plow

Sherratt (1981, p. 266–271) cited plows (ards) beginning in Europe, the Near East, and India in about the Fourth Millennium BC (about 5000 BC, in Sherratt, 1986b), and emphasized the importance of animal domestication for plow usage. Steenberg (1973), using a model of a 4000 BC 'spade' from the region of Denmark, demonstrated how it could have been used by two people to dig a furrow suitable for planting grains. Plowing greatly increased agricultural productivity over human planting of seeds and cultivation. The presence at about the same time of carts does not necessarily correlate with the plow and domesticated cattle and oxen, since carts have many uses besides those closely allied with the farm. Bogucki (1993) emphasized the potential importance of oxen in plowing over the use of hoes. Liverani (2006; Moorey, 1994, pp. 2, 3, citing the Third Millennium BC) briefly discussed the presence of seed plows and irrigated plant rows from Uruk. Crawford (2004, Figure 3.3) illustrated a Sumerian seed funnel and a plow.

Mateescu (1975) used bovine skeletal data to indicate that these animals were used in the Middle and Late Neolithic Danubian region for transport and plowing.

Water Resources

Summary

Water, so basic to life of animals, humans included, is available from bodies of water, still and running, and from the water derived from eating plants. With the advent of the Neolithic, there arose the need for reliable water sources for irrigation, plus various water storage devices. The more sophisticated water-resource devices such as lined wells, siphons, cisterns, pipes, aqueducts, and dams are largely of Bronze Age and more recent.

Wells

Tegel et al. (2012) provided a careful, extended description of the relatively complex wooden wall supports of some Early Neolithic German water wells. The carpentry of the well-wall supports, in oak, was done with stone tools, and involve mortise-and-tenon joints. Tube-like well linings using hollowed trunk sections were also used.

Jansen (1989; see also Marshall, 1931, for extensive information) described the Bronze Age Mohenjo-Daro brick-walled wells in some detail. Angelakis et al. (2007; Angelakis and Spyridakis, 1996) refer to earlier Bronze Age Minoan wells on Crete.

Crouch (1993) discussed Ancient Greek city wells in some detail, with emphasis on those in karst regions, and cited a 7000 BC well from Cyprus.

Mackay (1938) described the stone-lined wells at Mohenjo-Daro.

Siphons

Hodge (1992; see also Viollet, 2007, Figure 6.13 for a Roman siphon and header tank) briefly dealt with Greek and Roman siphons and their header tanks. Hodge (*in* Wikander, 2000, p. 77–87) provided a useful account of siphons.

Cisterns

Hodge (1992) briefly dealt with Greek and Roman cisterns. Crouch (1993) discussed Ancient Greek city cisterns in some detail. Hodge (2009) dealt in some detail with Roman cisterns.

Springs

Crouch (1993) discussed Ancient Greek city springs, particularly those in karst regions. Hodge (1992) dealt with Roman springs. Miller (1980) discussed the utility of large, porous clay jars for the storage of cool water.

Pipes

Marshall (1931) dealt with earthenware drain pipes from Bronze Age Mohenjo-Daro. Evans (1964, Figures 103, 104) illustrated Minoan terra-cotta water pipes. Crouch (1993) discussed pipes, chiefly made of ceramics of terra-cotta, in ancient Greek cities. Hodge (1992) dealt with Roman pipes with discussion of lead pipes. Peleg (1991; see also Hodge, *in* Wikander, 2000, Figure 11, Turkish example; p. 107–110 for Greek examples; and p. 112–120 for Roman examples) described some ancient stone pipelines from Israel, constructed of stone elements cemented together.

Water Wheels

Reynolds (1983) described the Late Iron Age development of the vertical water wheel, from approximately 1000 BC to their demise early in the Industrial Revolution with the development of steam power.

Aqueducts

Hodge (*in* Wikander, 2000, p. 39–65) provided an extensive treatment on aqueducts, particularly the Roman examples. We tend to think of aqueducts as a strictly Roman invention (see Viollet, 2007, for a discussion of Roman aqueducts), but Angelakis et al. (2007) make it clear that the earlier Bronze Age Minoans might have been the first to use aqueducts. Crouch (1993) discussed Ancient Greek city aqueducts. Jacobsen and Lloyd (1935) described Sennacherib's Assyrian aqueduct at Jerwan in Iraq, dated approximately 700 BC.

Dams

Hodge (1992, p. 79–92) described various Roman dams. Smith (1971) discussed ancient to present dams, beginning with an Egyptian dam from the Early Third Millennium BC; see also Viollet (2007) for a Third Millennium BC Egyptian dam at Sadd El Kafara, and Hathaway (1958). Viollet (2007) discussed a Fourth Millennium BC dam from Jordan that stored seasonal waters. Garbrecht (1987, 1991) edited two volumes dealing with dams, both recent and ancient, from many parts of the world; this is a useful source. Belli (1999) described Urartian dams from before 2700 BC. Schnitter (1979) discussed some Anatolian dams, one of which dated to the Second Millennium BC.

Irrigation Systems

Irrigation is, of course, strictly a function of the onset of agriculture in the Neolithic! Fagan (2011) provided an authoritative global account of evidence for irrigation from the historic past, including such things as the unique qanats of the Near East. Hodge (in Wikander, 2000, p. 35–38) provided a useful introduction to qanats. Beaumont et al. (1989) provided an extensive account of the characteristics and functioning of the qanat. Oleson (in Wikander, 2000, p. 183–215) provided a useful introduction to ancient irrigation, and Oleson (in Wikander, 2000, p. 217–302) dealt with questions involving the lifting of water, including that used for irrigation. Some traditional kingdoms in which the king was responsible for the wellbeing of the people, hence of the crops, hence of the irrigation system, have been termed 'Hydraulic Kingdoms'.

Egyptian

Butzer (1976) briefly reviewed early Egyptian irrigation system, which depended on impounding the flood waters of the Nile and using the shaduf to supplement that water where needed, with an early date of about 3000 BC. Also see Bonneau (1964).

Near Eastern

Bagg (2012) provided a comprehensive account of irrigation techniques in the Ancient Near East, with evidence dating from the Sixth Millennium BC. Included here were the use of the shaduf and, beginning in the First Millennium BC, the qanat type tunnel for bringing groundwater from elevated regions into the arid dryland at lower elevation. Pemberton et al. (1988) provided additional information from the Iraq region.

Garbrcht (1980) described the well-designed irrigation system at the Bronze Age site of Tuşpa in Armenia, a unique, self-contained example.

Ur (2005) provided the results of satellite imagery and aerial photography that revealed the extent of Sennacherib's Late Assyrian water works and canals.

Oates and Oates (1976) summarize information about Mesopotamian Bronze Age irrigation systems.

Burney (1972) discussed Urartian irrigation items, including canals and qanat-like features of the First Millennium BC.

Miller (1980) discussed the location of Natufian and earlier Neolithic settlements near water sources that could readily have been used for small-scale irrigation of wild and later domesticated crops.

Chinese

Li and Xu (2006) described the Dujiangyan irrigation scheme in Sichuan, dated 256 BC. Involved in this complex irrigation project is a short canal cut through solid granitic rock, using firesetting. This ancient irrigation work is still in use.

Southeast Asian

Higham (1996) discussed the use of Bronze and Iron Age 'moats' in parts of Southeast Asia as water-conservation devices during the dry season.

Human-Waste Disposal, Rubbish Disposal, Baths, and Drainage

The problem of human waste involves the fact that our bipedal erect posture results in the medial aspects of our buttocks being pressed together with the anus being almost an internal body opening. This is in contrast to the situation in the other higher anthropoids, gorillas, and chimpanzees, where the anus is in the open. Thus, humans have always had the problem of cleaning the anal area after defecation.

Disposal of human waste and rubbish was not a problem during the hunter-gatherer stage of evolution, any more than it is today where such people merely use the vicinity where they are staying for this purpose with relatively frequent movement (no pun intended), making the accumulation of waste and rubbish adjacent to living space not a problem. Beginning with the Neolithic, however, if not even earlier in the Natufian, more sedentary villages and larger groupings made the disposal of human waste a real problem. The disposal of human waste in privies and toilets of one type or another by means of running water is a largely Bronze Age and more recent business that has not been practiced universally by any means. The poor devil afflicted with diarrhea has been forced from the very beginning to get relief by washing with running water from nearby water courses or other sources, including water containers.

Stöllner et al. (2003, p. 150) cited very unhygienic conditions with the presence of feces from the Early Iron Age salt mines at Dürrnberg-bei-Hallein.

When one reviews the evidence of human-waste disposal, plus that for rubbish, one is impressed by the fact that humans living under village and city conditions have all too commonly been gross slobs who threw their human waste and rubbish into the common alley or street, resulting in stinking, vile conditions (the Scottish interjection 'gardyloo!' comes to mind; ditto Barbassa's quote about Nineteenth Century Rio de Janeiro, 2015, p. 133, 'The habit of throwing the contents of chamber pots out the window was also common, with results so disastrous that an 1831 municipal decree sought to regulate the activity. From then on, the waste could only be thrown into the street at night, and after three warnings of Água vai! Or 'here goes the water!' Failure to obey led to fines and hefty compensation for the victim.') The similar condition of many Medieval cities needs no discussion. Hobson (2009) dealt in some detail with the chamber-pot 'problem' in Roman times.

Humans do not, however, foul their own nests with human waste or rubbish, be they temporary shelters, permanent shelters, or homes in villages and cities.

Wilson (*in* Wikander, 2000, p. 151–179) provided a useful introduction to the overall issue of drainage and sanitation.

Angelakis et al. (2005; see also Angelakis and Spyridakis, 1996) described urban wastewater and stormwater in terms of sewers, and also discussed toilets from Second Millennium BC Minoan Crete, summarizing information for other parts of Greece up to the Iron Age for sewers and toilets, providing a picture of relatively 'modern-type' sewers and toilets using running water. Evans (1964, Figures 171a, 172) diagrams positions of Minoan latrines, baths, and drains.

Cahill et al. (1991) described several Seventh to Sixth Centuries BC Jerusalem-stone toilet seats situated over cesspits that yielded remains of whipworms and tapeworms.

Taylor (2015) considered the problems of waste removal, including human waste, from the city of Rome and from Medieval London in some detail, citing the many common problems and solutions. Hall and Kenward (2015) gave an account of waste disposal of all sorts in the City of York over a 2000-year interval from Roman times on, with the amusing information that during the Viking period large amounts of moss were provided as bottom wipes.

Charlier et al. (2012) discussed numerous methods, in addition to toilet paper, used for the cleaning of the human 'bottom' from the First Millennium BC to pre-present-day times, with the use of such things as leaves, grass, stones, corn cobs, animal furs, sticks, snow, seashells, hands, and sponges on sticks (see Koloski-Ostrow, 2015, p. 86–87), broken pottery fragments, and pebbles. Wikipedia (2015) illustrated wooden sticks (see also Matsui et al, 2003, p. 133, for Japanese use of flat sticks) and cited seaweed used for cleaning the human 'bottom' in Japan during the past.

Wright (1960) provided a brief account of pre-medieval bathrooms and toilets while giving an extensive account of medieval to present bathing facilities and toilet facilities.

Jansen et al. (eds., 2011) discussed a great variety of evidence concerning toilets from brief material on ancient Egyptian toilets to Roman times, with excellent illustrations for much of the material.

Hardy-Smith and Edwards (2004) provided a detailed account of Natufian evidence with rubbish disposal concentrated within dwellings, with larger items pushed to the periphery of the dwellings, transitioning in the Pre-Pottery Neolithic A to the Pre-Pottery Neolithic B situation in which rubbish was thrown into the spaces between the rectangular dwellings with the floors kept clean by sweeping. There is no evidence in these structures for areas of human waste, presumably because the inhabitants left the dwellings to relieve themselves at a respectable distance.

Miller (1980, p. 334) pointed out the utility of water-storage plaster or fired-clay containers that contain an alkaline solution provided by ashes from the hearth for washing dishes and doing laundry.

The smell of sewer gas from toilets was finally eliminated in 1775 by Alexander Cummings' patent for the sewer trap!

Mohenjo-Daro

Marshall (1931) discussed the privies, cesspits, and rubbish flues in Bronze Age Mohenjo-Daro. Jansen (1989; see also Marshall, 1931, for extensive information) described the Bronze Age Mohenjo-Daro street-level drains for sewage and the associated latrines. MacKay (1938) described the latrines with 'chutes' for human waste at Mohenjo-Daro, as well as the many drains. MacKay (1938) described the bathrooms, with chutes for water disposal, from Mohenjo-Daro.

Marshall (1931) described the bathrooms present in many Bronze Age Mohenjo-Daro homes and also a very large communal bath.

Minoan

Gray (1940) provided a fine introduction to the problem of human-waste disposal from Minoan times to the present with numerous examples, some amusing.

Greek and Roman

Neudecker (1994) provided a comprehensive account of Roman latrines and also some information about Classical Greek latrines plus extensive accounts of Roman baths with their commonly associated latrines. Manderscheid (*in* Wikander, 2000, p. 466–535) described Greek and Roman baths.

Greek

Owens (1983) described the problem of rubbish and the disposal of human-waste for Fourth- and Fifth-Century Athens, with the involvement of *Koprologoi*, who were employed in removing waste from the City proper, including cesspits. Crouch (1993) described Ancient Greek city wash basins ('louters').

Yegül (1992) described Greek baths in some detail.

Roman

Hobson (2009), together with Koloski-Ostrow (2015), provided extensive coverage of Roman toilets, private and communal, whether situated above cesspits or connected to running water, as well as those installed above street level with their lead piping. They cover not only Rome itself but also much of the Roman World.

It would be interesting to estimate the daily volume of human fecal matter and urine produced by 1,000,000–2,000,000 Romans and to then try to estimate how much of this material could have been handled by the sewers together with an estimate of how much ended up in cesspits used by individual households or as chamber-pot contents surreptitiously dumped at night into the streets below plus the small fraction carried away in carts during daytime.

Fagan (2011, Figure 10.9, p. 195; see also Hodge, *in* Wikander, 2000, Figure 13) illustrated a Roman public toilet where the human waste was carried away by water flowing beneath the seating arrangement of the latrine. The well-known *Cloaca Maxima* of ancient Rome was a sewer for the removal of waste, including human waste, although Koloski-Ostrow (2015, p. 70) emphasized that it served mostly for the removal of storm water. Jansen (1997) discussed the various toilets in Pompeii and their various methods of waste disposal. Scobie (1986) provides an extensive and appalling account of the poor condition of Roman sanitation, particularly in Rome itself, and also mentions that the situation in medieval cities was little better for most people; his description of the Roman baths suggests that they were probably places where one could pick up various diseases. Fagan (2000) discussed the probability that many Roman baths were not clean, with lots of dirty water, etc. They were not altogether very pleasant spots.

Hobson (2009) and Koloski-Ostrow (2015) describe the latrines associated with Roman baths in some detail, while Yügül (1992, 2010) provided an extensive description of the unique hot and cold Roman baths in Europe, North Africa, the Near East, Anatolia, and Northern Europe.

Behavioral–Technological Innovations

Thomsen (1836) in Denmark is generally credited with introducing the concept of a Stone Age followed by a Bronze Age and then an Iron Age, but Chang (1986, p. 45) noted that the Chinese anticipated him by a thousand years with the concept of stone, jade, bronze, and iron weapons and tools, and quoted Lucretius, 98–55 BC), with the concept of stone, bronze, and iron weapons.

Wescott (ed., 2001) discussed 'primitive technologies' that are easily duplicated today, including trapping, containers made of gourds and bark, tanning, working of hides, scapular saws, sandals, dugout manufacture, lamps (including clay lamps), bow drill, cordage, wooden mortar and pestle, snares, and deadfalls.

Major innovations with global distribution in time pose serious questions. One can think here of such major innovations as the beginnings of metal production and use (the Chalcolithic and Bronze Ages followed by the Iron Age), and the global beginnings of the 19th Century Industrial Revolution based on the steam engine. These major innovations are relatively instantaneous in time and global in distribution. For the earlier metal ages, one can conceive of their being initiated in one limited area and then spreading globally in a very short time, with the implication that the global spread was aided by language. The same is obvious for the global initiation of the Industrial Revolution. 'Good ideas travel well.'

Tools and Materials

Summary

One of *Homo*'s most basic behaviors is the making and using of various tools, Homo faber. Chimpanzees have a limited tool-making and tool-using behavior, including the use of twigs to secure termites from their nests, the use of wadded leaves to soak up water for drinking, and the breaking of nutshells with stones and wooden hammers. Crane (1999, p. 30) cited chimpanzee use of sticks, including pointed sticks, to obtain honey from a tree-stump hive, but these limited items pale when compared with the complex human repertoire.

Here is a list of some of the materials from which humans make their tools: stone, wood, plant materials, bone, horn, leather and skin, ceramics, metal, and ultimately plastics. The recognition and persistence of these materials in the 'record' ranges from the strong persistence of stone tools – with wood a close second – to the very poor persistence of leather and skin. As a result, the first 'appearance' of objects made from these various materials in the archaeological–anthropological record is badly skewed and must not be taken as their 'true' first usage. Previously, I briefly discussed the human skeletal characteristics that make the use and fabrication of tools practical.

Rosen (1996) reviewed the question of the replacement of flint tools by metal, bronze first, followed by iron; he makes a strong case that the replacement was not simple but involved many factors, some unrelated to tools, and that it should not be regarded as a simple, linear process.

Stone and Bone Tools

Generalities

Without question, until the Neolithic the record of stone tools is the most widespread source of information about early man and his changing technologies. Of course, there are many caveats concerning the stone-tool record, including regional differences, differences engendered by different stone types, with the determination of 'absolute' dates among the most difficult to deal with.

Barham and Mitchell (2008) provide an extensive account of the manufacture and use of Oldowan-type and Acheulian tools, including both stone and bone tools. They raise the possibility that use of the sharp flake as a tool may have been as an unintentional byproduct of nut-cracking.

Hallos (2005) suggested that Middle Pleistocene knappers were not just making tools for immediate use at a site but rather were also involved with planning for future use at other sites. Roche (2005) usefully described the manufacture of several types of stone tools from 2.6 million years ago to the Middle Pleistocene.

Schick (1998) summarized the Paleolithic record, with the Oldowan technology first recorded at about 2.5 million years BP in East Africa. Quade et al. (2004) provided a very thoughtful account of the sedimentological–environmental evidence accompanying Oldowan artifacts from Ethiopia and their relations to underlying beds with *Australopithecus* and overlying beds with Acheulian artifacts. Plummer et al. (2001) described Late Pliocene Oldowan material from Kenya, followed by the Acheulean technology beginning at about 1.5 million years BP, and then the Middle Paleolithic, Mousterian (with McBrearty and Brooks, 2000, Figure 5, for examples), and widespread appearance of the Levallois technology. Clark (1992) summarized knowledge about the Oldowan and Acheulian of the Maghreb in North Africa, and also discussed scattered data from other North African regions. Tchernov (1998) discussed the 'dating' problem as well as the possible migration paths followed by various human groups, with the implication that many aspects of both questions are far from being settled. Conroy (1990, p. 351–357) summarized much of the data relating to cultural activity, placing the use of tools back in the 2.0–2.8-million-year interval.

De Heinzelin et al. (1999) discussed some Late Pliocene evidence from Ethiopia of hominid use of prey for meat and marrow, in conjunction with Oldowan tools known from the region.

Sieveking and Newcomer (1987; see also Potts, 2003) summarized the literature on human uses of flint, including chert, back into the Paleolithic.

Volman (1984) discussed the Oldowan and Acheulian artifacts of the southern African Early Stone Age and also the Middle and Late Stone Age types.

Hughes et al. (2021) discuss all aspects of jade, archaic to modern, with a bibliography of 1280 references.

Oldowan

The earliest stone tools thought to have been made by hominins belong to the African Oldowan Industry, 1.6–2.6 million years BP (Plummer, 2005, *in* Stahl, ed.), and are thought to have been made by some species of *Homo,* or possibly by an advanced australopithecine, or by members of many of the contemporary species. Susman (1991) discussed the difficulties in deciding which hominid(s) made the Oldowan tools, and concluded that both *Homo* and *Paranthropus* may have been capable. Semaw et al. (1997) provided evidence of Early-Pleistocene–type stone tools from Ethiopia that are dated at 2.5 million years. Panger et al. (2002) raised the possibility that pre-Oldowan hominids might also have been using stone tools, despite the present lack of positive evidence.

Steele (1999) commented on the presence in Kenya of flint-flake tools made by flint knapping techniques that are approximately 2.5 million years old; they are the oldest known evidence for real hominid-type tool-making behavior-technology, but he also mentioned slightly older reliable evidence from Ethiopia (Quade et al., 2004; Semaw et al., 1997). Roche et al. (1999) discussed the Kenyan evidence in detail. Bishop et al. (2006) discussed an Oldowan site of Pliocene age in western Kenya.

Rogers and Semaw (2009) reviewed the known Pliocene Oldowan sites in Africa and commented on their climatic and cultural context; sites may go back even to 2.9 million years.

Delagnes and Roche (2005; see also Ludwig and Harris, 1998) provided an informative account of Late Pliocene Oldowan hominid knapping skills, making it clear they were far more complex than one might have thought.

Semaw et al. (2003) discussed Oldowan artifacts and associated skeletal materials from Gona, Afar, Ethiopia. Plummer et al. (2001) discussed Oldowan material from the Late Pliocene of Kenya. Of interest is the account of Toth et al. (1993) of primitive tool-making capabilities of a bonobo at about the Oldowan-industry stage.

Kimbel et al. (1996) described Oldowan tools and associated remains of *Homo* at 2.33 million years BP from the Kada Hadar Member of the Hadar Formation, Ethiopia. Dennell et al. (1988) described an Oldowan-type occurrence from Northern Pakistan that is 2 million years old, the oldest known from Asia east of the Levant.

Acheulian

Marean and Assefa *in* Stahl, ed. (1995), discussed the lithic and other evidence for the change from Acheulian to Middle Stone Age tools and bone artifacts. Lycett and von Cramon-Taubadel (2008) discussed a possible dispersal model of Acheulian technology out of Africa.

Clark (2001b) discussed later Acheulian African stone-implement technology. Mishra et al. (1995) discussed the earliest Acheulian industry from Peninsular India. Allchin and Allchin (1982, p. 37–38) illustrate and refer to an Acheulian hand axe from Bhimbetka Hill in central India. Misra et al. (1977) briefly discussed Acheulian stone tools from Bhimbetka.

Yerkes et al. (2003) discussed Early Stone Age tools used for woodworking, determined by microwear analysis, with a discrimination of tools used for 'finer' woodworking, such as adzes and chisels, from axes used for such things as tree felling. Dominguez-Rodrigo et al. (2001b) described plant phytoliths from Acheulian stone tools from Peninj, Tanzania; these were unlike the plant phytoliths in associated sediment and, with plant fibers as well, this was taken as evidence for use on wood. Binneman and Beaumont (1992) described microscopic evidence of plant material, possibly wood and/or sedges, from two Acheulian handaxes from the Northern Cape.

Lepre et al. (2011) described a Kenyan site that yielded Acheulian-type stone tools reliably dated 1.76 million years, the oldest known to date, while remarking that similar tools are unknown in Eurasia until much later and that the Kenyan material is associated with Oldowan-type tools.

Santonja and Villa (2006) discussed the distribution of Acheulian artifacts, including those in southwestern Europe, England, Germany, southeastern Europe, and western Asia from Georgia to Israel, and the Arabian Peninsula.

Garrod and Bate (1937) illustrated various Acheulian handaxes from the Mount Carmel area, Israel.

Clark (2001) discussed Acheulian tools from Kalambo in East Africa.

Potts et al. (2004) described remains from Olorgesailie, Kenya, assigned to *Homo erectus*, that are associated with Acheulian artifacts with an age of 0.97 to 0.90 million years. Asfaw et al. (2002) described *Homo erectus* remains from Bouri, Middle Awash, Ethiopia dated at 1 million years BP, associated with 'early Acheulian' stone tools.

Middle Paleolithic

The account of Brown et al. (2012) of 71,000-year-old South African microliths does support a Middle Stone Age occurrence of 'modern' technology far in advance of that known from Europe.

Brooks et al. (1995) and Yellen et al. (1995) described worked bone tools, including barbed points, with minimum ages of approximely 90,000 years BP, which they ascribed to the Middle Stone Age as a minimum estimate of very complex tools and associated hunting activities.

Jacobs et al. (2008) suggested that Middle Stone Age evidence from Southern Africa, approximately 65,000 to 70,000 years old, suggests that modern humans were making stone tools and using beads and ocher.

Barham et al. (2002) described some bone tools from Broken Hill (Kabwe) Cave, Zambia, and commented on their evolutionary significance as an early instance of manufacture and use of bone tools. Henshilwood et al. (2001) described Middle Stone Age bone tools from Blombos Cave, 70,000 years BP.

Upper Paleolithic

De Beaune (1993) discussed Aurignacian stone tools of various types and their many uses. Glover (1981) discussed preliminary data of Upper Paleolithic age involving flint implements from South Sulawesi.

Natufian

Natufian bone and stone tools are discussed in Bar-Yosef and Valla (eds., 1991), including hammer stones and the bow drill.

Neolithic

Chalcolithic

Yerkes and Barkai, 2004) discussed information for woodworking with some Chalcolithic tools as determined by microwear.

Bronze Age

Marshall (1931) dealt with awls from Mohenjo-Daro.

Iron Age

Stöllner et al. (2003) briefly described the iron-headed picks and axes used in Austrian rock-salt mines.

Tool Use

Keeley (1977) described the various types of 'polish' preserved on stone tools that indicate their use on wood, meat, leather, hides, and other materials—a very insightful paper. Campana (1989) discussed bone and antler tools, their manufacture and use, and the microscopic surface evidence of use from the Natufian of Israel and the Protoneolithic of the Zagros.

Crabtree (1972) provided an extensive account of stone-tool manufacturing.

Stone (see also *Mining and Quarrying*, below)

Man, *Homo*, is a very observant creature who has always paid close attention to his environment in all of its complexity. For example, today city dwellers, urbanites, pay close attention to traffic at all times, to pet-dog droppings, to the possibility of pickpockets, muggers, and similar criminals, to cracks in the pavement that can trip a pedestrian, and so forth.

There is every reason to believe that from our earliest ancestors in the Pleistocene close attention has been paid to all features of the environment, including such things as stone suitable for implement making, edible plants, stones and objects suitable for ornament making, the habits of potentially dangerous carnivores and of potential prey animals, and so forth. Saul (2013) argues that the Mesopotamians had attempted to inventory the world.

Stone suitable for tool making has occupied humans since the inception of the Oldowan. Such stone, including flint, obsidian and other 'flinty' volcanic rocks, diorite, and many granitic rocks most commonly occur as obvious exposures in high-relief regions such as cliffs, talus slopes in steep terrains, coastal beaches, and some stream banks in high-relief terrains, to name just a few.

In an important paper, Barkai and Gopher (2009; see also Barkai and LaPorta, 2006) made clear from the Sede Ilan site in Israel that Late Early Paleolithic and Early Middle Paleolithic humans had a well-developed sense of what we consider 'raw materials' and 'resources'. At this site they used basalt quarrying tools from a nearby exposure, and developed extensive debris piles that were well organized relative to the quarrying face from which their flint was derived. They made a considerable impact on the surface environment, landscape alteration, with their selective quarrying of a flint-rich horizon and the locations of their adjacent, extensive debris piles. This represents long-term, sustained activity rather than some sort of random activity very early in time. A somewhat similar situation is discussed by Nadel et al. (2011) from Mt. Carmel, Israel.

The earliest evidence of the use of flint for 'cutting' and other uses is from the Oldowan, with present information suggesting that loose pieces of surface stone were used. Evidence from the oldest Oldowan site, at Gona, Ethiopia (Stout et al., 2005), indicates that a distinct level of selectivity was already exercised from among the various available cobble lithologies, i.e., that this was not some sort of random selection from available cobbles.

Klemm and Klemm (2008) provided a very comprehensive account of Egyptian stone resources used during the various dynasties. Moorey (1994) extensively discussed ancient Mesopotamian stones used for various purposes. Also see A. Lucas (1962).

Stiles et al. (1974; see also Stiles, 1998, for the evidence against this factory-site being some kind of random aggregation brought together by various agencies) discussed an Olduvai Gorge Oldowan factory site in Tanzania with chert from beds not situated at the site, i.e., brought in from elsewhere. The cherts are magadiite pseudomorphs from remnants of a nearby hypersaline lake. It is clear that there was a lot of sorting of desired chert during the manufacturing process, i.e., it does not represent random acquisition.

Paddayya et al. (1999, 2002) discussed a Karnataka, Peninsular India, factory-type site of Acheulian age that employed local siliceous limestone for implements. Underground mining is not required, although some surface work at exposures may have been carried out here and there to free suitable materials. Cobbles of suitable stone from streams were used.

Mining and Quarrying

The earliest evidence of quarrying is from the Acheulian (see examples below), but this limited information tells us little more than the obvious, i.e., that early *Homo* was well aware of the possibilities for obtaining desirable stone by surface quarrying of a primitive sort. It is not until the Mesolithic that there is any recognized evidence of actual digging down beneath the surface to pry out desirable stone, apparently using antlers. In this sense quarrying and mining are among the earliest of recognizable human behavior–technologies. Underground mining is a strictly Early Neolithic feature, with mining soft chalk beds for enclosed flint nodules in parts of Europe, widely beginning with Copper and Bronze Age searches for deposits. Heavy-duty underground mining of anything harder than chalk evidently had to wait until the latest Neolithic–Bronze Age when the use of firesetting (see below) was employed to break free waste gangue in order to get at desirable ore. Why the recognition and use of metals waited so long is unclear.

Boaretto et al. (2009; see also Verri et al., 2004, 2005) provided a geochemical method using beryllium isotopes, whose ratios depend on cosmic radiation that affects surface materials, to discriminate between flint secured from surface exposures and flint derived from 'deep' sources. For a comprehensive overview of early mining, see Poss (1975).

Acheulian

Sampson (2006) discussed Acheulian quarrying of hornfels from the Upper Karoo, South Africa.

Petraglia et al. (1999) described Acheulian quarrying and artifacts from Peninsular India.

Barkai et al. (2002) described some Late Acheulian–Early Mousterian Mt. Pua, Israel, localities where surface exposures of limestone containing flint nodules were exploited with hammer-stones to break off the 'barren' limestone to obtain the flint nodules, with debris heaps left behind – a primitive type of quarrying of surface materials.

Boaretto et al. (2009) used beryllium-isotope data to indicate that stone tools from the 400,000-years BP Qesem Cave, Israel, were made from deeply buried materials.

Mesolithic and Late Paleolithic

Schild (1976; see also Schild, 1983) provided a detailed account of flint mining in central Poland during the Late Paleolithic and Mesolithic involving the so-called chocolate flint, with shaft depths of several meters.

Vermeersch et al. (1984; see also Vermeersch et al., 1990) described Upper Paleolithic 33,000-year-old chert mining of riverine cobbles in the Nile Valley. Vermeersch and Paulissen (1997; see also Vermeersch et al., 1995) described occurrences in the Qena Area, Nile Valley, Egypt, of pits up to about a meter or so in depth from which chert cobbles for implement manufacture were obtained from conglomeratic horizons.

Meszaros and Verts (1955) described an Upper Paleolithic paint mine from near Lovas (Hungary, County Veszprem).

Bronze Age

Klemm and Klemm (2008) illustrated many Classical Egyptian quarry sites. Harrell and Storemyr (2009) described and cited Egyptian stone quarries from the pre–Dynastic Period to the Greco-Roman Period. Nicholson and Shaw (2000, Tables 2.1 and 2.2) provided an extensive list of ancient Egyptian quarries for both stone and other products. Ward-Perkins (1971) discussed Egyptian, Greek, and Roman quarrying techniques. Waelkens et al. (1992) described various Egyptian and Classical Period quarries and something of the quarrying techniques employed.

Biagi and Cremaschi (1991) described some Harappan flint quarries from the Rohri Hill, Pakistan. Maggi et al. (1996) described Copper Age – Early Bronze Age quarrying of bedded Tithonian radiolarite cherts from Liguria.

Variscite and Cinnabar Mines

Schild and Sulgostowska (eds., 1997) include a paper on a Barcelona-region Neolithic variscite mine, the mineral presumably destined for personal adornments. Rosell et al. (1993) discussed the Neolithic variscite mines of the Gava area south of Barcelona. Dominguez-Bella (2004, 2012) reviewed Spanish variscite sources, including those of northeastern and northwestern (Huelva) Spain. Domínguez-Bella and Céspedes (1995) discussed variscite beads from the Neogene dolmen at Alberite near Cadiz, and also cited cinnabar and ocher from this burial, as well as discussing the distribution of variscite deposits in Iberia. Sharpless (1908) described a Bronze Age cinnabar mine from Koniah, Turkey. Jovanović (1980, p. 152) cited cinnabar from near Vinča close to Belgrade.

Amazonite, Amethyst, and Emerald Quarries and Turquoise Mines in Egypt

Harrell and Osman (2007) described a pegmatite quarry from which amazonite suitable for jewelry and other purposes is present, associated with 'pounders' of the type used elsewhere in ancient Egypt for quarrying. However, this quarry and others nearby appear to be inadequate as a major source of amazonite for ancient Egyptian usages.

Harrell and Sidebotham (2004) described an Eastern Desert Ptolemaic quarry for amethyst from quartz veins cutting Precambrian basement. Amethyst was available as far back as the Late Predynastic in Egypt with uncertain sources.

Harrell (2004) described the Egyptian emerald mines as being the earliest source of these gems. Giuliani et al. (2000) discussed the isotopic composition of Old World and Colombian emeralds, and something of their archaeological ages.

Nicholson and Shaw (2000, pp. 62, 63; see also Tallet, 2012) briefly discussed turquoise mines in the Sinai that date back to the 'Early Dynastic Period', with malachite also involved. Moorey (1994, p. 101–103) provided an extensive account of turquoise in the ancient Near East.

Also see A. Lucas (1962).

Mining of Flint and Obsidian

Neolithic and Middle Paleolithic Flint and Other Tool-Materials

Weisgerber, Slotta and Weiner (eds., 1999) provided a fairly comprehensive account of European flint mining, including the various tools employed, with emphasis on both flint and deer-antler types, with accounts of Dutch, German, English, Belgian, French, Hungarian, Swedish, Austrian, Swiss, Danish, Italian, Portuguese, and Polish mines and deposits, chiefly Neolithic. Becker (1959) dealt with flint mining in Neolithic Denmark.

O'Brien (2015, p. 206–213) discussed and illustrated tools used in Bronze Age mining, including hammers, mauls, antler picks, and wedges.

Schild and Sulgostowska (eds., 1997) edited a volume that includes references to flint mining, mostly Neolithic, in Switzerland, Pakistan (possibly pre-Neolithic), Poland, Slovakia (including possible pre-Neolithic items), Germany, England, Spain, Egypt (Middle Paleolithic), and Italy.

Schmid (1972) discussed a Mousterian silex mine and dwelling place in the Swiss Jura.

Russell (2000; see also Barber et al., 1999, with dates back to the Fourth Millennium BC) extensively described British Neolithic flint mines in Cretaceous chalk, including mining techniques and the antler tools (see Clutton-Brock, 1984, for details) used in the mining. Allard et al. (eds., 2008) include papers dealing with Neolithic European flint mining from Tertiary and Cretaceous source rocks in Scotland, the Netherlands, Belgium, France, Poland, Bulgaria, and Spain. Shepherd (1980) provided information about Neolithic flint-mining techniques from an engineer's viewpoint, dealing chiefly with English deposits plus some data on European occurrences.

Although not mining in the strict sense, the shallow 'quarries' into Ordovician tuff from Great Langdale and Scafell Pike 'axe factories' in the English Lake District (Claris and Quartermaine, 1989) are mentioned here since they involved large-volume Neolithic activity.

Bronze Age Obsidian Quarry

Broodbank (2000) provided some data about the volume of obsidian involved in the Melos source during the Bronze Age. Torrence (1986) described the obsidian quarries on Melos and considered just how they were exploited.

Williams-Thorpe (1995) emphasized the reliability of using the varied, but characteristic, compositions of various obsidians from the Mediterranean, central and eastern Europe, Anatolia, and the Near East to 'trademark' them and to use this information to decipher the routes used for trade and the like.

Metal, and Metal Mining and Smelting

Özdoğan and Özdoğan (1999; see also Yalçin and Pernicka, 1999, and Esin, 1999) discussed Prepottery Neolithic copper artifacts, mostly beads, from Çayönü Tepesi and Aşikli, Turkey, that were made from native copper by hammering and possibly by heating as well, i.e., no real metallurgy involved.

Craddock (1995) provided a comprehensive summary of various mining techniques practiced early on, including such things as drainage, ventilation, tools, and smelting.

Craddock (2001) reviewed the evidence of metal smelting in the transition from the Chalcolithic to the Bronze Age. He emphasized that the initial use of hearths to fabricate native copper resulted in items lacking any significant iron content, but that this changed with the Bronze Age. With the Bronze Age there is evidence of early wind-powered 'furnaces' suitably situated on hilltops to catch the prevailing winds, with charcoal as fuel. This was followed by furnaces using blowpipes and eventually with bellows and tuyere tips. He pointed out that this type of smelting had nothing to do with kilns used for the making of pottery or for the making of plasters; it is an independent process. Craddock (1999) concluded that the iron content of Bronze Age materials began to rise significantly toward the Middle Bronze Age, indicating the presence of reducing conditions with slagging, whereas earlier conditions were relatively oxidizing and non slagging, and thus unsuitable for the reduction of iron compounds to metal.

Golden et al. (2001) described Chalcolithic remnants and slags from small crucibles from Shiqmim in the northern Negev that were used for copper smelting. Bourgarit et al. (2003) described Chalcolithic fahlore slags from Cabrières in Southern France that developed under oxidizing conditions with copper prills present.

Glumac and Todd (1991) discussed the evidence of Chalcolithic smelting of copper in the Middle Danube Basin. Glumac and Tringham (1990) discussed some copper-bearing slags from Selevac in former Yugoslavia that are of Chalcolithic age.

Smith and Williams (2012) described evidence for Late Bronze Age copper-smelting activity in the Great Orme, Llandudno, Conwy area in terms of slag with copper prills. The nature of the actual furnace(s) was not determined. Chapman and Chapman (2013) described a primitive type of pit furnace from the area and showed a blowpipe that might have been used that is similar to one that might have been used in the Bronze Age pit furnaces.

Firesetting

Weisgerber and Willies (2000; see also Craddock, 1995, p. 33–37) provided an excellent account of firesetting in the Neolithic. Firesetting is the early mining practice of setting a fire against the rock face in a mine or quarry to open cracks, which can then be dealt with using primitive wedges, hammer-stones, and picks, the last of which commonly involved deer antlers. Firesetting was the early equivalent to the use of explosives for the same purpose. Following firesetting, water is thrown on the heated rock. Craddock (1995) indicated that the 'water' did not help to shatter the rock face, but rather might have been used to put out any embers left from the burned fuel used for firesetting. Crew and Crew (eds., 1990) described evidence of firesetting from many British Bronze Age copper mines, chiefly Irish and Welsh, and also described modern firesetting experiments. Penhallurick (1986, p. 71) suggested that firesetting might have been suggested to early humans who observed how stones adjacent to a hot fire became cracked.

Heldal et al. (2005) emphasized the extensive use of firesetting in Egyptian stone-quarrying activities that did not involve metal mining.

Mining and Quarrying Tools

From their earliest evidence, mining and quarrying involved the use of tools. Early mining and quarrying was very labor intensive. The tools used chiefly were such things as hammers made of oblong-shaped stone hafted with vines or other cordage-type materials and of hand-held, loose, rounded stones, 'pounders' (see Figure 16 *in* Harrell and Storemyr, 2009, for Egyptian examples) used as hammers, and of 'picks' and 'wedges' made from such things as deer antlers. Firesetting was employed to open up cracks in bedrock faces, which could then be further loosened by means of picks and wedges wedged into place. It is only with the Bronze Age and the subsequent Iron Age that more effective picks and wedges became available. Firesetting was finally made obsolescent only with the development of explosives and of modern-type drilling equipment.

Craddock (1995, p. 37–46) discussed the stone hammers used by the earlier miners as well as antler and bone picks, with metal tools coming in much later than the ubiquitous stone hammers.

Jovanović (1995) made the obvious comment that pre-Chalcolithic shallow mining of desired types of stone far back in time facilitated the transition to metal mining. Early Neolithic underground mining for flint nodules from the Chalk come to mind here.

Neolithic and Younger Mines

Weisgerber and Willies (2000) canvassed a number of Neolithic mines, some dating back to the Copper Age, involving hard-rock mining for copper and tin with references to locales from Central Asia to Western Europe.

Many of these ancient mines have yielded hammer-stones and more than a few deer antlers used as picks, both useful for loosening ore-bearing material after firesetting. Occurrence of hammer-stones in many Old World mines testifies to their widespread use in the actual mining process as well as in beneficiating ores. Crew and Crew (eds., 1990) described a number of Irish and Welsh Bronze Age hammer-stones and just what methods were used to haft them.

Penhallurick (1986, p. 61) suggested that copper smelting might have been suggested by what occurred during high-temperature reducing conditions used in pottery making.

Craddock (1995) dealt extensively with the 'development' of copper smelting in the Copper and Bronze Ages, emphasizing that the point or points of this development in the Old World are still uncertain, with a Sixth Millennium BC age indicated for the first evidence from the Near East, with extensive evidence beginning in the Fourth Millennium BC.

Čhernych (1992, p. 3) cited the oldest known metal occurrences in the Near Eastern Sixth Millennium BC with small copper and lead objects from Anatolia in the Seventh Millennium, more widespread copper and lead ornaments by the Sixth and Fifth Millennia BC, and copper implements, all in Mesopotamia, Anatolia, and possibly Iran. Giles and Kuijpers (1974) described a 2800 BC copper mining site at Koslu in Anatolia with wooden mining props in evidence.

Čhernych (1992) has the second half of the Fifth Millennium BC to the start of the Fourth Millennium BC for the Copper Age; the middle of the Fourth to the middle of the second half of the Third Millennium BC for the Early Bronze age; the second half of the Third Millennium to about the Sixteenth Century BC for the Middle Bronze Age; and from the Sixteenth to the Tenth or Ninth Centuries BC for the Late Bronze Age.

Muhly (1986) suggested that rare Near Eastern Seventh Millennium BC copper objects were made from native copper, with copper smelting dating from the Fifth Millennium BC.

Anthony (2007, p. 162–163) mentioned the appearance of copper about 5000 BC, while (Čhernych, 1992, Figure 5) cited Seventh millennium BC. Anthony (2007, p. 124–125) discussed the timing of Bronze Age beginnings in Europe.

Čhernych (1992, Figure 2) made it clear that widespread copper usage preceded arsenical bronze, which in turn was followed by tin bronze, followed by iron.

Craddock, ed. (1980) includes papers that deal with copper mining in the Copper Age and Bronze Age in several Irish counties, one Spanish locality (Huelva), and a Yugoslav locality, and gives references to Afghan, Iranian, Israeli, and Omani localities, plus mining of lead and silver in the Aegean.

Ancient Copper Mines

Europe

O'Brien (2015) gives a comprehensive account of European Copper Age and Bronze Age copper mining, from which much of the following material is abstracted. He discussed, described, and illustrated mining tools, including stone hammers, antler implements used as picks and wedges, and ancient adits, shafts, and stopes.

Kienlin (2012) reviewed Chalcolithic metallurgy in Central Europe and the Carpathian Basin.

Southeast Europe

Jovanović (1979) reviewed the Ai Bunar and Rudna Glava mines and the tools used by the miners.

Serbia

Jovanović (1982; see also, 1978, 1988) described a Serbian Bronze Age copper mine at Rudna Glava, complete with hammer-stones. O'Brien (2015, p. 45) indicated 5400–5300 BC for initial copper mining here. Jovanović and Ottaway (1976) described Chalcolithic (Vinča Culture) copper objects (beads, etc.) and details of the Rudna Glava deposit. Borić (2009) dealt extensively with the dating of the Vinča Culture beginning in the Chalcolithic and also provided details about the Rudna Glava mines.

Bulgaria

Čhernych (1978; see also O'Brien, 2015, p. 34–53) described the Bulgarian shallow shafts and surface trenches worked for oxidized copper ore (malachite and azurite) in the Fourth Millennium BC. Čhernych (1978) described the shallow shafts and opencast copper workings, Fourth Millennium BC, Copper Age from Aibunar, Bulgaria; these are among the oldest known from Europe.

Eastern and Central Mediterranean

Greece and the Aegean Islands

Dill et al. (2008) briefly commented in Early Bronze Age copper mining on Seriphos in the Cyclades. O'Brien (2015) summarized the scattered evidence of potential Early Bronze Age copper mining in the Aegean Islands (his Figure 1), and commented on other potential sites.

Stos-Gale (1989) discussed Cycladic copper metallurgy, particularly on Kythnos, with the thought that local ores were later replaced by those from Laurium.

Cyprus

O'Brien (2015, p. 58–66) summarized the extensive evidence of Early Bronze Age and younger copper mining and smelting.

Sardinia and Corsica

O'Brien (2015, p. 67) summarized the evidence of Bronze Age copper mining on Sardinia and possible sites on Corsica.

Italy

Maggi and Pearce (2005) reviewed the Bronze Age and Copper Age Italian copper mines of Liguria, somewhat younger than the oldest European copper mines of Bulgaria. O'Brien (2015, p. 67–76) reviewed the extensive evidence of Bronze Age copper mining in various parts of Italy.

Iberia and the Western Mediterranean

O'Brien (2015, p. 75–76) summarized the possibility that Spanish copper mining and smelting developed independently.

Southwest Spain

Rothenberg and Freijeiro (1980) described Chalcolithic and younger copper mining at Chinflon, Huelva, but O'Brien (2015, p. 85–87) made it clear that Late Bronze Age is more likely; he also commented on numerous other copper localities in the same region.

Portugal

O'Brien (2015, p. 89–90) summarized the limited evidence of Chalcolithic and Bronze Age copper mining in Portugal.

Southeast and Central-East Spain

O'Brien (2015, p. 90–92) summarized the limited evidence of Chalcolithic and Bronze Age copper mining in the Balearic Islands. In Southeastern Spain there is no recognized evidence of mining, but the presence of Chalcolithic copper metallurgy and slags strongly suggests that mines are present (Hook et al., 1991).

Northern Spain

O'Brien (2015, p. 92–103) discussed the evidence of Chalcolithic and Bronze Age copper mining, with emphasis on locales in León and Asturias.

France and the Western Alps

O'Brien (2015, p. 105) cited evidence of Late Bronze Age copper smelting plus the potential presence of mine sites in Graubunden, Switzerland, and discussed the Cabrières area, where there are Third Millennium BC copper mines in the northeast Alps and the Provence–Alpes–Côte-d'Azur regions. Gattiglia and Rossi (1995) dealt with the Saint-Véran (Hautes-Alpes) copper mine and nearby smelting site with remains of tuyeres and slags of Chalcolithic or Early Bronze Age.

Other Mines in Southern France

O'Brien (2015, p. 115–117) discussed Early Bronze Age copper mines along the southwestern margin of the Massif Central and in the western Pyrenees.

The French Alps

O'Brien (2015, p. 117–121) briefly discussed evidence for Chalcolithic and Bronze Age mining in the Rhône-Alps and the Provence–Alpes–Côte-d'Azur.

Northern Europe

O'Brien (1961) reviewed the nature of various Irish and English Bronze Age copper mines, mostly very shallow but with a few exceptions, and cited the hammer-stones and antler tools found with them, plus evidence for firesetting. Crew and Crew (eds., 1990) discussed and described many British Bronze Age copper mines, chiefly Irish and Welsh.

Ireland

Jackson (1968) described many aspects of the County Cork, southwest Ireland, Bronze Age copper mines.

Jackson (1980) tabulated the Early Bronze Age Irish copper mines in Counties Cork and Kerry and referred to some of the actual workings, together with the stone mauls found associated with them. O'Brien (2015, p. 125–137) discussed the Chalcolithic and Bronze Age copper mines of Ireland.

Britain

O'Brien (2015, p. 138–150) discussed British Chalcolithic and Bronze Age copper mines, chiefly Welsh plus a few English localities.

Central and Eastern Europe

The Austrian Mines

Stöllner et al. (2006) reviewed Middle Bronze Age copper mining in the Austrian Mitterberg region, with mines reaching depths of 200 meters. Rieser and Schrattenthaler (1998; see also Höppner et al., 2005) provided a survey of the Bronze Age copper working of the Schwaz-Brixlegg, Tyrol region, with an extensive account of stone hammers found in the deposits and evidences of firesetting. O'Brien (2015, p. 161–184) extensively reviewed the various Austrian Bronze Age copper mines with their complex ores.

Slovakia

O'Brien (2015, p. 186–187) reviewed the evidence of Chalcolithic and Bronze Age copper mining in Slovakia.

Southern Urals

Koryakova and Epimakhov (2007, p. 28, Figure 1.3) referred to Middle Bronze Age copper mines in the Southern Trans-Urals. O'Brien (2015, p. 187–193) reviewed the extensive Kargaly Late Bronze Age deposits.

Near East

Comment: In view of the extensive, relatively detailed material provided by O'Brien (2015) for the European Chalcolithic and Bronze Age, it is obvious that much work of a similar nature remains to be done for the Near East and Asia. The following materials should be considered as a statement of the present condition of our limited knowledge.

Holzer and Momenzadeh (1971) described the Early Bronze Age copper mines in the Veshnoveh Area of West-Central Iran, with evidence of firesetting and stone mauls, giving an Early Bronze Age date of approximately 3200 BC.

Weisgerber and Hauptmann (1988) reviewed the evidence of Bronze Age copper mining in the Sinai and areas to the north of the Dead Sea. Weisgerber (2006) described the Feinan and Timna deposits in some detail. Hauptmann (1987) discussed the Timna and Feinan deposits in Sinai and Jordan with Chalcolithic and Bronze Age mining, and the presence of trade in these ores into adjacent Palestinian and Egyptian sites for smelting.

Jovanović (1982; see also, 1978, 1988) also cited Bronze Age copper mining in Iran and Anatolia. De Jesus (1980, Part i) discussed ancient copper mining in Anatolia. Jovanović (1980, p. 164) cited Kozlu in central Turkey as an Early Bronze Age site with shafts 50 meters deep.

Weisgerber (1978) described the evidence for Third Millennium BC copper mining at Samad, Oman. Goertler et al. (1976) cited a number of Oman Bronze Age copper mining sites as well as slags. Hauptmann et al. (1988) discussed the Third Millennium BC copper-rich slags from Oman and made clear that Oman had produced considerable copper for export to Ur and Sumer.

Asia

Černych (1992, Figure 3) pointed out that in northern Eurasia mining and metallurgy centers were concentrated in the Caucasus, the Urals, Kazakhstan, former Soviet Central Asia, the Sayan–Altai, and Transbaikalia, with extensive trade in metals from these regions.

Černych (1992, p. 276) briefly cited copper mining in Transcaucasia and the Main Caucasus Range.

Mudzhiri and Kvirikadze (1979) described a Bronze Age copper working from Abkhazia with associated hammer-stones.

Zhou Baoquan et al. (1988) described end-Second-Millennium BC copper mining from Hubei Province with well-preserved tools and equipment. Du Faqing and Gao Wuxun (1980) briefly reviewed Bronze Age and other intervals for nonferrous mining activity in China. Wagner (1986) discussed the Tonglüshan copper mining site near Wuhan, and Vogel (1982) discussed some details of the mine, as did Liu et al. (1993), who also provided details concerning some wooden structures still present, as well as some of the bronze and wooden tools, plus reference to additional Chinese copper mining elsewhere.

Chakrabarti (1988) discussed ancient copper mines from Eastern India.

Although there do not appear to be any comprehensive studies of the evidence for Bronze Age copper mining and mines in Southeast Asia, there are enough slag heaps, together with limited evidence for mine openings and the like, to deduce vigorous Bronze Age copper mining in the region.

Bennett (1988; see also Bennett, 1990, 1989, and Villiers, 1984) discussed occurrences of copper ore in Thailand associated with smelting activities, which one assumes would correspond to Bronze Age and younger mining activities at many Thai sites. Pigott and Natapintu (1988) briefly discussed some First Millennium BC copper mining sites in northeastern Thailand. Pigott and Weisgerber (1999) discussed the copper-mining complex at Phu Lon in northeast Thailand, probably from the Second Millennium BC. Pigott et al. (1997) discussed copper smelting and mining activities from the Khao Wong Prachan Valley area of Central Thailand in the First Millennium BC. Natapintu (1988) cited various Thai copper mining sites, with some information about age.

Ancient Tin Deposits

Old World

Andráš (2008) provided a useful summary of many occurrences of tin and copper mines from the Mediterranean to Central Asia.

Yener and Vandiver (1993; see also the 'Appendix' to this paper by Willies; see also Yener, 1995, 1998; Yener et al., 1989; Kaptan, 1995) discussed Early Bronze age tin deposits of the Taurus,

Turkey, particularly those of the Kestel mining complex, while dealing with and refuting previously published skeptical comments, and also discussed the nearby beneficiating and smelting at Göltepe. Yener (1994) discussed the actual mines, the evidence for firesetting, and the possible use of children in the mines.

Nezafati et al. (2006) discussed an ancient, probably First Millennium BC, tin deposit in the Zagros.

Sabet et al. (1976) described some placer tin deposits in Egypt's Eastern Desert.

Parzinger and Boroffka (2003; see also Boroffka et al., 2002, for associated Central Asian tin deposits) described occurrences of tin ore in Central Asia, with their associated human communities in the Zeravshan Valley region, reporting that the actual workings were of the order to 5 to 15 meters in length and that granite and vein quartz sites, i.e., hard rock, were handled by using firesetting at the face to crack the rocks, making the ore-containing material available for processing with hammer stones. Alimov et al. (1998) discussed the ancient tin deposits of Uzbekistan and adjacent Tadjikistan in some detail, including the deposits at Karnab and Musiston.

Higham (1996) cited the various Bronze Age (relatively young, several thousand years old) deposits of tin and copper of Southeast Asia.

Penhallurick (1986, p. 173–224) indicated that all of the Bronze Age Cornish tin came from alluvial and colluvial deposits, worked back to about 2000 BC; hard-rock tin mining in Cornwall is a far more recent development of the post–Roman occupation of Britain.

The virtual absence of tin deposits in North America is a good explanation for the absence of a local Bronze Age; native copper from Michigan and elsewhere was used by the Native Americans for various objects. It is worth noting that the elements are not all equally distributed on the modern continents; tungsten is also very poorly represented in North America.

South America

Lechtman (1980) briefly discussed the presence of tin bronze from Andean locales associated with the Incas, a contrast with North America, which lacks any evidence for bronze owing to the almost complete absence of tin sources from that continent. Numerous copper deposits are available in both continents.

Ancient Lead, Silver, Gold, and Zinc Deposits

Craddock (1995) dealt with the recovery of silver from diverse ores, including the fabled deposits at Laurium in Greece. Hopper (1968) dealt at some length with the Laurium mines, including the complexity of the deposits; their general geology; and various other aspects of the process of getting the silver from the argentiferous galena ore. Conophagos (1980) went into detail about the many various aspects of the Laurium Mines: the actual mining techniques; ore beneficiation; and smelting, which involved the recovery of silver from the lead. He then dealt with the reduction of the litharge (red lead monoxide) back to lead, and the coining process of the famed Athenian silver tetradrachms with the Owl of Athens on the reverse side of the coin.

Wagner et al. (1980) described Early Bronze Age lead–silver mining on Siphnos in the Aegean.

Craddock (1995, p. 125) cited the earliest appearance of lead in an artifact of the Sixth millennium BC.

Jones (1980) described a number of remnants left at Rio Tinto, Spain, from Roman mining, chiefly for silver, with subsidiary gold and copper, with evidence of earlier Seventh Century BC Punic work. Craddock (1995; see also Willies et al., 1984 for lead and zinc deposits at Zawar) briefly described the lead–silver deposit at Dariba, Rajasthan, India, which was probably worked at least back to the late Second Millennium BC, with workings that penetrated to 100 meters below the water table, a real record for ancient mining, with systems of bailing undoubtedly used in ancient mines to keep the mine-water level within limits.

Gold Mining

Klemm et al. (2001) described Egyptian gold mining from Pre-Dynastic times on in some detail. The Egyptian hard-rock gold mining is exceptional since most early gold mining was undoubtedly restricted to the recovery of alluvial gold by one technique or another from many widespread sources. Allchin (1962) provided an older account of Peninsular India gold-mining activity, including firesetting and some stone and metal tools from approximately 2000 years BP. Lewis and Jones (1970) described Roman alluvial gold deposits in Northwestern Spain that were largely exploited hydraulically, employing aqueducts and dams, among other things, as well as some hard-rock mining.

Salt Mining and Production

Salt is necessary for life! Pre-Neolithic humans undoubtedly obtained the bulk of their salt from meat, whereas agriculturalists, depending on grain, legumes, and other agricultural products, needed outside sources of salt for health reasons and also as a preservative for storing food such as meat, and for pickling. Weller (2002) briefly described Neolithic salt obtained from a Catalonian salt dome, while referring to another salt dome in Romania. Underground salt mines are a post-Neolithic development. Weller and Dumitroaia (2005) described a Romanian Neolithic locality featuring salt springs and associated 'briquetages' (crude pottery used to produce salt from salt brine with the use of fires, found as debris at salt production sites). Weller (2000) cited a number of European locales, from Romania to France, where briquetage occurs with age extending back to the Sixth Millennium BC, and illustrates some of these vessels. The briquetage occurs as fragments left after the salt cake was freed from the ceramic material. Monah (2002; see also Andronic, 1989; Ursulescu, 1995, provided information about Moldavian localities) added more information about eastern Carpathian salt springs and Neolithic salt production, including some briquetage occurrences in Romania. Tasić (2000) pointed out the obvious when commenting that the Early Neolithic Çatal Hüyük site is relatively close to a salt lake that evaporates during the summer and that Early Neolithic Jericho is close to the Dead Sea and its shoreline salts. Similar situations were probably present at many other Neolithic sites! Nikolov and Bacvarov (eds., 2012) provided a series of papers dealing with European Neolithic sites, some fairly Early Neolithic, where brines were boiled or heated enough to remove water by evaporation in ovens, but not enough to actually boil down to salt in crude pottery vessels (briquetage; see p. 226, Figure 1, for diagram showing the form and age distribution in Europe of briquetage vessels) that were broken away from the salt block. Many of the papers provide details of the basic brine boiling process and a good discussion of the localities where briquetage fragments occur, and evidence of Bronze Age rock-salt mining (p. 190–195) in Transylvania and Maramureş.

Flad (2007; see also Flad et al., 2005) described salt production with associated briquetage from a Late Neolithic, Third Millennium BC Chinese site at Zhongba, northeast of Chongqing (Chungking). Li and von Falkenhausen (eds., 2006, 2010, 2013) provided an extensive account of salt production in the Upper Yangtze Basin, South China, dating back to the Late Neolithic, about Fifth Millennium BC, with extensive description of briquetage vessels, among other things, emphasizing the Zhongba Site. Wagner (2001) described salt making from marine water in Shandong (Shantong), employing iron vessels for boiling.

Habu (2004) cited Japanese Late and Final Jomon salt-evaporating pots.

Nitta (1996) briefly described salt manufacture in the late First Millennium BC in the Isan area of the Khorat Plateau, eastern Thailand. The salt was derived from underlying Cretaceous salt deposits overlain by younger strata through which water percolated to the surface. This brine was then boiled in ceramic pots similar to those used in Europe, with their characteristic sherds present at the sites. The pots were broken after the water had been evaporated to release the salt.

Barth (1982) provided an introduction to the First Millennium BC Hallstatt salt mines in Austria, with comments about the mining practices. Stöllner et al. (2003) extensively described the earlier Iron Age Dürrnberg-bei-Hallein Austrian rock-salt mines from Permian and Early Triassic beds, with both children and adults involved in the mining.

Ocher; Hematite Mining

McBrearty and Brooks (2000, p. 526; see also Boshier and Beaumont, 1972, for some details) summarized evidence of hematite mining for pigment in Southern Africa as far back as the Middle Stone Age. Bar-Yosef and Valla (eds., 1991) discussed Natufian ocher occurrences. Bordes (1961) cited lumps of manganese oxide, some in 'pencil' form, from the French Mousterian.

Brick

Brick for construction purposes does not occur naturally. The oldest bricks for construction purposes appear in the Neolithic as variously shaped mud brick (Kenyon, 1981, cited and illustrated examples from the Neolithic of Jericho), and Schmandt-Besserat (1977) cited earlier Neolithic examples from Tell Aswad, Syria. Vat (1940) commented on the use of fired brick at Bronze Age Harappa, additional to mud brick, while Marshall (1931) discussed the burnt brick present at Mohenjo-Daro. Fired brick first appears for widespread construction purposes with the Romans. Moorey (1994, p. 302–329) dealt extensively with brick making in ancient Mesopotamia.

Wood and Reed (see also Hafting, Below)

Owing to poor preservation, there is very little information concerning wood use by hominids. Movius (1950) described a German Acheulian sharpened yew hunting spear found between the ribs of a straight-tusked elephant. Dennell (1997) and Thieme (1997; see also Balter, 2014) discussed the technological–cultural significance of a wooden throwing spear from a 400,000-year-old Paleolithic locality in Germany and its implications for the presence of an advanced hunting technology far earlier than previously assumed. Thieme (2005) described the occurrence at Schöningen, Lower Saxony, of a number of wooden hunting spears with the skeletons of numerous horses and a few hearths plus one stick that might have been a spit over which horse meat was smoked. Tyldesley and Bahn (1983) summarized much of the available data: postholes from the Mousterian and even the Lower Paleolithic indicate use for structural framework. Some of the Late Paleolithic cave art is present high enough on cave walls to have required the use of ladders of some sort. Charcoal from fires and for use as a pigment is also involved. A few fire-hardened spears are recorded, and the presence of spear throwers certainly indicates the use of wooden spears. Bone hafts and split bone bases of points are indicated. Polish on some Oldowan stone tools (Keeley and Toth, 1981; Keeley, 1977) indicates that they were used on wood. Dominguez-Rodrigo et al. (2001) found plant residues, probably from wood, on Acheulian stone tools from Tanzania. It is clear, however, that our knowledge of prehistoric wood use is very limited.

Fagan and Van Noten (1966) described Late Stone Age wooden implements from Zambia at a hot-springs site, including wooden posts, possibly part of a windbreak, wooden arrowheads associated with a plant poison, and digging sticks. Bordes (1961) described a Mousterian posthole cast (occupied originally by a pole) and some worked lumps of manganese oxide presumably used as a colorant. Deacon and Deacon (1999, p. 145–147, Figure 8.6; see Ouzman, 1997, for an extended account) described digging sticks and the bored stones used to weight them ('digging-stick weights') while being used. He shows a rock-art depiction of them, and with women carrying them; the interpretation of these 'bored stones' in the 'absence', i.e., non-survival, of their wooden digging sticks from various sites elsewhere of various ages is easily explained in terms of these South African examples.

Bergman (1993) discussed European Middle Paleolithic bows and arrows. Lombard (2011) considered the possibility of bow-and-arrow technology in the Middle Stone Age of South Africa.

Yerkes et al. (2003) discussed Early Stone Age tools used for woodworking, determined by microwear analysis, discriminating tools such as adzes and chisels used for 'finer' woodworking from axes used for such things as tree felling. Dominguez-Rodrigo et al. (2001b) described plant phytoliths from Acheulian stone tools from Peninj, Tanzania, unlike the plant phytoliths in associated sediment, and with plant fibers as well, as evidence for use on wood. Binneman and Beaumont (1992) described microscopic evidence of plant material, possibly wood and/or sedges, from two Acheulian handaxes from the Northern Cape.

Momber et al. (eds., 2011) made the important point that the techniques of tree felling and making planks and possibly dugout boats with stone tools are significantly different from those employing metal tools, beginning with the Bronze Age.

Belitzky et al. (1991) described a Middle Pleistocene wooden plank showing evidence of man-made polish. Dugout boats are known, but their ages are all less than 10,000 years BP, obviously a minimum age.

Rudenko (1970) described some small wooden tables with detachable legs (lathe-turned in some cases) and a shallow concavity suggesting that they were used for the serving of food from the Iron Age of the Pazyryk burials.

Lacquerware

Webb (2000) provided a good description of lacquer and lacquerware. The oldest known lacquerware is from the Hemudu Culture of Zhejiang Province, China, about 7000 years BP, and younger examples are known from India to Japan (Habu, 2004, cited Early Jomon). Zhou Bao Zhong (1988) described Hemudu lacquerware. Chang (1986, p. 298) cited Shang Dynasty lacquerware, and also pre-Shang Hsia-age lacquerware.

Bone

D'Errico and Backwell (2003) discussed the characteristics of bone modified by men as tools with emphasis placed on Early Stone Age samples from Swartkrans, South Africa. Backwell and d'Errico (2005) concluded that bone tools were used by *Homo erectus* in the Oldowan and earlier Acheulian but not to any extent later, nor do they indicate a very early stage in tool sophistication. Mania and Vlček (1987) described bone and ivory tools from the *Homo erectus* site at Bilzingsleben, eastern Germany. Brain and Shipman (1993) made a reasonable case for Swartkrans bone tools having been used to dig for tubers and corms.

Henshilwood et al. (2001) described bone tools in the Middle Stone Age (70,000 years BP) at Blombos Cave, South Africa. Yellen et al. (1995) described barbed and unbarbed points from a Middle Stone Age site (Katanda) in Zaire with a minimum age of 90,000 years.

Knecht (1993) discussed Early Upper Paleolithic European bone and antler split-based projectiles. Belfer-Cohen and Bar-Yosef (1999) briefly cited some Levantine bone 'objects' from the Aurignaican of Hayonim Cave.

Yellen (1998) discussed and described Saharan and sub-Saharan barbed bone points, presumably used mostly for fishing, as both harpoons and as hafted points, with very few from Europe (a Magdalenian example is cited), with ages from Middle Stone Age and younger in Africa.

Ivory

Mania and Vlček (1987) described ivory tools from the *Homo erectus* site at Bilzlingen, eastern Germany. Einwögerer et al. (2006) described an infant burial with some associated ivory beads from the Austrian Upper Paleolithic, about 27,000 years BP. Hahn (1986) cited South German Upper Paleolithic ivory figurines of animals and humans. Soffer (1985) cited Aurignacian-type Central Russian Plains items inscribed onto ivory. White (1993) described Aurignacian Eurasian mammoth-ivory beads. Bader and Lavrushin (eds., 1998) described numerous mammoth-ivory beads from a Gravettian site near Vladimir, east of Moscow. Conard et al. (2009) described some German Aurignacian mammoth ivory flute fragments. Moorey (1994) discussed ivory in the Near East, Southern Asia, and North Africa chiefly from the archaeological literature, i.e., relatively recent. The Oriental Ceramic Society (1984) produced an elegantly illustrated 'Catalogue' of carved Chinese ivories dating back to the Shang Dynasty.

Antler

Mania and Vlček (1987) described tools made from deer antlers from the *Homo erectus* site at Bilzingsleben, eastern Germany.

Brain (1988) interpreted a number of antler and long-bone artifacts with evidence of wear and scratches at their tips as digging sticks for obtaining vegetable foods from the Early Stone Age of Swartkrans Cave, with the inference that a robust australopithecene was involved.

Knecht (1993) discussed Early Upper Paleolithic European bone and antler split-based projectiles that clearly were hafted, as well as lozenge-shaped and spindle-shaped points.

Knecht (1983b) considered the use of antlers in making split-based, hafted projectiles of Aurignacian age. Cattelain (1999) described a Solutrian spear thrower made out of reindeer antler from the Dordogne. Belfer-Cohen and Bar-Yosef (1999) cited some Levantine, Aurignacian antler 'objects' from Hayonim Cave.

Campana (1989) discussed bone and antler tools, their manufacture, use, and microscopic surface evidence of use from the Natufian of Israel and the Proto-Neolithic of the Zagros.

Bogucki and Grygiel (1983) cited and illustrated some Polish earlier Neolithic spoons made from wood and red-deer antler.

Russell (2000; see also Barber et al. (1999), with dates back to the Fourth Millennium BC) extensively described British Neolithic flint mines in Cretaceous chalk, including mining techniques and the antler tools used in the mining (see Clutton-Brock, 1984, for details).

Leather and Rawhide

Leather is commonly made with tannin-producing vegetation, as opposed to rawhide, with the latter being 'untreated'. The oldest leather appears to be represented in the Old World by a Fourth Millennium BC shoe found in Armenia (Pinhasi et al., 2010). Leather easily decomposes under natural conditions, which accounts for its rarity and the minimum age cited here. Bar-Adon (1980) figured and discussed leather sandals from the Chalcolithic of Israel.

Egg and Spindler (2009) described the footwear of the Ötztaler Alpine 'iceman', including a leather sandal, dried grass, and a cord covering, a 5000-year-old Copper Age example.

Sherratt (1981, p. 282–283; see also Barber, 1991, 2001) indicated the presence of leather in the Neolithic with buttons. Griaznov and Boulgakov (1958; see also Rudenko, 1970) illustrated a number of leather objects from the Pazyryk graves dated approximately 2400 years BP.

Garrod and Bate (1937) illustrate and cite lissoirs presumed to have been used in 'smoothing' leather from the Mount Carmel Natufian.

Stöllner (1999) briefly described leather articles (shoes and backpacks) from the Iron Age Dürrnberg salt mines.

Pottery and Ceramics (See Also Kilns, Pottery Wheels, Below)

Clay deposits suitable for pottery and ceramic-making are widespread wherever the weathering of suitable rocks has occurred.

Vandiver et al. (1989) made the important distinction between 'pottery' and fired ceramic materials with pottery appearing significantly later than ceramics.

Ceramics

Farbstein et al. (2012; see also Einwögerer, 2,000, for some Austrian material from Krems) described some Upper Paleolithic ceramic figurines from Croatia, dated at 15,000–17,500 years BP. Soffer et al. (1992; see also Soffer et al., 1998, for Gravettian ceramics) described the Late Paleolithic Moravian ceramics and both hearths and kilns where the various ceramic objects were recovered, while making clear that pottery appears significantly later than ceramics. Frierman (1971) suggested that during the Pre-Pottery Neolithic B the use of lime plaster for various purposes would have led to the observation that the interiors of lime kilns included clay that had been transformed into ceramic-like material, and that fireplaces in continual use might have led to a similar observation. This is better understood combined with the fact that humans were modeling in clay as far back as the earlier Paleolithic.

Rollefson et al. (1992) cited early Yarmoukian Pottery Neolithic ceramics from Ain Ghazal. Moorey (1994) cited Mesopotamian ceramics dating from approximately 6000 BC.

Pottery

Barnett and Hoopes (eds., 1995) discussed the inceptions of pottery in various parts of the world, emphasizing that it did not necessarily coincide with the beginnings of agriculture.

Wu et al. (2012; see also Chang, 1986, p. 105) described 20,000-year-old Chinese pottery, among the oldest yet known, and pre–East Asian agriculture, with the implication that the pottery was used for cooking.

Kuzmin (2006) reviewed the earliest pottery from East Asia and estimates a 13,700–13,300 years BP age. Habu (2004) describes Jomon pottery from Japan dated at 16,500 BP and from Maritime Russia possibly at 16,000 BP, while Aikens (1995) described early Jomon Japanese pottery from about 12,700 BP.

Anthony (2007, p. 148–149) noted 7000–6500 BC as the oldest European pottery, i.e., near the beginnings of the Neolithic if that interval is defined as the first appearance of pottery, or recognizing that the pottery-Neolithic extended from about 8700 BC to approximately 7000 BC (the Pre-Pottery Neolithic B).

Moorey (1994, p. 141–215) dealt extensively with Mesopotamian pottery, faience, and glass making. Moore (1995, p. 40) cited 8300 BP from Çatal Hüyük.

Barham and Mitchell (2008) discussed the mid-Neolithic and younger evidences of pottery in various parts of Africa. Close (1995) discussed the late Tenth to earlier Ninth Millennium BC appearance of pottery across much of northern Africa and remarked on the relatively small quantities except in the Nile Valley (see Jesse, 2000, for some possible examples from Sixth Millennium BC northwest Sudan). Breunig et al. (1996) discussed pottery from the Lake Chad shoreline in northeastern Nigeria, dated approximately 6000 years BP. Sadr and Sampson (2006; see also Mazel, 1992) concluded that pottery in southern Africa preceded the local Iron Age by only a few centuries.

Haaland (2007) makes the important observation that pottery appears in the southern Sahara and Sahel in the Tenth Millennium BC, several thousand years before it appears in the Near East, with the implication that in Africa it was used for stewing of fish and for other aquatic materials, as well as porridges and possibly for beer.

Tegel et al. (2012) found Early Neolithic German pottery involving broken pots that had been repaired with birch tar.

Kilns

Ellis (1984) discussed various kiln types, earlier Neolithic to post-Neolithic, from Eastern Europe. Kilns are, of course, an essential part of the pottery-making process. Moorey (1994) dealt briefly with Near Eastern kilns, chiefly from the more recent past. Alizadeh (1985) described in some detail some Chalcolithic, 6800 BC, kilns from Chogha Mish, western Iran, and made comparisons with older kilns from elsewhere.

Pottery Wheels

The earliest wheel-built pottery, using a Late Chalcolithic end-Fifth Millennium BC or first-half Fourth Millennium BC slow wheel (tournette), is described and discussed by Roux and de Mirodchedji (2009) from the Southern Levant; they make it clear that such wheels used a coiled clay base that was 'finished' on the tournette with the tournette being incapable of dealing with large lumps of clay as opposed to the capabilities of the later-appearing fast wheel. Moorey (1994) described a number of Near Eastern wheels, chiefly from historical sites.

Wheels appear to have been used for pottery-making before they were used for transport, a matter to be verified.

Plaster and Bitumen

Limestone suitable for burning is widespread in many parts of the World, but gypsum suitable for gypsum plasters has a far more restricted occurrence.

Kuijt (1994) provides evidence that Pre-Pottery Neolithic A floors were earthen, followed by plastered floors in the Pre-Pottery Neolithic B. Several papers in Kuijt, ed. (2000) cited the presence of lime plaster on the floors and 'plastered' skulls (see Hodder, 2006, Figure 7, for a good example) at many Neolithic Pre-Pottery sites. The burning of limestone for lime plaster was a substantial advance. Kingery et al. (1988) discussed the lime-plaster on such skulls.

Garfinkel (1987) considered the lime floor-plastering of the Pre-Pottery Neolithic B to late Pre-Pottery Neolithic in the Near East, pointing out that the production of *tons* of burnt lime was necessary for this plastering and that the labor involved a number of workers and the use of lime kilns. Rollefson et al. (1992) mentioned lime-plastered floors painted with red ocher.

Kingery et al.'s (1988) important treatment of pyrotechnology cites a lime-burning Natufian hearth with a chimney from the Tenth Millennium BC and discusses the addition of tempering material to the plaster, including mineral grains and plant fibers. They pointed out that architectural use of lime plaster dates back to the Natufian, with floors in the Pre-Pottery Neolithic B, emphasizing that small villages still had earth floors, with the thick plaster floors characteristic of large villages. Kingery et al. (1988) cited work in China that described similar lime plastering there at a somewhat later date. They also cited lime plaster in the Epipaleolithic, 12,000 years BP.

Kingery et al. (1988) also discussed the likelihood that pre—Late Paleolithic people were familiar from far back in time with the fact that limestone became calcined while exposed to fire in hearths, and that when water was added, the slaked product became plastic, but advantage of this fact could not be taken until the Natufian advent of dwellings, and similarly with gypsum plaster. They also suggest that the high temperature of the lime calcining process could also have led to the observation that metal-bearing minerals would produce metal at high temperatures. Needless to say, the production of lime plaster required a large supply of competent manpower, i.e., larger villages where some people could specialize.

Hodder (2006, p. 60–61) discussed 'soft' plasters used in the earlier Neolithic, which involved more frequent replastering of both walls and floors.

Kingery et al. (1988) discussed the lime-burning process for lime plaster and the allied process for gypsum plaster, and also the burnishing process, as well as the addition of tempering materials. They also showed (their Figure 14) that Near Eastern localities characterized by lime plaster are geographically very distinct from those characterized by gypsum plaster.

Hauptmann and Yalcin (2000) pointed out that the use of raw materials for lime burning that contained enough 'contaminants', including silica and alkali, could have produced superior, harder, pozzolanic-type results. Malinowski (1982) described a number of Classical-interval mortars and concretes and their uses in locales from ancient Crete to Rome.

Moorey (1994, p. 332–335) discussed the use of bitumen in early Mesopotamia as a plastering material.

Forbes (1936) and Figuier (1996 [reprint of 1876]) provide overviews on the use of bitumen and related substances in antiquity.

Metal

Gold

Presumably, primitive man early on became aware of gold as small, bright flakes and nuggets in streams, as these folks paid close attention to all aspects of their environment. Gold early on was obtained as alluvial gold, there being no evidence of hard-rock gold mining until the Bronze Age. Anomalously, there are no occurrences of gold in any form linked to humans until the Copper Age of Bulgaria (Chapman et al., 2006), when various types of gold ornaments turn up in Copper Age graves at Varna and in the Chalcolithic occurrence cited below. We have no explanation for this phenomenon. Hauptmann and Klein (2009) provide a few details about Fourth-Millennium BC underground gold mining in Georgia. Gopher et al. (1990) and Gopher and Tsuk (1991) described some gold artifacts of Chalcolithic Age from Nahal Qanah Cave in the Samaria Hills, with the gold coming from hard rock, not an alluvial source, and with the objects smelted at 1064°C and then cast in molds. Mascetti and Triossi (1999) cited gold earrings from the Levant at 5000 years BP. Hartmann (1970) described Bronze Age and younger gold artifacts from Europe that further underline the fact that pre-Chalcolithic gold artifacts are unknown. This is a mystery that remains unsolved at present. There are Pre-Pottery Neolithic hammered objects made from native copper, so why are there no known pre-Chalcolithic gold artifacts?

Copper and Bronze

Černych (1992, Figure 2) made it clear that widespread copper use preceded arsenical bronze, which in turn was followed by tin bronze followed by iron.

Anthony (2007, p. 124–125) discussed the timing of Bronze Age beginnings in Europe. Anthony (2007, p. 162–163) mentioned copper appearing approximately 5000 BC, while Černych (1992, Figure 5; see also Maddin et al., 1991, for native copper artifacts from Turkey) cited Seventh Millennium BC. Černych (1992, p. 3) cited the oldest known metal occurrences in the Near Eastern Sixth Millennium BC, with copper implements and small copper and lead objects from Anatolia in the Seventh Millennia BC, with more widespread copper and lead ornaments by the Sixth and Fifth millennia BC, all in Mesopotamia, Anatolia, and possibly Iran. Yener (2000, p. 1) states the early presence in Anatolia of copper objects as Eighth Millennium BC, and extensively documents the Anatolian development of metal mining, including tin, and the processing of ores from very early. Tylecote (1976, Tables 5, 6) compiled information on smelted copper artifacts and copper metallurgy in the Near East and Southeastern Europe. Tylecote (1976) also surveyed early occurrences of bronze from Europe, the Near East, and Asia, and also described smelting techniques for bronze and iron. De Jesus (1980, part i) devoted considerable attention to copper smelting in antiquity, while casting doubt on its origin as having anything to do with campfires, and also (1980, part ii) surveyed Anatolian occurrences. Bartelheim et al. (2003) described smelting of fahlerz-type copper ores at Brixlegg, Austria, dated to the second half of the Fifth Millennium BC.

Steuer and Zimmermann (eds., 1993) reviewed many of the European Bronze Age to Medieval mining districts in terms of actual mining activities and also ore processing and smelting.

Muhly (1986) suggested that rare Near Eastern Seventh Millennium BC copper objects were made from native copper, with copper-smelting dating from the Fifth Millennium BC. Stech (1990) described copper artifacts made of hammered native copper from Çayönü, southeastern Turkey, with an age of 9000 years BP, well before evidence of copper smelting. Giumlia-Mair (2005, p. 276) cited an awl probably made of very pure native copper by hammering from the Italian Middle or Late Neolithic.

Hook et al. (1991) discussed evidence for Chalcolithic copper smelting from Southeastern Spain.

Marshall (1931) dealt with bronze arrowheads, knives, saws, chisels, and daggers from the Bronze Age of Mohenjo-Daro.

Tylecote (1987) provided an extensive account of mineral dressing, smelting, and metal fabrication as applied in the distant past.

Iron

Wertime and Muhly (eds, 1980; see also Pigott, 1996) provided numerous discussions of the 'gradual' introduction of iron into the record during the later Bronze Age, particularly with data for the Old World, leading up to the Iron Age proper.

Wagner (1993) suggested that the earliest occurrence of Chinese iron is at about 500 BC. Wagner (2001) described several aspects of the Han Age Chinese iron industry, including blast furnaces, cupula furnaces, and fining hearths. Hartwell (1967) reviewed the development of the iron and coal industries in northeast China during the Sung Dynasty and even somewhat earlier, with the replacement of charcoal by coal for iron smelting; this is far earlier than similar developments in Europe.

Waldbaum (1980) reviewed the appearance of iron in the archaeological record, concluding that it appeared very rarely before 3000 BC, and that its real first appearance for practical purposes is slightly before the beginning of the First Millennium BC. Kosambi (1963) and Chakrabarti (1977) discussed the appearance iron in India and agreed that the First Millennium BC is reasonable and also noted the widespread occurrence of iron ores.

Burke (1986, Ch. 7) gives worldwide usage of meteoritic iron and lists (p. 230) objects presumed to be of meteoritic iron found in Near Eastern localities dated prior to 1000 BC, and gives an exhaustive bibliography on such matters.

Glass

Henderson (2000) reviewed the evidence and concluded that Near East glass of the earlier Second Millennium BC and possible occurrences in the Third Millennium might be the earliest. Nicholson and Shaw (2000) reviewed Egyptian glassmaking in some detail, while also dealing with faience with the thought that the glassy surface of faience is linked to the later development of glass. Faience consists of finely ground quartz with small amounts of an alkali, including lime, that is allowed to fuse but not become liquid (glass). Moorey (1994) dealt extensively with Near Eastern faience and glass. Marshall (1931) discussed Bronze Age faience from Mohenjo-Daro. Evans (1964, p. 486–523) described Minoan faience including the beautiful little 'snake goddesses'.

Oppenheim (1970) surveys cuneiform texts relative to glass and glassmaking in ancient Mesopotamia. A prime resource for research on glass is the Rakow Research Library at the Corning Museum of Glass.

Abrasives

Abrasives have been used to shape and polish stone. The commonest and most readily available abrasive is quartz sand. Also commonly available are abrasive siliceous plant stems such as those provided by many grasses and reeds and by *Equisetum*. Volcanic pumice and hematite-based red ocher have been available for fine polishing. In Classical Antiquity in the Eastern Mediterranean, emery from Naxos was available. Emery consists of naturally occurring, fine-grained corundum and hematite, a product of the regional metamorphism of bauxite. Moorey (1994, p. 82) discussed various Bronze Age sources of emery.

Lu et al. (2005) demonstrated the presence of fine-grained corundum on Neolithic stone tools from China, c. 4000–3500 BC; they inferred that powdered diamond might have been involved in some cases, though the evidence is circumstantial.

Containers

Containers for liquids, water particularly, are an obvious necessity for humans from the beginning, with gourds, skin containers, and ostrich-shell containers as obvious candidates, with cords for carrying. Vencl (1981) discussed Paleolithic and Mesolithic containers. Oshibkina (1985, p. 406–408) illustrated and briefly discussed birch-bark containers for flints from the Lake Onega Mesolithic.

Utensils

Once it is admitted that *Homo* from the very beginning had the ability to use stone boiling for making soups and stews as well as for baking underground plant-organs, it is clear that various types of 'spoons' and 'scoops' would have been necessary, with wood as the most easily available material. Bogucki and Grygiel (1983) cited and illustrated some Polish earlier Neolithic spoons made from wood and red-deer antler. Marshall (1931) dealt with ladles from Mohenjo-Daro.

Rope, String, Cords

Adovasio et al. (2001; see also Soffer et al., 1998) provided details on Gravettian plant-fiber-based textiles, basketry, and cordage from Moravia. White (1986, Figure 48) illustrated an impression of a piece of Magdalenian rope. Nadel (1994) described some cord fragments from

Ohalo II, in Israel, about 23,000 years BP, and suggested that they might have been used for nets or fish traps or other in other ways.

The presence of beads (see below for discussion of **Beads**) from at least the Acheulian indicates the presence of cords or strings, whether of plant or animal materials.

Hafting

Pramankij et al. (1997) provided an excellent well-illustrated account about how stone tools can be hafted by means of woody vines, thongs, and other materials.

Middle Stone Age

Wadley et al. (2009; see also Wadley et al., 2004; Lombard, 2006, 2008) detailed experiments with natural plant gums mixed with ochers to produce adhesives similar to remnants found in a hafting area for stone tools from the South African Middle Stone Age; their data, involving Sibudu Cave samples, 70,000 years old, is impressive and suggests a high level of technical sophistication. Mazza et al. (2006) described several Mid-Pleistocene Paleolithic Italian tar-hafted stone implements. Paleolithic birch-bark pitch (Grünberg, 2002) was used for hafting both flint and wooden implements. Boëda et al. (1996; see also Boëda et al., 1998) described bitumen used as a hafting material with flint tools from a Syrian Middle Paleolithic Mousterian site. Aveling and Heron (1998) described birch-bark tar from an English Mesolithic site.

Lombard (2004, 2005) found evidence from Sibudu Cave of vegetable hafting-material on the proximal ends of Middle Stone Age points with faunal residues present on the distal ends, i.e., evidence of hafting. Gibson et al. (2004) described 60,000–68,000-year-old evidence of hafting from Rose Cottage Cave, South Africa, with ocher on or near the backed edges of tools.

Wilkins et al. (2012) described good evidence for Middle Stone Age, about 500,000 years BP, South African stone points that were hafted and used as spears for hunting, with the implication that this is near the time where *H. neanderthalensis* and modern humans first diverge.

Late Stone Age

Knecht (1983b) considered the use of antlers in making split-based, hafted projectiles of Aurignacian age.

Knecht (1993) discussed Early Upper Paleolithic European bone and antler split-based projectiles that clearly were hafted, as well as lozenge-shaped and spindle-shaped points. Peterkin (1993) dealt with European Upper Paleolithic lithic-point morphologies for points that were probably hafted, and with harpoons made from bone.

Neolithic

Kingery et al. (1988) cited Epipaleolithic lime-plaster hafting material on microliths from the Twelfth Millennium BC.

Boëda et al. (1996) cited the use of collagen as a hafting material from a Neolithic site. Clark et al.'s (1974) descriptions of ancient Egyptian bows and arrows provides a useful background against which to consider Paleolithic hafting possibilities, including the use of microlithic blades as barbs on arrows.

Post-Neolithic

Forbes (1936) surveyed evidence concerning occurrences and usages of bitumen during the post-Neolithic.

Warfare and Interpersonal Violence (See Also Hafting, Above)

Summary

Keeley (1996; see also Larsen, 1997, p. 119–151; Golitko and Keeley, 2007; Keeley, 1997) makes a convincing case, using modern ethnographic data and some archaeologic data from the Neolithic, that humans have been involved in, and have an inclination for, warfare and intraspecific violence, which 'undoubtedly' dates from the 'beginning' of the genus *Homo*. Further support for this position is provided by Lahr et al.'s (2016) description of 10,000-year-old hunter-gatherer group violence in Kenya, with the reason for the violence unknown, i.e., whether it was an argument over women or property rights (hunting grounds, sources of vegetable foods, etc.); this the first clear occurrence of pre-Neolithic group violence, as opposed to many older reports of individuals killed by others, with uncertainty whether this was isolated violence or not.

Diamond (2012, p. 131) pointed out that war can be defined 'as recurrent violence between groups belonging to rival political units, and sanctioned by the units'. Otterbein (2004) goes over the same ground and essentially agrees that warfare has been present from the very beginning, as does Gat (2006), with the latter relying chiefly on ethnographic information for the initial appearance of war with *H. erectus*.

Morris (2014) documented the fact that interpersonal violence, including warfare, has significantly decreased as a percentage of total human population per unit of time from the hunter-gatherer stage to the present: despite the horrendous deaths in modern warfare up to the present day, the overall percentage of violent deaths per total population has decreased very significantly.

Warfare in the conventional sense involves organized violence between opposing groups, not just violence between isolated family groups, clans, or individuals. In this sense warfare became possible only when humans first began to live in relatively permanent groups that included far more than the members of a single family or clan; this occurs only with the very Late Paleolithic, including the Natufian and then the Neolithic. Before this time, interfamily and interclan violence between hunter-gatherers manifests itself in the record as isolated 'broken' skulls and projectile-caused wounds; this is not warfare in the conventional sense.

The problem of trying to recognize evidence of warfare in the archaeological–anthropological record is complex. First, evidence of intraspecific violence is easy to recognize, based on clearly fatal damage to skulls and the presence of potentially fatal wounds affecting post-cranial skeletal material made by weapons. But warfare is another matter. Warfare by definition is organized group-vs.-group violence and cannot be recognized solely on the basis of isolated fatal wounds to an individual. Truly positive evidence of warfare in this sense begins to be recognized only in the Neolithic with the first evidence for fortified sites accompanied by clearly massacred groups of people.

Several of the contributions in Carman and Harding (eds., 1999) made the important distinction between recognizing evidence of battles as contrasted with sieges. Sieges first appear on the scene with the Neolithic, while evidence of battles is more elusive since isolated remains of fatal violence are not adequate.

Pre-Neolithic violent encounters between individual humans with fatal results do not by themselves qualify as evidence of warfare. So, we are left with the perception that intraspecific fatal violence has a very lengthy history dating far back into the Paleolithic, but that truly organized warfare is a more modern phenomenon.

Intraspecific Violence

Wrangham and Peterson (1996) reviewed the evidence for intraspecific violence as characteristic of groups of modern chimpanzees (*Pan*) and then went on to comment about the ethnographic evidence of similar intraspecific violence present in various groups of modern hunter-gatherers. They also considered that the ancestor of the chimpanzee–hominid lineage, present some millions of years ago (see Morris, 2014, Figure 6.5, for a family tree illustrating these phylogenetic relations), was possibly also characterized by intraspecific violence. The eventual appearance of Australopithecines, following the 'split' of the chimpanzee lineage from that leading eventually to the homininae, probably also was characterized by intraspecific violence.

Boaz and Ciochon (2004) present evidence for the possible presence of intraspecific violence involving some specimens of *H. erectus* from China, which, if correct, would indicate intraspecific violence from the very beginning of the genus.

Courville (1950) discussed and described some Neandertal skull injuries suggesting blunt trauma, presumably caused by their fellows, with the implicit message that such lethal behavior within our genus has been with us for a long time; he also gives more numerous Cro-Magnon examples. Zollikofer et al. (2002) described evidence for a weapon-produced injury on a French Neandertal skull and reviewed other possible examples of Neandertal intraspecific injuries caused by weapons, including Trinkaus's (1983) consideration of evidence from Shanidar Cave in Iraq. In view of what we now know, it is possible that intraspecific violence using weapons by hominids against members of their own species is hard wired, as contrasted with the interspecific aggressive behaviors of the other higher anthropoids. Walker (2001; see also Martin and Frayer, 1997) reviewed the evidence of violent human behavior through time with the inference that it has been continuous. D'Errico et al. (2012) detailed the evidence for weapons tipped with poison from far back. Roper (1969) reviewed the evidence for Pleistocene 'intrahuman' killing and found it supportive, although evidence for war during this interval is not convincing. Péquart et al. (1937), while describing the Late Mesolithic burials at Téviec, an island off the south coast of Brittany, illustrated an arrowhead preserved in a vertebra. Coleman (1977, Plate 94, figure j) illustrated a Late Neolithic skull from Keos, in the Cyclades, that showed evidence of a healed projectile wound. Wendorf and Schild (1986) include a description by T.D. Stewart et al. of a male skeleton from Wadi Kubbaniya, Egypt, of Late Paleolithic age, that had been involved in earlier violence and who died from later human-inflicted wounds.

Vencl (1999; see also Vencl, 1991, for brief accounts of European Mesolithic and younger skeletal wounds caused by weapons) summarized various evidence for 'violence' present on Paleolithic and Mesolithic skeletal materials. His 'evidence' does not indicate whether or not warfare, rather than individual encounters, was involved. He has a more detailed account of more organized earlier Neolithic and Eneolithic violence, including those associated with villages having associated ditches into which victims of violence were thrown. Bachechi et al. (1997) described an Italian late Upper Paleolithic female pelvis with a flint projectile still embedded, with evidence that the victim survived for a time; their Table 1 tabulates similar occurrences from elsewhere for both humans and animals. Bocquentin and Bar-Yosef (2004) described an Early Natufian example of violence involving a flint projectile embedded in a spinal vertebra.

Cordier (1990) compiled numerous occurrences of weapon-induced injuries to humans and animals from the Paleolithic to the present, with the human injuries, including many that were probably fatal, presenting evidence for the presence of violence far back in time, as well as during more recent times.

Egg and Spindler (2009) described the shoulder wound caused by a flint-tipped arrow that caused the death of the Ötztaler Alpine 'iceman' during the Copper Age approximately 5000 years BP. The arrowhead remained in the wound.

Campillo et al. (1993) described a later Neolithic Spanish burial with two skeletons showing evidence of crushed skulls, with one of the individuals having suffered instant death caused by an arrow that penetrated the abdominal region before becoming lodged in a vertebra. No evidence is provided whether warfare was involved. Armendariz et al. (1994) described four cases of Spanish Neolithic – Bronze Age arrow wounds, one (from a hypogeum) undoubtedly fatal, with no discussion of whether warfare was involved. Etxberria and Vegas (1992) summarized a number of Spanish later Neolithic and Bronze Age arrow-wound examples.

Evidence of Warfare

The convincing documentation assembled by Golitko and Keeley (2007; Keeley, 1997) indicating defensive structures dating from the beginnings of the Neolithic make it unreasonable to conclude that the warlike nature of *Homo* had not been present from far earlier, far back in time, possibly from the appearance of the genus.

Rowlands (1972) assembled a comprehensive account of various defensive structures used by relatively small groups of people for various purposes under differing conditions and with different opponents, including animals.

Gabriel and Metz (1991) summarized the problems and capabilities of ancient armies from Sumer to Rome; this is a comprehensive brief treatment dealing with weapons and armor and their capabilities, the incidence of wounds, disease, tactics, and more.

Wrangham and Glowacki (2012) and Fry and Soderberg (2013) take opposing views on the potential antiquity of warfare, based on their analyses of nomadic hunter-gatherer groups (see Bower, 2013, for comment), but Fry and Soberberg's (2013) Panglossian view of human nature is thoroughly refuted by Keeley's (1996) data and the dramatic depiction of a 7000-year-old Neolithic massacre (Teschler-Nicola et al., 1999) at Schletz, Lower Austria, in which a large number of inhabitants of a fortified village were massacred with the bodies left unburied where they were scavenged, possibly by wolves, and with a 'deficit' of young females suggesting their abduction. If the Schletz massacre were not enough to make the point, the German Talheim Neolithic massacre (Wahl and König, 1987), using hafted stone weapons and arrows, further underlines the nasty proclivities of our species. Frayer (1997) discussed the Late Mesolithic skull 'nests' from Ofnet, Bavaria, as evidence of a massacre (note his Figure 1, which makes it clear that most of the victims were not adult males of fighting age, i.e., the massacre probably occurred while such males were occupied elsewhere), with Orscheidt (2005) discussing the evidence without making any definitive statement about 'cause'.

Milner (2005) made the telling point that wounds involving parts of the skeleton, commonly vertebrae, seriously underestimate the number of weapon-caused injuries, including deaths, owing to the significant percentages of soft tissues also affected. He used data from Nineteenth Century arrow wounds as the basis for his conclusions, a very perceptive item!

Neolithic and Younger

Anthony (2007; see also Kuzmina, 2001, for similar settlements in the region) summarized the evidence from Sintashta, just East of the Urals, for a heavily fortified walled settlement with an outer ditch, and for evidence of a battle in the area of the Abashevo settlements west of the Urals, all about 4000 years BP.

Osgood et al. (2000) summarized evidence concerning warfare in Europe during the Bronze Age, including their conclusion that archery was most important during the earlier Bronze Age, following which spears, daggers, and swords played a more important role.

Of interest is Vencl's (1991, Figure 2) compilation of 'wounds' from the Paleolithic to the Middle Ages of 'wounds' (mostly skeletal damage) to males and females with a much higher percentage for males.

Anthony (2007, p. 224) suggested that real cavalry, organized into fighting units, appeared only in the Iron Age along with shorter bows and more standardized arrows.

Dolukhanov (1999) summarized data from Eastern European Mesolithic and Neolithic sites indicating the presence of warfare.

Keeley (1996; see also Golitko and Keeley, 2007) makes a very good case for the widespread presence of warfare in various groups, small to large, in the Neolithic.

Mercer (1999) described British Neolithic evidence of defensive structures associated with indications of arrowhead concentrations at weak point[s], indicating serious violence, with several associated skeletons with arrowhead wounds.

Clare et al. (2008, Figure 5) described potential evidence of widespread warfare involving Neolithic small fortified villages in Southwestern Turkey, including extensive fires, unburied, burned bodies, and slingstone caches. Marshall dealt with sling balls from Bronze Age Mohenjo-Daro, but whether they were used in warfare is not discussed.

Bouville (1982) described French Neolithic, Fifth Millennium BC, evidence of a massacre while also citing a Neandertal example of a violent death.

Wendorf (1968) described what appears to have been a massacre of a large group of adult men and women, together with children, from the Late Paleolithic or Neolithic of Nubia, with flint weapons involved.

Louwe Kooijmans (1993) described a Middle Bronze Age Dutch burial site that was concluded to represent the results of a violent conflict.

Anglim et al. (2002) provide detailed discussions of Egyptian, Assyrian, Greek, Carthaginian, and Roman tactics, armor, weapons, and various battles. Gabriel and Metz (1991) provided additional information about these topics for the same time interval.

Fortification

Keeley et al. (2007) comprehensively reviewed and summarized those aspects of military architecture that have the potential for being preserved in the prehistoric record, and also cited the earliest examples known, with Jericho in the Pre-Pottery Neolithic A and Pre-Pottery Neolithic B cited for a ditch backed by a massive rock wall. Their treatment is an elegant example of what can be obtained from the prehistoric record.

Renfrew (1972) discussed Late Neolithic evidence of fortification in the eastern Mediterranean. Anthony (1970, pp. 227, 230) cited a 4500–4000 BC wooden palisade from the Balkans and another with ditches and earthen banks. Vencl (1999) provided a summary of many Neolithic and Eneolithic fortifications and the criteria used in recognizing them. Anthony (2007) provided information about many of the fortified sites in Late Bronze Age Soviet Asia. Barber (1987) discussed a number of Bronze Age fortifications in the Cyclades. Golitko and Keeley (2007) summarized various evidence for Neolithic Linearbandkeramik fortifications in Europe. Hrala et al. (2000) provided accounts of Middle and Late Bronze Age fortifications in Bohemia. Danti and Zettler (1998) cited a Third Millennium BC fortified city from Syria without describing any details. Leisner and Schubart (1966) described the remaining parts of a Portuguese Chalcolithic fortification at Pedra do Ouro, and also discussed other Iberian and western Mediterranean sites, together with an extensive account of accompanying artifacts.

Projectiles

McBrearty and Tryon (2006) made it clear that the Acheulian is characterized by handaxes, with projectile-type weapons coming in with the Middle Stone Age. Shea (2009; see also Brooks et al., 2005, for Africa) indicated that projectile weapons came in with the Middle Stone Age. Mourre et al. (2010) described the presence of pressure flaking in Middle Stone Age points from Blombos Cave, South Africa, well before the Upper Paleolithic , Solutrian age of pressure-flaked tools from Europe.

Shea (2009) reviewed questions regarding the introduction of projectiles. He pointed out that present evidence suggests the earliest introduction in Africa approximately 50,000–100,000 years BP. European evidence appears later. The use of thrown spears is less likely than of spear throwers (atlatls) and the bow and arrow, with the evidence for both being slim owing to poor preservation of wood in the record. Shea (2006) concluded that evidence for hafting goes back into the Middle Paleolithic – Middle Stone Age, although whether this involved projectiles versus non-projectile weapons is unclear. Lombard (2011) considered the possibility of bow-and-arrow technology in the Middle Stone Age of South Africa.

Peers and McBride (1990) illustrated crossbowmen in China in the Fourth or Third Century BC. For crossbows, the Wikipedia entry is detailed.

Hughes (1998) discussed the nature of the 'points' used in weapons, including atlatl, plus bow and arrow, using North American data. Bergman (1993) discussed European Middle Paleolithic bows and arrows.

Knecht (1993) described Early Upper Paleolithic European bone and antler projectiles used for hunting.

Farmer (1994) reviewed the evidence of atlatl and bow-and-arrow presence in the Upper Paleolithic, Mesolithic, and Late Stone Age. Cattelain (1999) described a Solutrian spear thrower made out of reindeer antler from the Dordogne. Garrod (1955) described a variety of Paleolithic spear throwers, and White (1986) illustrated various Magdalenian spear throwers.

Clark et al. (1974; see also Clark, 1963, for additional information about bows and arrows, and Friis-Hansen, 1990; also Cattelain, 1997, who emphasized the difficulty of discriminating between points employing spear throwers or bows) described ancient Egyptian bows and arrows and provided a useful background against which to consider Paleolithic weapon possibilities. Zutterman (2003) reviewed Near East bows in depth, dating from the Second Millennium BC to much later.

Barham (2002b) discussed the presence of backed tools in the Middle Stone Age of south-central Africa.

Bar-Yosef and Valla (eds., 1991) illustrated various Natufian shaft straighteners from the Near East. Shaft straighteners are commonly used for straightening the shafts of arrows and spears; the grooved stone straightener is strongly heated, and the arrow or spear shaft pressed against the very hot stone groove to 'straighten' the shaft.

Yellen (1998) discussed and described Saharan and sub-Saharan barbed bone points, presumably used mostly for fishing, as both harpoons and as hafted points, with very few from Europe (a Magdalenian example is cited), with ages from Middle Stone Age and younger in Africa. Yellen et al. (1995) described barbed and unbarbed points from a Middle Stone Age site (Katanda) in Zaire with a minimum age of 90,000 years BP.

Crane (1999, p. 540–544) cited the use of bees as weapons from Roman times to the present.

Morris (2014, p. 168–169) suggested that artillery, or guns, first appear in 1248 on a Manchurian battlefield.

Armor, Including Shields and Weapons

Stoddart (1990, p. 219) observed that primates cannot kill with their teeth, which helps to explain the importance of weapons for humans. The largely herbivorous habit of other primates leaves them largely uninvolved with killing for food with weapons, despite their occasional violent, weaponless killing of conspecifics and other taxa.

Harding (1999; see also Osgood et al., 2000) reviewed information concerning European Bronze Age armor and shields. Harding (1999) and Kristiansen (1999) described the introduction of the sword in the Late Bronze Age. MacKay (1938) described swords from Mohenjo-Daro. Higham (1996) cited and briefly described Bronze Age armor from Southeast Asia.

Born and Seidl (1995) described and illustrated in detail pointed helmets of bronze and a few with iron and bronze together, with bronze shoulder armor from Assyria and Urartu. Born (1991, 2001) described and illustrated in beautiful detail various specimens of Bronze Age armor, chiefly European and Near Eastern, including helmets and breast plates, and weapons including various sword types, axes, spear points, and the like. Oakeshott (1960) dealt extensively with Bronze Age to Medieval swords, with passing reference to spears and other weapon types, shields and axes, plus limited information on armor.

Hafting (see above)

Hafting by itself is not necessarily evidence for the use of weapons in intrahuman conflict or warfare, since hafting of tools for many non-violent purposes is widespread.

Textiles, Clothing, Basketry, Matting, Wooden Containers

Soffer et al. (2000; see also Adovasio et al., 1997) described widespread evidence from the European Upper Paleolithic (approximately 22,000–28,000 years ago) for the presence of plant-fiber-based textiles and basketry.

Soffer (1985) suggested, on the basis of abundances of bones of fur-bearing animals, that the Paleolithic inhabitants of the Central Russian Plain used animal skins for clothing.

Habu (2004) cited Japanese Jomon fabrics and baskets.

Jørgensen (1992) provided a comprehensive account of European textiles, Neogene to 1000 A.D., covering materials, weave types, and occurrences in some detail; this is the basic reference for Bronze Age to Iron Age European textiles and their manufacture.

Kittler et al. (2003) used molecular-clock data to conclude that the evolution of *Pediculus humanus*, the body louse, restricted to wearers of clothing, originated approximately 72,000 years plus or minus 42,000 years ago.

Neolithic

Barber (1991; see also Winiger, 1995; Labriola, 2008) provided an extensive account of Neolithic textiles, the oldest dated 7000 years BC, to more modern textiles and their manufacture. Renfrew (1972, Figure 17.3) illustrated spindles and indicated that they occur in the Early Neolithic. Kuijt (1994) cited Pre-Pottery Neolithic A loom weights. Stordeur et al. (1996) cited basketwork and 'coarse woven material' from a Mureybetian (Pre-Pottery Neolithic A) site on the Middle Euphrates. Sherratt (1981, p. 282–283; see also Barber, 1991, 2001) indicated the presence of wool at about the late Third Millennium, linen in the Fourth Millennium BC, preceded by leather with buttons in the Neolithic. Vogt (1947) discussed and described a variety of European later Neolithic and Bronze Age fabrics, weaving types, and basketry, with comments on conditions of preservation. Barber (1991, p. 10–11) cited flax from the Sixth Millennium at Çatal Hüyük. Schick (1988) described cordage, basketry, and fabrics in some detail from the Pre-Pottery Neolithic B of Nahal Hemar Cave, Israel, with the use of reeds, rushes, and grasses including flax, and examples treated with bitumen presumably for waterproofing. Forbes (1964) discussed a variety of items involved in Mediterranean and Near Eastern cloth manufacture, spinning, sewing, basketry, weaving, dyeing, fulling, and the various equipment and techniques involved, with emphasis on Classical sources.

Barber (2001) summarized what is known about textile-based clothing including 'bands, caps, and string skirts for women' by about 20,000 BC, blanket wraps as cloaks, blankets for sleep and even shrouds, and kilts in the Fifteenth Century BC, pants about 1200 BC. Hodder (ed., 2007) discussed and illustrated many Pre-Pottery Neolithic B examples of basketry and mats from Çatal Hüyük. Helbaek (1963) briefly described textiles from the Early Neolithic of Çatal Hüyük, with plant and animal fibers implicated. Garfinkel and Miller (2002, Figure 2.28) illustrated some Yarmukian clay spindle whorls from Israel. Schmandt-Besserat (1977) cited earlier Neolithic spindles from Tell Aswad in Syria. Adovasio (1975) described a number of textile and basketry impressions in clay from earlier Neolithic Jarmo, Iraq. Mellaart (1964, Figure 34) described baskets, from Çatal Hüyük, that were used for burials of children, skulls, and adults, and also described wooden vessels, including bowls, dishes, and boxes, as well as textile fragments plus various bead types and bone implements, including ladles, as well as various objects in obsidian, including weapon points, all from about the Sixth Millennium BC. Stordeur et al. (1997) cited basketry and possible textiles from a Syrian Pre-Pottery Neolithic A site. Barker (2006, p. 162; Moulherat et al., 2002) cited cotton from the Neolithic of Mehrgarh, Baluchistan, Pakistan.

Tapa is cited by Bell (1983) at about 4300 BC. Van Heekeren, 1972, p. 165, Figure 39b, briefly discussed and illustrated a stone bark-cloth beater of 'early Neolithic' age from 'west Borneo'.

Copper Age

Jarrige (1981, p. 99) cited cotton from the Mehrgarh Early Chalcolithic – Bronze Age.

Egg and Spindler (2009) described the complex clothing of the Ötztaler Alpine 'iceman', a 5000-year-old Copper Age example from the Alps. Bar-Adon (1980) figured and briefly discussed woven wool fragments from the Chalcolithic of Israel, together with remains of woven linen.

Bronze Age

Higham (1996) cited spindles from the Bronze Age of Southeast Asia. MacKay (1938; see also Marshall, 1931) described spindles from Bronze Age Mohenjo-Daro. Marshall (1931, pp. vi, 194) discussed cotton textiles from Bronze Age Mohenjo-Daro. The great number of objects found at Troy and identified as spindle whorls suggests that they may have had additional purposes, perhaps as talismans.

Iron Age

Schlingloff (1974) quoted Herodotus for the First Millennium BC presence of cotton in India. Hundt (1967; see also Stöllner, Figure 21, p. 153) described a number of fragments of woolen cloth from the Hallstadt salt mines. Polosmak (2000; see also Griaznov and Boulgakov, 1958) described felt articles from Pazyryk in the Altai, about the oldest known, with an age of about 2400 years BP. Gervers and Gervers (1974) provided an excellent description of the felt-making process, including the 'painting' of felts. Rudenko (1953, 1970) are the basic documents for the contents of the Gorny Altai graves, Kurgans, including those from Pazyryk, and include extensive descriptions of clothing worn by both men and women.

Griaznov and Boulgakov (1958; see also Rubinson, 1990; Rudenko, 1970, p. 298–304) illustrated articles of Chinese silk and knotted wool rugs, the oldest carpets known, from Pazyryk, and Böhmer and Thompson (1991) discussed technical details of their weaving and dyeing and their place of origin. Ryder (1990) described various wool types with emphasis on those from Pazryk. Sheng (2010) discussed Chinese silk fabrics from Pazyryk, and cited felt from Xinjiang at about 1800–1500 BC, plus silkworm cocoons from the Chinese Neolithic Yangshao Culture dated 5000–3000 BC. Jakes and Sibley (2009) demonstrated the presence of silk fabric wrapped around a Shang Dynasty (c.1300 BC) bronze halberd, with the silk replaced, 'pseudomorphed', by copper compounds, including malachite. Zhao Feng (ed., 2002) provided an elegant treatment of Chinese textiles, beginning with Shang Dynasty woolen items from Xinjiang, and emphasizing more recent materials. Zhao Feng (ed. 1999) produced a relatively comprehensive, historical account of Chinese silk textiles, with an account beginning in the Neolithic, also including a helpful glossary.

Ryder (1990b) described remains of woolen textiles from Iron Age Hallstatt with sheep and goat wools of brown and black colors. Hundt (1960) described the various weaving styles and colors present in wool fragments from the Hallstatt salt deposit, and Hundt (1987; see also Stöllner, 1999, 2003) described additional Hallstatt textile fragments in more detail. Von Kurzynski (1996) discussed Hallstatt textile weaving types and also dyes. Banck-Burgess (1999) provided an extensive account of Iron Age textiles, weaving types, and dyes, with excellent illustrative materials.

Dyeing Materials

Biochemical evidence is provided from the 'interiors of Canaanite jars' (McGovern and Michel, 1990) for evidence of Tyrian Purple, the 'royal purple' of Classical Antiquity, obtained from a specific gastropod (*Murex*). Karmon and Spanier (1987) described a number of Israeli sites with good evidence of the purple-dye industry dating to near the beginnings of the First Millennium BC. Kardara (1961) described a Classical Greek dyeworks at Rachi and commented (p. 263) about the various plants used in dyeing, as well as the purple from *Murex,* and also discussed aspects of the dyeing process. Roy (1978), in a consideration of Indian plant dyes, cited a purple-colored cotton from Mohenjo-Daro that he indicated was dyed with madder, *Rubia cordifolia,* and Marshall described a Bronze Age dyeing trough from Mohenjo-Daro. Stöllner et al. (2003, Figure 20, p. 150) cited Mediterranean red dye (*Kermes vermilio*) from the Early Iron Age salt mines at Dürrnberg-bei-Hallein from textile fragments. Rogers *in* Banck-Burgess (1999), provided a very helpful account of the Iron Age red and blue textile dyes in the Hochdorf textiles.

Bedding

Stone Age

Wadley et al. (2011) described bedding (77,000 years BP) from South Africa, with associated plant materials possessing aromatic anti-mosquito and larvicidal properties. Opperman and Heydenrych (1990) described heaps of grasses presumed to be vestiges of bedding material from a Middle Stone Age cave in the Northeastern Cape, South Africa. Opperman (1996) interpreted patches of grass remains as bedding from a late Middle Stone Age northeastern Cape Province cave. Beaumont and Vogel (2006) discuss what they consider to be grass bedding material from the Middle Stone Age of South Africa. Henderson (1992) suggested that the Middle Stone Age Klasies River Shelter 1B features use of fire for cooking food, including shellfish, well away from what is interpreted as bedding grass.

Upper Paleolithic

Nadel et al. (2004) described the presence of bedding materials from an Israeli site dated at about 23,000 years BP, where highly perishable material was preserved under very unusual conditions.

Neolithic

Hodder (2006, p. 186–188; see also Mellaart, 1967) referred to earlier Neolithic floor matting, basketry, and wooden containers from Anatolia.

Property (see also Trading and Transport of Materials)

The concept of property is almost exclusively a *Homo* characteristic. Exceptions in the animal world include harems that some animals guard and maintain, as do humans. The common regard of daughters, wives, and other women as chattel also requires mention. The presence in some cultures of the requirement for a 'bride price' or for a dowry is also suggestive of a property concept. The presence of tools requiring a high level of skill in their manufacture certainly began in the Acheulian with their beautifully crafted stone axes, which, whether made by the 'owner' or obtained from a specialist, implies ownership. [For Saul (2013, 2019), such axes, many of which show no signs of wear, were originally conceived as 'cosmic tokens' or talismans, rather than tools. – Eds.]

It is unclear whether ownership or holding for personal use pertained for the far simpler Oldowan implements, since they might have been made at a use-site and then discarded rather than retained. Tools can of course include weapons for offense and defense, and implements for hunting, woodworking, and other possibilities. Tools also include those used in food preparation such as stone mortar and pestle, and metate, which would not be readily portable.

Property also applies to trade in the broad sense, since trade normally involves the exchange of property of one kind or another with profit of one kind or another involved, as well as 'gift' giving, with something of value being provided by the recipient. Property is also probably involved with human ornaments such as beads.

Beginning with the Neolithic the concept of property is obviously involved with the planting, tilling, and harvesting of crops. Also involved are such things as fruit orchards and olive groves (Joffe, 1998, p. 300; olive trees require about 10–12 years before bearing fruit!). The presence of settlements occupied by individual family groups implies property.

Caves used for shelter and living space, whether seasonal or on a permanent basis, were possibly considered as property by the families or clans that occupied them. It is also likely that in the hunter-gatherer stage one would not be surprised if family groups or clans regarded favorite hunting or gathering areas and water holes as property.

Trading and Transport of Materials

Musonda (1987) provided evidence that neither trade, nor barter, nor theft need always be involved in the distribution of materials. In his case, Later Stone Age Zambian hunter-gatherers may well have scavenged ceramic material from abandoned contemporary Iron Age sites.

Merrick et al. (1994) pointed out that distinguishing between 'movement' of materials by sequential trade, as contrasted with long-distance travel from sites of origin, is hard to distinguish.

Sherratt (1999) provided a very perceptive view of 'trade' in the general sense, emphasizing that trade in various commodities, including foodstuffs and other perishable materials, need to be more seriously considered rather than the all-too-common emphasis on non-perishable items such as stone-weapon materials, seashells, and robust eggshells. Sherratt (1977) provided an extensive account of trade, chiefly European, featuring sections on Neolithic flint and other lithic resources, plus an extended discussion of copper and bronze trade.

Renfrew (1986) reviewed many of the concepts of 'value' as applied to archaeological–anthropological materials, indicating that this is an area of great difficulty to analyze.

Diamond (2012, p. 74) pointed out that trade can involve advancing political or social goals.

Larsen (2015), using evidence from Bronze Age Anatolia, provided extensive insight into the complexities of trade; this is a classic account of somewhat modern-type trading organizations, trade routes, and the use of money, from far back in time, that rivals our knowledge of similar activities from Classical Antiquity in Greece and Rome; these were not simple people merely exchanging goods at an entrepot but were merchants in the medieval to modern sense.

Postgate (1992, p. 206–222) discussed Early Mesopotamian foreign trade.

Hopkins (1983) discussed the motivations involved with trade in Classical Antiquity from Greek and Roman sources, a complicated story. Millett (1983) discussed the evidence of how loans were arranged in Fourth Century BC Athens, a peek into the world of ancient finance.

Snodgrass (1991) emphasized that the 'exchange' of materials and objects may involve the concept of merchants, profit, and the like, or simple gift-giving exchange unrelated to profit.

Mauss (1954) discussed gifts, with the inference that gift-giving is normally carried out with the expectation of a return gift from the recipient.

Oldowan

Toth and Schick (1993) quoted data concerning Oldowan transport of lithic materials more than 10 km from source. Braun et al. (2008) analyzed Oldowan tools from the Kanjera Formation of Kenya and concluded that transport of some of the materials involved distances greater than 10 km.

Early Stone Age

Negash et al. (2006) provided geochemical evidence that obsidian in some Early Stone Age artifacts from some Ethiopian sites had a nearby source only 10 km away.

Middle Stone Age/Middle Paleolithic

Merrick et al. (1994) provided some data on potential Middle Stone Age 'movement' of obsidian from various Kenyan and northern Tanzanian sites, with little evidence from the Early Stone Age.

Féblot-Augustins (1993) provided a view of lithic exchange in the late Middle Paleolithic of Central Europe and Western Europe that shows most lithic 'movement' occurring very close to home, but with a significant percentage of medium-range and even long-range transport.

Connan (1999) provided information concerning routes involving Near East bitumens during the Mousterian.

Jochim (2000, p. 189–190) cited trading during the Late Mesolithic of Western Mediterranean seashells into Central Europe and even far down the Danube.

Upper Paleolithic

Clark (1948) cited the presence in northernmost Scandinavia of flint from southern Scandinavia. Schild and Sulgostowska (eds., 1997) included information about Polish chocolate-flint trade during the Final Paleolithic.

Anikovich et al. (2007) concluded that early Upper Paleolithic shells, presumably used as ornaments, at the Kostenki site on the Don River came from the Black Sea region approximately 500 km to the south.

Soffer (1985, p. 438) cited Upper Paleolithic amber from sites in the Kiev region, with the source(s) of the amber uncertain since Baltic amber is similar to that recovered at these sites. The literature on amber and amber trade is extensive.

Natufian

Bar-Yosef and Valla (eds., 1991; see also D.E. Lieberman, 1993, p. 613) discussed the probable Natufian trade of varieties of marine seashells from the Red Sea and Mediterranean to various Levantine sites.

Neolithic

Renfrew (1972, Figure 20.1) indicated the Neolithic distribution of obsidian from Melos in the Aegean region. Broodbank (2000, p. 44) cited the presence at the Franchthi Cave in the Argolid of obsidian from Melos, with implication of oversea transport about 7000 years BC, and dealt in depth with the various possibilities for trade within the Aegean and eastern Mediterannean during the Bronze Age. Cauvin et al. (1998) described in some detail the many Near East and Middle East sources of obsidian and the many archaeological sites for the latest Upper Paleolithic, Natufian, and Neolithic with obsidian from relatively nearby sources.

Bogucki (1988) reviewed the evidence for trade in north-central Europe during the Neolithic that includes such items as flint, seashells, copper, and salt, concluding that much of the trade was relatively local, with exception of such things as marine seashells and amber.

Connan (1999) provided information concerning the routes involving Near Eastern bitumens within the Near East during the Neolithic.

Scarre (2005, p. 414–415; see also Petrequin et al., 2008) cited Neolithic trade in jadeite originating from the western Alps and reaching as far north as Scotland, as well as elsewhere in Britain and also in Brittany in northwestern France. Dominguez-Bella et al. (2015) report the discovery of a Neolithic Alpine jade axehead from the second half of the Fifth Millennium BC in Aroche, in the far southwest of Spain, which revives the question of long-distance exchange between the Iberian Peninsula and the rest of Europe. Many such axeheads show no signs of use or wear.

Reese (2005) discussed occurrences at Çatalhöyük of Neolithic marine and nonmarine shells, with evidence of trade in marine shells from both the Mediterranean and the Red Sea.

Carter et al. (2006) discussed the distribution of obsidian from two South Capadocian sites to use-sites elsewhere in the Near East and Cyprus through the Pre-Ceramic Neolithic and later. There was clearly an active trade in this resource.

Monah (1991), while reviewing salt-spring occurrences associated with salt production in the later Neolithic of the Eastern Carpathians, cited the probable exchange of salt for obsidian obtained elsewhere. For Nikolov and Bacvarov (eds., 2012) the obviously large production of salt at Provadia-Solnitsata in the Fifth Millennium BC, was surely too large to have been used locally, i.e., trade must have been involved, but the problem with trying to link salt production to trade in salt is that the traded product leaves no trace. In England, some of the Roman roads had originally been salt tracks. The salt trade contrasts with trade in easily recognized things such as obsidian, flint, other mineral products, amber, unique seashells, and even fossils, recognizable far from their original source and enabling a case to be made for trade from a particular source to the distant locality.

Wen and Jing (1992) reviewed the Neolithic nephrite jade occurrences of China with no certain locations for the source(s) of the nephrite [some of which have since been discovered – Eds.]. Habu (2004) cited Early Jomon jade articles from Japan.

Domínguez-Bella (2012, Figure 6) illustrated and discussed the Neolithic trade in Spanish variscite from various localities to both Iberian and French localities.

Bronze Age

Gale (ed., 1991) includes a series of papers on Bronze Age trade in the Mediterranean, chiefly from Sardinia to the Near East, some of which discuss just what the evidence of trade consists of, including trading for profit by merchants, exchanges of goods between rulers, tribute, and other possibilities; clearly, this is not a simple business. These papers make very clear the inherent difficulties in trying to understand Bronze Age trade in metals – copper, tin, lead and silver, in particular – even with modern isotopic techniques... *caveat emptor.*

Snodgrass (1983) discussed the marine transport of 'heavy freight' in Archaic Greece, making it clear that large, oar-propelled galleys were entirely capable of transporting loads in the 'ton' range, including heavy stone statuary and the like.

Broodbank (2000) summarized a mass of information about Bronze Age trading involving the Aegean region, the Cyclades in particular. Wachsmann (1998) considered many aspects of Bronze Age trade in the eastern Mediterranean, as well as warfare and piracy.

Moorey (1994, p. 10) emphasized that transport in the nonmarine environment was really feasible only by water, i.e., riverine.

Knapp (1991) discussed the trade in organic materials during the Bronze Age in the Eastern Mediterranean, covering an extensive range of goods.

Sherratt and Sherratt (1991) provided a very useful summary of Bronze Age trade in the eastern Mediterranean and the Near East. They emphasized the overall importance of waterborne transport, both riverine and marine. Also significant is the fact that most of the early transport was of high-value products, with later lower-value items such as wine and olive oil plus ceramics and textiles entering into the mix, but bulk agricultural products not playing a large role at this time. Sherratt and Sherratt (1991) also reviewed the various thoughts about the motivations behind Bronze Age trading in the Eastern Mediterranean.

Sherratt (1997, p. 320–332) reviewed the extensive evidence of Neolithic to Bronze Age trade, largely in stone, in Central Europe, noting that the presence of copper objects in European regions lacking sources of copper is clear evidence of trade during the Copper and Bronze Ages.

Muckelroy (1981) described what he interpreted as scrap bronze at two presumed Middle Bronze Age shipwreck sites on the south coast of England, with the metal deduced to have been imported from adjacent France and worked up locally in southern England.

Broodbank (1989) considered the possible use of Early Bronze Age longboats in Cyclades-region trade.

Very long-distance trade in the Late Bronze Age is evidenced by the presence of lapis lazuli from Badakhshan Province, northeastern Afghanistan, in both Mesopotamia and Egypt. Herrmann (1968) reviewed the evidence of lapis lazuli trade, beginning by at least 4500 BC, between Mesopotamia and Badakhstan, a distance of 1500 miles! Even earlier trade involving lapis lazuli is cited from the Early Neolithic of Mehrgarh in Baluchistan with the occurrence of lapis lazuli beads! Pinnock (1988) discussed local Near East trade in lapis lazuli during the Third Millennium BC. Moorey (1994, p. 75) noted Mesopotamian use of lapis lazuli by the Fifth to Fourth Millennium BC and provided (p. 85–92) an extensive account of its use and presence in various ancient locales, and related topics.

Delmas and Casanova (1990) suggested that lapis lazuli might also have come from the Pamirs and the Chagai Hills in Pakistan.

Larsen (1987) discussed Second Millennium BC trade in the Near East between the various local centers.

Catling (1991) pointed out that the presence of Baltic amber at various Late Bronze Age Mediterranean locales is good evidence for long-distance trade.

Bass (1986; see also Pulak, 1998; Haldane, 1993, for plant remains including grains, fruits, spices, and resin) described the rather exotic 'mixed' cargo of a Late Bronze Age shipwreck from off the southeast Turkish coast, indicating a complicated commercial trade in the Eastern Mediterranean. Bass (1986; see also Bass, ed., and C. Pulak *in* Bass, 2005, for discussion and illustrations) described a Bronze Age shipwreck from the Eastern Mediterranean that included large numbers of 'oxhide' bronze ingots, remnants of tin ingots, various tools and weapons, pottery, plus miscellaneous items, including basketry fragments. They also describe other Bronze Age wrecks from the same region.

Algaze (1995) discussed Fourth Millennium BC trade in the upper reaches of the Tigris and Euphrates, including trade between the Anatolian and Zagros highlands with the plains-dwellers below.

Clark (1952) provided extensive older information about European Neolithic and younger trade in various materials, including, flint, amber, copper, and some finished products.

Beck and Bouzek (1993, eds.) include a variety of papers covering Bronze Age sites, and some Neolithic, from various parts of Europe and the Near East plus Egypt with amber, probably mostly of Baltic origin.

Artzy (1994) discussed Bronze Age trade in the Eastern Mediterranean and also across Arabia by camel and from Somalia.

Oates et al. (1977) made a case for Bronze Age transport of Sumerian Ubaid-type pottery down the western side of the Persian Gulf, Arabia, possibly by fishermen or pearl fishers.

Rice (1994) provided information concerning the Persian Gulf region during the Bronze Age with emphasis on Bahrein ('Dilmun' during that time interval) but is somewhat disorganized as regards trade.

Inizan (1993) discussed in some detail the probable trade in carnelian (cornelian) between Indian sources in the Deccan and Mesopotamia, particularly for the Third Millennium BC. Rao (1963) discussed the trade significance of various types of seals, including one of probable Bahreini type, between the Persian Gulf region and Western India, including Gujarat.

Iron Age and Younger

Veraprasert (1992) discussed evidence of probable trade involving beads and other items from Khlong Thom, on the western coast of southern Thailand, to India and points west during a time interval that might have extended back into the Iron Age, as shown by the presence of glass and Roman coins.

Miller (1969) extensively reviewed the many spices of Indian and southeast Asian origin involved with Roman trade.

Transportation

Summary

Transportation of materials from the very beginning has been carried out by humans, followed in time by the use of other animals and techniques.

Movement by Water and 'Boats'

Humans from the very beginning undoubtedly crossed small bodies of water by means of simply constructed rafts and other means that have left no record.

Evidence concerning human waterborne movements is of two basic types. The first is information from well-dated tombstones, monuments, rock art, painted ceramics, and the like that depict actual boats, sails, or other means of over-water transport; Casson (1994) provides a good sample of this class of information. The second is information of an inferential type from the presence of well-dated artifacts and other cultural data on isolated landmasses and islands that could have been reached only by over-water transport. McGrail (2001) compiled the most extensive account to date of various building techniques for water transport and related equipment.

Inferential Data

Bednarik (1999, 2002, 2003; see also McGrail, 2007) summarized a mass of circumstantial evidence that points out the presence of significant marine navigation in Indonesia east of the Wallace Line, dating back about a million years, with younger evidence from many other East Asian island areas. These sources also summarize information for a variety of Mediterranean island-sites going back to the Paleolithic. This is convincing evidence, although the actual means of transport is unknown (Bednarik discusses rafting). McGrail (1991, 2001) summarized similar evidence and discussed the various types of transporters that might have been used.

Sea level was lower at times in the past.

Indonesia and Nearby Islands

See Morwood et al. (1998) for 880,000- and 800,000-year-old stone tools on Flores, and see Antón and Swisher (2004) for *H. erectus* type tools on Timor. Sondaar et al. (1994) discussed 730,000-year-old artifacts from Flores. Van den Bergh et al. (2016) described stone artifacts from Sulawesi dated at 'before 200 thousand years ago until about 100 thousand years ago' that further document the presence of early man in this island region between Southeast Asia and Sahul, with Late Paleolithic graphic cave art present in the same area on Sulawesi; (**see Graphic Art**).

Australia and New Guinea (Sahul)

For the presence of our species in Australia, 40,000 years BP or possibly even 55,000 years BP, see Spriggs (1997). See Roberts et al. (1990) for a 50,000-year-old human-occupation thermoluminescence date from northern Australia; Roberts et al. (1994) for 53,000 and 60,000-year measurements; and Groube et al. (1986) for a 40,000-year-old human occupation site on Papua, New Guinea. O'Connell and Allen (1998) expressed skepticism about the very early Australian dates. For another perspective on the problem, see Groves (1996). O'Connell and Allen (2004) considered the available dates from New Guinea and Australia and suggest a 42,000–45,000 BP date for the earliest arrivals. Cosgrove (1989) described 30,000 years BP for human occupation of Tasmania and possibly intermittent Australian connections back to 50,000 years BP.

Islands East and North of Australia–New Guinea

Wickler and Spriggs (1988) discussed evidence of 28,000 years BP for human settlement on Buka Island, northernmost Solomons. Allen et al. (1988; see also Gosden and Robertson, 1991; Leavesley et al., 2002) discussed the evidence of human presence on New Ireland by 33,000 BP. Pavlides and Gosden (1994) described a site on New Britain dated at 35,000 years BP. Roe (1992) described a site on northwest Guadalcanal at 6000 years BP. Gosden et al. (1989) discussed the ages of various Lapita sites in the Bismarck Archipelago. Fredericksen et al. (1993) cited a minimum of 14,000 years BP for human occupation of Manus Island, to the north of Papua New Guinea, which involves several hundred kilometers of over-water transport. Fox (1970) discussed evidence from the Tabon Caves on Palawan with ages extending back past 30,000 years BP, possibly to 45,000–50,000 years BP.

Bay of Bengal

Ray (1990) reviewed the evidence suggesting that seafaring across the Bay of Bengal dates from the First Millennium BC.

Mediterranean Region

Strasser et al. (2010) argue for seafaring capability by at least 130,000 years BP and refer to Mesolithic and Lower Paleolithic artifacts at Plakias, Crete. Sondaar et al. (1995) discussed the evidence for human colonization of Sardinia by about 20,000 years BP. Broodbank (2000) provides a detailed analysis concerning trade routes in the Bronze Age Aegean, including Crete, with inferences about the type of transporters involved. Broodbank (2000, p. 11) commented on the presence of Middle Paleolithic tools on the island of Cephalonia, western Greece, which implies reasonably early waterborne transport. Vigne (2008, p. 191) commented on the presence of domesticated cattle on Cyprus by 8300 BC with what it implies about sea transport of such large animals, young or adult. Cherry (1990) listed numerous radiocarbon dates for human occupation of Mediterranean islands: Table 1, Cyprus 10,000 years BP; Table 3, Cyclades 9000 years BP; Table 4, Corsica 8500 years BP, Sardinia 13,590 years BP.

Actual Data

McGrail (2001, p. 431) dated the earliest log-boat (dugout) in the Eighth Millennium BC, and the earliest plank boat in the Third Millennium BC (see also Tallet, 2012, Egyptian boats); these are obviously minimum dates, and he commented that there are no excavated examples of hide, bark, bundle, basket, or pot boats, or of bundled or buoyed rafts, and only five known log rafts, the earliest being Roman. Habu (2004) cited Early and Late Jomon Japanese dugouts.

'Boats' have been suggested from 7000-year-old remains from Kuwait (Lawler, 2002), from a Dutch dugout of 7600–8200 years BP (van der Heide, 1975), and from another Dutch dugout of approximately 5500–6700 years BC (van Zeist, 1957). Niblett (2001) described an early Fourth Millennium BC Hertfordshire dugout that was probably used as a cremation site for the human bones within. Robinson et al. (1999) discussed several Irish dugouts, dated approximately 3900 BP and their construction. Lanting and Brindley (1996) provided an account of radiocarbon dates for Irish dugouts (their 'logboats'), indicating ages from Late Bronze Age almost to the present, with a summary of similar data for Britain and Scandinavia. Breunig et al. (1996) discussed a Chad Basin dugout dating at about 8000 BC. These are obviously minimum ages. Casson (1994) discussed and illustrated ceramic 'pot' boats, inflated skin-rafts, coracles, reed rafts with primitive sails, wooden boats, and more advanced sails, paddles, etc. from the Eastern Mediterranean, Egypt, and the Near East, with some of these categories dated as early as the Third Millennium BC (see also Broodbank, 2000, pp. 98, 142, Figure 115, Minoan sails). Broodbank (2000) emphasized the importance of harbors needed for sailing vessels as contrasted with the earlier canoe-type vessels that could land just about anywhere. Wachsmann (1998) summarized a mass of eastern Mediterranean and Egyptian Bronze Age data concerning ships and allied matters such as ship construction, rigging, rudders, anchors, sails, rowers, and paddlers. Broodbank (1989) considered the use of Early Bronze Age longboats in the Cyclades, manned by large crews of paddlers, concluding that only a few 'centers' within these islands could have provided the necessary manpower. Bass (ed., 2005) discussed the construction of Bronze Age vessels found in the Eastern Mediterranean, as well as younger wrecks from other parts of the Mediterranean and many other parts of the world.

McGrail (1991) briefly discussed various shipbuilding methods used in the Bronze Age and related questions about equipment and sailing practices.

Wright (1990) described several Bronze Age wooden boats found on the bank of the Humber, eastern England, with extensive discussion of their construction method. Fenwick (ed., 1978) described the remains of the Ninth Century A.D. Graveney Boat from Kent, providing some detail about its construction and comparisons with similar vessels from elsewhere in the region.

Ellmers (1984) described some Late Paleolithic Scandinavian carvings of skin boats on rocks with dates of approximately 8000 and 2000 BC.

Movement on Land

Horse

Anthony (2007, p. 460) suggested that horseback riding began in the Pontic–Caspian steppes by about 4200–4000 BC, for which Levine (1999) suggested Second Millennium BC.

Carts, Wagons, Wheels, Axles

Sherratt (1981, p. 263–264; see also Anthony 2007, for early Asian carts; Anthony and Vinogradov, 1995) cited carts with solid wheels pulled by oxen at 3200–2800 BC from Sumer, and 3400–3100 BC (pp. 265, 271) or Fourth Millennium BC from Europe, the Near East, and India (p. 266), as well as carts with spoked wheels (p. 273) in the Third Millennium BC. Using pictograms, Bakker et al. (1999) made a good case that wheeled vehicles were present in Europe and the Near East during the Fourth Millennium BC. Bogucki (1993) emphasized the potential importance of oxen for use in transport back into the Neolithic, without, however, discussing the actual evidence of them as labor-saving sources. Forrer (1932) argued that the first carts were for cultic use rather than transport and 'not the other way around'.

Gasser (2003; see also Sherratt, 1986, for later Neolithic and Early Bronze Age European wheels, and Oates, 2001, for some from the Near East; also Anthony, 2007) described a wooden wheel from Slovenia dated at 5100–5350 years BP, with an accompanying axle. Bakker et al. (1999) discussed Fourth Millennium BC wheels and axles from Europe and the Near East. Piggott (1983, 1968) gave detailed consideration to wheel construction and occurrences, including in the Caucasus, and discussed wheel morphologies at some length, together with associated horse furniture, describing numerous Third Millennium BC wheels and later spoked wheels from Europe, with a useful account of wheel types.

Anthony (2007, p. 397–405; see also Piggott, 1983; Levine, 1999; and Anthony and Vinogradov, 1995) discussed Late Bronze Age spoked chariots and the cheek pieces of their horses dating from 1900 to 1800 BC. Chang (1986, p. 323) cited Chinese two-wheeled, spoked, horse-drawn chariots.

Rudenko (1970) illustrated and briefly described an Iron Age carriage with spoked wheels from the Pazyryk burials.

Roads

Knowledge of ancient roads is limited. Trails used by human carriers undoubtedly existed from the very beginnings of humankind without leaving traces. Kelany et al. (2009) described the ancient Egyptian roads used in the Aswan region to transport quarried stone and boulders down to the Nile for riverine transport. Shaw (2006) provided additional information about Egyptian roads used to transport heavy stone loads from quarry to the Nile, and the use of sledges drawn by oxen in this work. Heldal et al. (2005) also discuss the roads used by the Egyptians to transport stone from quarrying sites to the Nile. See also Harrell and Storemyr, 2009 for more road information. There is no information, however, about how the stone once at the water's edge was transferred to Nile vessels or for its offloading. One possibility would have canals dug close to the Nile occupied by large barges, with the stone then dragged across the canal onto the barge on its sled *if* the barge was kept low in the water by the weight of smaller stone blocks of ballast, which could be removed once the load was centered on the barge. McGrail (2001, p. 44–45) has a useful discussion of just how massive stone might have been transported down the Nile in Classical times.

Jager (1985) described a Netherland cart track of Funnel Beaker Age and cited similar cart tracks from northwestern Europe elsewhere.

Roads in the modern sense appear to have been a Roman innovation used for the movement of troops and goods over long distances. Alexander the Great was not a road-builder.

Heavy Transport

Burford (1960) addressed the question of transport in Classical Antiquity. He made a strong case that heavy transport of stone for buildings, city walls, heavy timbers, and the like was carried out by oxen, which required that suitable roads be made available, that the transport be done in the summer when roads would be dry, not muddy, and that special wagons adequate to carry the heavy loads be employed. Horses were used where speed was required as in war chariots, but oxen were used wherever heavy loads were involved. Since oxen were commonly used for plowing, the source of the oxen would have been relatively local. Shaw (2006; see also Harrell and Storemyr, 2009, for more road information) provided information about ancient Egyptian roads used to transport heavy stone loads from quarries to the Nile with the use of sledges drawn by oxen in this work.

Heavy transport would not have concerned early (or later) hunter-gatherers. With the Neolithic, however, the necessity of building strong city walls as well as massive religious structures of one kind or another ('monumental architecture') would have posed a problem, with the availability of oxen used for plowing as the at-hand solution.

Dayyah and Sami (2001) discussed some Roman quarrying techniques and methods for the transportation of stone.

MacRae (1988) reviewed questions whether handaxe transport and flint sources involved belts, shoulder bags, or baskets.

Fire (refer to Diet also, above)

Summary

The first use of fire by humans is a contentious issue. Positive evidence in terms of hearths and the like are only of Acheulian age. The nature of Oldowan-age deposits, reworked depositional materials, is only negative evidence. The most convincing positive evidence is from Wrangham (2009), who points out that human skeletal anatomy and digestive physiology require the use of fire in food preparation, as do data on infant nutrition.

Evidence

James (1989) critically reviewed many of the alleged early evidences of fire and is skeptical about the overall reliability or interpretation of most. Clark and Harris (1985) reviewed many of the Lower Paleolithic sites where man-made fires might be expected, but with mostly ambiguous results. Berna et al. (2012) provided the oldest currently available reliable evidence for the use of fire from the early Acheulian of Wonderwerk Cave, Northern Cape Province, South Africa, with a date of 1.0 million years.

Villa et al. (2002) reviewed the use of wood and bone as fuel, as well as lignite, in the European Paleolithic. Théry-Parisot and Meignen (2000) discussed the use of both wood and lignite at a French Mousterian shelter as judged from the charcoal remains.

Petraglia (2002) points out the potential usefulness of recognizing and paying attention to concentrations of fire-cracked rock as an indication of ancient hearths, a feature that does not appear to have been used to any extent in earlier archaeological work.

Once one accepts the evidence for the use of fire by *Homo* from the beginning, it becomes a question of how fire was made. Stapert and Johansen (1999, 1999b) reviewed the evidence, indicating that striking sparks with flint and a lump of pyrite is one possibility, with the use of a fire drill as the other major possibility. The use of a fire drill, unfortunately, has little chance of being preserved in the fossil record. There are many variants of the fire drill, such as the bow drill. The use of the bow drill is also probable in the manufacture of beads made of very hard materials such as carnelian.

Lower Paleolithic

The use of fire during the Oldowan is subject to inference. The basic inference is the small mouths of humans (Wrangham, et al., 1999) and the need for cooked foods from the beginning. Cooked foods are essential for the feeding of weaned infants (see **Infant Nutrition,** above) needing a high-protein diet to keep pace with their rapidly developing brains. The absence of any positive evidence for the use of fire during the Oldowan can be explained taphonomically. Oldowan sites consist of both tools and skeletal remains in what were open-air sites subject to possibly significant water transportation that would have destroyed any evidence of fire, as well as the ever-present confounding possibility of wildfires unrelated to human activities.

Wrangham (2009) inferred the presence of fire beginning with *Homo erectus,* an inference based on his conclusions regarding skeletal characteristics of *Homo* that indicated use of cooked food (see under **food** and **taxonomy)**.

Monnier et al. (1994) described Lower Paleolithic evidence from Brittany of the use of fire, 350,000 to 500,000 years BP.

Brain and Sillen (1988; see also Brain, 1993) made a case for fire-use as far back as 1.6–1.8 million years from a South African cave, and Berna et al. (2012) made the case for human use of fire at 1.0 million years (see above) from a South African cave site.

Barham and Michell (2008) reviewed the evidence for the use of fire in the Lower Paleolithic.

Rowlett (2000) found evidence of fire-use by *Homo erectus* in East Africa at Koobi Fora in a 1.6-million-year-old site, and from China at Zhoukoudian (see Weiner, 1998, for evaluation of the Zhoukoudian evidence). Rowlett (1999; see also Rowlett, 1999b) argued that *Homo erectus* used fire at various African sites.

Bellomo (1994) described the evidence for the presence of human-caused fire at Koobi Fora in the Early Pleistocene.

Gowlett et al. (1981) described possible evidence for the presence of fire, Oldowan artifacts, and Australopithine skull remains from Chesowanja, Kenya.

Acheulian

Clark (2001, p. 316) cited a possible hearth with burnt wood and charcoal from the Acheulian at Kalambo Falls (Zambia–Tanzania border).

Alperson-Afil (2008) and Alperson-Afil et al. (2007) discussed the use of fire by hominins at Gesher Benot Ya'aqov, Israel, an Acheulian site. Alperson-Afil (2008) described the evidence for Acheulian use of fire at a site in Israel of Early and early Middle Pleistocene age.

Mason (1993) described evidence for the use of fire at a Transvaal cave during the Acheulian.

Goren-Inbar et al. (2004) discussed Israeli evidence for human use of fire from the Acheulian more than 790,000 years ago.

Meignen et al. (2001) described many Near East Paleolithic hearths and commented on evidence of hearths from the Acheulian as well.

Preece et al. (2006) discussed the very positive evidence for the use of fire during the Acheulian at Beeches Pit, Suffolk, dated at about 400,000 years BP, and commented on other Acheulian-age sites where the use of fire was implicated.

Shahack-Gross et al. (2014) described a very large hearth from the Acheulo-Yabrudian Cultural Complex at Qesem Cave, Israel dated at 300,000 years BP.

Middle Paleolithic

Roebroeks and Villa (2011) summarized the evidence for the habitual use of fire by Neandertals in Europe, reliably dating to 300,000–400,000 years ago, including its use in hafting stone tools.

Speth (2006) described the presence of Middle Paleolithic Neandertal-style hearths in Israel. Opperman and Heydenrych (1990) described Middle Stone Age hearths from the northeastern Cape, South Africa. Henderson (1992) discussed evidence for hearths at the Middle Stone Age Klasies River Shelter. Barham (1996; see also Barham, 2000) described Middle Stone Age hearths from a Zambian cave, and the presence of a windbreak, postholes, and a 'clean' area within the windbreak.

Théry et al. (1996) discussed the presence of lignite at a Mousterian and at a Mesolithic site in the Massif Central with nearby Jurassic brown coal as a possible source. This occurrence is further evidence of the ability, or habit, of very ancient humans for noticing and taking advantage of locally available resources.

Brown et al. (2009) discussed the use of fire at 164,000 years BP to improve the flaking qualities of silcrete at Pinnacle Point, coastal South Africa.

Upper Paleolithic

Soffer (1985) cited the many hearths associated with Upper Paleolithic sites on the Russian Plains.

Karkanas et al. (2004) described some Aurignacian clay hearths, probably used for cooking, from southern Greece. Klima (1956) provided evidence of coal fires being used in the earlier Upper Paleolithic of Silesia.

Neolithic and Later

Gabriel (1987) described the widespread Saharan 'stone circles' that represent Neolithic fireplaces made by wandering pastoralists during the times when the Sahara still supported adequate forage.

Castaways on Tromelin Island in the Indian Ocean maintained a fire during 15 years in the 18th Century A.D.

Architecture

Shelter and Buildings

Acheulian

De Lumley (1979) described an Acheulian, 300,000-years-BP site at Nice where a hut had been constructed, complete with postholes, a hearth, numerous animal bones, and remains of some marine shellfish, plus a little ocher; this is about the oldest known evidence of a living structure, albeit a temporary campsite type.

De Lumley (1969, 1969b) described a Paleolithic, Acheulian cave site from Nice, about 130,000 years BP, that includes what appears to have been a shelter inside the cave, which might have been constructed as a tent-like feature, in which there is evidence of two small areas where fires had been present, as well as stone tools and animal bones, some of the latter having been subject to fire. Also see de Lumley (1979).

Earlier Paleolithic

Mania and Vlček (1987) described three circular dwelling places with hearths and adjacent workshops from the *Homo erectus* site at Bilzingsleben, eastern Germany.

Middle Paleolithic

Kolen (1999) discussed Middle Paleolithic European living spaces, concluding that they were not dwellings in the usual sense but rather were places where various activities were carried out from which 'debris' was pushed out centrifically. Barham (1996) described a possible Middle Stone Age windbreak, suggested by postholes and absence of debris in the 'interior', plus hearths outside a Zambian cave.

Upper Paleolithic

Soffer (1985) discussed the unique Paleolithic mammoth-bone shelters so widespread on the Central Russian Plain, where many are associated with storage pits. Habu (2004) cited Jomon storage pits from Japan. Fagan and Van Noten (1971) described possible windbreaks from Gwisho, Zambia.

Natufian and Neolithic

Goring-Morris and Belfer-Cohen (2008) describe the Natufian to Neolithic transition from hut-type circular structures to the rectangular structures of the Neolithic and also mention the sociological implications of the change.

Bar-Yosef and Valla (eds., 1991) cited round building 'hut'-foundations in the Natufian.

Goring-Morris and Belfer-Cohen (2008) described the transition from the Natufian circular dwellings to the rectangular Pre-Pottery Neolithic B structures with a number of examples.

Weiss et al. (2008) described an Epipaleolithic Israeli brush-hut site with a concentration of plant-food seeds and a grinding stone at one side of the hut.

Perrot (1966) described Natufian circular dwellings from Israel.

Cauvin (2000) reviewed the presence in the Pre-Pottery Neolithic of the Levant of rectangular buildings as contrasted with earlier circular dwellings. This appears to be a major change in style. Stordeur et al. (1996) figured a half-sunken semi-round house from a Mureybetian (Pre-Pottery Neolithic A) site on the Middle Euphrates.

Yan Hu et al. (2007, p. 31) cited rectangular house outlines from the 8000-years-BP Xinglongguo Site.

Peters et al. (2005, Figure 5; see also Bellwood, 2005, p. 54; Kuijit, ed., 2000; see also Kuijit, 1994) indicated the presence in the Pre-Pottery Neolithic A of round buildings, followed in the earlier Pre-Pottery Neolithic B by rectangular buildings, a trend seen in several Near East areas.

Mellaart (1967) described in some detail the rectangular buildings from the Early Neolithic of Çatal Hüyük with their unique roof top entrances and evidence of using ladders to get to the roof.

Darvill and Thomas (1996) described a large number of Neolithic house 'outlines' from Northwest Europe, largely British, with relatively simple timber construction, in contrast to the far more substantial Near East types.

Door Locks

Feriolo and Fiandra (1993) described the nature of door locks from the Late Bronze Age at Arslentepe, which makes it clear that an inclination for 'bad' behavior has been with us for a long, long time.

Measurements and Numbers

Rudgley (1999, p. 86–105) surveyed many of the attempts to interpret various 'systematic' incised 'points' etc. on bone, antler, and ivory of the Paleolithic and post-Paleolithic 'pre-writing' stage as evidence of counting systems, measurements, calendars, astronomical observations, and the like. Marshack (1972) interprets such marks as calendar counts or lunar counts.

Renfrew (1972) cited evidence for Late Bronze Age measures and weights as well as the use of numbers. Rosenberg and Redding (2000, *in* Kuijt, ed.) cited and briefly mentioned Natufian tallies from Anatolia, with uncertainty about what was being tallied.

Scales, Balances, Weights

Vandenabeele and Olivier (1979) illustrated a variety of Mycenaean Age scales from the Eastern Mediterranean. Barber (1987) illustrated and discussed a number of lead weights from the Bronze Age of the Cyclades. Pulak (1998) describes weights recovered from a Late Bronze Age Eastern Mediterranean shipwreck, and Bass (1967; see also Bass, ed., 2005) described weights recovered from another Bronze Age Eastern Mediterranean wreck and also discussed weights found elsewhere in the Eastern Mediterranean and Near East. Vat (1940) discussed a number of weights found at Bronze Age Harappa. MacKay (1938; see also Marshall, 1931) described stone weights and scales from Mohenjo-Daro.

Many Sumerian or Old Babylonian weight stones are in the form of ducks with the head turned back (M.A. Powell, 1979).

Religion

Diamond (2012, p. 323–368) provided a good account of the many, varied characteristics of religions. Boyer (2001, p. 5) made a relatively comprehensive list of the reasons that have been put forward for the origins of religion.

Hodder (ed., 2010; ed., 2014) considered many aspects of the potential religious significance of the Çatalhöyük sites.

Kramer (1963) provided an account of Sumerian religious thought, the basis for many aspects of younger Near Eastern and Western religious concepts. Postgate (1992) dealt with many aspects of Early Mesopotamian religion.

Saul (2013, 2019) treats religion as an initial condition of our 'humanness', a communal effort to contend with death, with implications for the origin of language.

Shrines

Rollefson and Schmidt (eds., 2005) edited an account of Early Neolithic, Southeast Asia 'ritual centers'. These are defined as non-residential structures, plus a few occurrences of monumental megalithic, Stonehenge-like features. Their functions are not very well understood but may include religious and other uses.

Mellaart (1967) described a number of unique shrine-like rooms with numerous bucrania from the early Neolithic of Çatal Hüyük.

Cairns, only some of which were for burials, require further investigation.

Divinities

J. Cauvin (1987, 2000) suggested that a divinity in the form of a Mother Goddess and an associated Bull God first appear near the beginnings of the Neolithic in the Levant. He referred to sanctuaries and statuettes to support this conclusion. He regarded the earlier Late Paleolithic cave art as not involving any divinities.

Schmandt-Besserat (1998b) discussed the significance of a pre-pottery Neolithic statue of a woman from Ain Ghazal, Jordan, holding her breasts. This is a common theme in Near Eastern iconography and involves the woman as a fertility symbol and includes Inanna, Ishtar, Asherath, Astarte, and Tanit later on, with Aphrodite and Venus still later.

Burials, Afterlife, Memorials, and Monuments

Summary

Humans are the only animals to practice burial of their dead. Grieving is apparently present in some other mammals, such as dogs, chimpanzees, and elephants, but burials are unique to humans, as are the associated memorials and monuments, as well as the concept of an afterlife. The provision of grave goods from just about the earliest records of human burials attests to the afterlife concept. Burials, memorials, and monuments continue as common practice to the present day.

Morris (1992) dealt with the significance of grave goods, when present, and the use of cremation as contrasted with inhumation; this is a basic treatment. Chapman et al. (eds., 1981) reviewed some of the information provided by burials, including such things as graves with abundant, rich grave goods that may represent higher-rank individuals in a population, with details concerning the various burials themselves that include boat burials, monumental stone burials, cairns, and so forth; there is clearly very great diversity in human burials. Many cairns are of uncertain purpose. The original purpose of the inuksuit (singular inuksuk) of Greenland and the North American Arctic is unclear.

Riel-Salvatore and Clark (2001; see also Belfer-Cohen and Hovers, 1992) provided a useful discussion of the question whether there are Middle Paleolithic as contrasted with Upper Paleolithic burials. Overall, the evidence appears to indicate burials in the Middle Paleolithic, mostly involving Neandertals.

Harrold (1980) considered the significance of Eurasian Paleolithic burials. None are recognized from the Lower Paleolithic as yet, which may be an artifact caused by erosion, but it is too early to draw conclusions.

Lower Paleolithic

Ullrich (1995) surveyed mortuary evidence from the Paleolithic, including a few burials, suggesting that with the small sample, particularly for the Lower Paleolithic, it is too early to draw hard-and-fast conclusions.

Middle Stone Age

Smirnov (1989) provided a useful summary dealing with many aspects of Mousterian burials in Europe, the Near East, and Crimea, while commenting that no validated pre-Mousterian burials are known.

Solecki (1971, 1975) described an Iraqi Neandertal burial with associated pollen of various native plants (Leroi-Gourhan, 1975) that suggested that flowers were used with the burial. Bonifay (1964) discussed some limited information from a Dordogne Mousterian burial site.

Rak et al. (1994) described Neandertal infant remains from the Amud Cave burial, Israel, which Valladas et al. (1999) dated at 50,000–70,000 years BP.

Movius (1953) provided a translation and comments on a Soviet excavation of a Mousterian site in southeastern Uzbekistan that included the burial of a child surrounded by a ring of Siberian Mountain Goat horns, covered with earth, with the implications involved.

Russell (1987) provided evidence that the Krapina Neandertal burials represent bones that were defleshed before secondary burial.

Vandermeersch (1970) briefly described a Mousterian child's burial with an associated deer's head and probable ostrich-eggshell fragments from an Israeli cave. Rendu et al. (2013) thoroughly reviewed the La Chapelle-aux-Saints Neandertal burial in the Bouffia Bonneval (bouffia is a local patois for cave), concluding that this is a real burial, with all that implies about human behavior. Peyrony (1934) described various Mousterian burials from La Ferrassie with associated worked flints and other objects.

Defleur (1993) reviewed the Mousterian burials known as of 1993 in some detail, a very useful account.

Molleson (1981) pointed out that evidence for formal burials and ceremonials indicated a level of social organization larger than the individual family unit and that the first well documented burials date from about 70,000 years BP for Neandertals of northwest Europe.

Chapman (1981) discussed various European burials from the Middle Paleolithic and younger sites.

Few burials are known from the Mesolithic, mostly of Neandertals.

Late Paleolithic

The relatively abundant Late Paleolithic burials include a large number without significant grave goods, as contrasted with the small number having such goods, with burials featuring significant grave goods being largely of males, although not entirely. Harrold suggests that only 'important' individuals had burials that feature significant grave goods. This is, of course, the same for the Neolithic-to-present situation, with Egypt presenting the ultimate expression.

Riel-Salvatore and Gravel-Miguel (2013, *in* Stutz and Tarlow, eds.) summarized data concerning Gravettian Upper Paleolithic Eurasian burials, pointing out that ocher was commonly employed and that beads were most commonly located on the upper torso and head regions. Rogachev (1955) described a Gravettian-age burial from Kostenki, in the Don River region.

Einwögerer et al. (2006) discussed an Upper Paleolithic Austrian infant burial (about 27,000 years BP), in which one burial included two side-by-side newborns, skulls facing north, covered with a thick layer of red ocher, one infant associated with some ivory beads near its pelvis, and the whole burial then covered with the scapula of a mammoth; this unique burial is consistent with ceremonial procedures and all this implies about human behavior.

Fagan and Van Noten (1971) briefly described a number of burials from Gwisho, Zambia, with no evidence of grave goods, except for one infant burial containing numerous shell beads.

Natufian and Neolithic

Lewis-Williams and Pearce (2005) briefly described many of the earlier Neolithic massive stone circles and graves from Western Europe that have given rise to various thoughts about possible rituals, religious, astronomical, and other possibilities, with Stonehenge, Avebury, Gavrinis, and Newgrange as notable examples. The main site in the Orkneys may be the oldest, erected almost a thousand years before Stonehenge.

Mellaart (1963, 1964) described the unique Neogene Çatal Hüyük shrines with their bull heads, female figurines, and other evidence of a complex nature. Schmidt (1998) described the curious, animal-ornamented, almost menhir-like stone columns from the preceramic Neolithic of several Upper Mesopotamian localities and their possible religious significance. Schmidt (2001) discussed the megalithic structures present at an Early Neolithic site at Göbekli

Tepe, Southeastern Turkey. Sherratt (1990) discussed the significance of the megaliths of West Europe.

Baumgarten (ed., 2005) edited a series of short papers discussing the potential significance of Late Natufian and Early Neolithic ritual centers. Needless to say, the authors found it difficult to really understand just what these ritual centers involve, which is not surprising in view of the complexities of modern rituals and religious thoughts.

Campbell and Green (eds., 1995) included a number of Neolithic and younger accounts of burial practices recorded from the Near Eastern record, plus a few Natufian items. The very distinctive items make it clear that there were a great many approaches to burial 'styles' in the past. All of this undoubtedly reflects the human preoccupation with death and the afterlife during the past, and its presence as well even today; styles change in time and place, but the initial preoccupation remains fixed.

Kuijt (1996) discussed the common Late Natufian and Pre-Pottery Neolithic practice of skull removal (see also Belfer-Cohen, 1988, for Late Natufian skull removal) and secondary reburial in the Near East. Kingery et al. (1988, p. 231–232) cited 'shell eyes', with 'a fine skin plastered coat that was colored pink', with attention to the beard and hair, i.e., very lifelike.

Grosman et al. (2008) described a Natufian burial from Israel with associated grave goods.

Neolithic

Wyse Jackson and Connolly (2002) briefly described a Neolithic Irish passage grave in which possibly cremated remains were associated with numerous marine invertebrate fossils.

Hershkovitz and Zohar (1995; see also Segal et al., 1995; Hershkovitz et al., 1995) considered the 'plastered human skulls' of the Early Neolithic during the Pre-Ceramic phase of the Pre-Pottery Neolithic B about 8500 years BP of the Near East, with their significance being unclear. Rollefson et al. (1992) cited plastered skulls from the Pre-Pottery Neolithic B at Ain Ghazal. Goren et al. (2001; also see Goring-Morris *in* Kuijt, 2000) described additional Pre-Pottery Neolithic B skulls from this region, and also considered something of their possible significance. Bienert (1991) reviewed the plastered-skull evidence, as well as 'severed' skulls, from the Pre-Pottery Neolithic B, concluding that ancestor worship is highly probable. Molleson et al. (1992) described plastered skulls from a Pre-Pottery Neolithic B burial at Abu Hureyra, northern Syria, that included the use of cinnabar as a bright-red pigment, with the source of the pigment unknown. Strouhal described a number of plastered skulls from the Pre-Pottery Neolithic B of Jericho.

Hauptmann (2002) considered in some detail the characteristics of the Pre-Pottery Neolithic in central Anatolia and its relations to nearby regions. Hodder (ed., 2007; see also Helbaek, 1963), while describing many of the Pre-Pottery Çatalhöyük burials, discussed some infant burials in baskets, one with fancy bead anklets and wristlets, and with ocher also involved, and discussed and illustrated the many burials present beneath the living quarters.

Zhang and Kuen (2005) described a number of Early Neolithic burials from Central China that include flutes and various grave goods.

Yang Hu et al. (2007, p. 36–39) discussed 'residential burials' at the 8000-years-BP Xinglongwa site, with some continuing to be occupied after the burials were made.

Wadley (1997) discussed various South African Neolithic to modern burial types.

Potter et al. (2014) discussed two Central Alaskan infant burials with grave goods and ocher, and an associated burial of a cremated child, all at 11,500 BP.

Iron Age

Rudenko (1970) discussed many aspects of what can be ascertained about religion from the Pazyryk burials.

Egyptian

Lloyd (2014) provides a brief introduction to the complexities of Egyptian burials and insight into some of the thought processes involved.

Ur

Wooley (1934) provided insight into the totally different, complex burial behaviors brought to light in Ur.

Saul (2013) has carts and 'toy' carts in burials as models of the ever-turning heavens, *Ursa Major*, in particular.

Bronze Age, Minoan, Greek, and Roman

Morris (1992) examined and discussed the evidence to be obtained from burials of various types in the Greek and Roman World. He raises the questions about what burials might tell us about rights of passage and beliefs in an afterlife. There is no reason to think that other, far older cultures did not also have some very curious burial customs about which we have no specific clues. The point here is that different cultures commonly have very distinctive burial practices but that the common thread of human burial practices displaying a concern with the hereafter is present.

Evans (1964) described and illustrated many of the unique Minoan ritual sites and also their goddesses.

Vat (1940) discussed the pot burials from Harappa, including some with cremated remains. Higham (1996) cited the unique Bronze and Iron Age jar-burials present in Southeast Asia.

Cemeteries

Harrold (1980) summarized information about the limited number of Middle Paleolithic burials, with somewhat more from the Upper Paleolithic and none known from the Early Paleolithic, with grave goods present in some. Klima (1988) described a unique Upper Paleolithic triple burial from Czechoslovakia.

Coleman (1977) described several small Late Neolithic cemeteries from Kephala on Keos in the Cyclades that contain stone-lined burial chambers and slab coverings, with infants buried in jars within the tombs. Renfrew (1972) discussed a number of Bronze Age grave goods from Crete and Mycenae, and discussed various burials of Bronze Age. Broodbank (1989) mentioned that most Cycladic burials involve single skeletons, a few with grave goods. Horwitz and Goring-Morris (2004) described a Levantine Pre-Pottery Neolithic B – Early Neolithic burial ground from Israel that showed evidence of animal remains involved with burial rituals.

Goring-Morris (2000, *in* Kuijt, ed.; see also Table 1) discussed the headless bodies and separate skull practices of the Pre-Ceramic Neolithic.

P. Lieberman (1991) discussed the possible conception of burials, including those with grave goods, and language capability as well as belief in an afterlife and religion in the broad sense.

Cremations

McKinley (2013, *in* Stutz and Tarlow, eds.) reviewed the significance and presence of cremated remains, and Williams (2013, *in* Stutz and Tarlow, eds.) discussed some Iron Age examples. This practice may have been far more widespread than commonly realized, since it is very difficult to identify them without burial of the cremated ashes in urns.

Gregariousness

Stoddart (1990, p. 222) pointed out that humans are unique in forming small groups including both sexes, in which pair bonds are the rule, as contrasted with most other gregarious mammal groups in which harems are the rule and where the newborns are not long dependent on their parents.

Slavery and Prostitution

Pre-Bronze Age evidence of slavery consists chiefly of Greek and Roman accounts, as is the case with prostitution. Yet extensive pre-Neolithic use of slaves or of prostitutes remains a distinct possibility. Klees (1998; see also Garlan, 1982) provided a comprehensive account of the widespread slavery in Classical Greece that deals with a wide variety of evidence. Postgate (1992, p. 106–108) briefly discussed some aspects of slavery in Old Babylonia. 'Temple prostitution' appears to be a separate phenomenon.

Mass Deportations

Oded (1979) provided details of the many mass deportations practiced by Assyrian kings for various purposes such as enlisting soldiers in Assyrian armies, punishment of rebellious groups, populating outlying districts, etc. Alexander the Great did much the same.

Human Disease

All organisms, plants and animals, present and extinct, are or have been subject to disease. Diseases can have genetic causes, or be due to deficiency or 'organism' mechanisms. The past record of disease is very fragmentary. Why? The bulk of the evidence is provided by hard-part skeletal materials. One needs to keep the proper perspective toward skeletal paleopathology in view of Tanke and Rothschild's (1997) observation that less than 1% of currently recognized human diseases leave any traces on bones! Mitchell (2015b) has outlined many of the factors involved in trying to understand the impact of disease on human populations living under various conditions, hunter-gatherers and younger.

Roberts and Manchester (2005) provided an extensive account of human disease in antiquity that covers a number of factors, particularly skeletal symptoms, which receive far less treatment by others.

Our basic problem when considering disease in *Homo* is trying to determine from very limited evidence whether they are relatively recent or of considerable antiquity, even present from pre-*Homo* times. We would like to determine whether the diseases characteristic of hunter-gatherer populations differ significantly from those of urbanized Neolithic and younger populations. Some diseases will thrive under conditions of easy human communication, as in urban environments, while others will not. For example, it is reasonable to suggest that insect vectors such as mosquitoes or tsetse flies can affect very small human populations whereas more 'urban'-type diseases such as measles or pulmonary tuberculosis require a large, communicating population. The possibilities inherent in disease 'switching' ['spillover'] from non-*Homo* hosts to humans should always be kept in mind [written by the author in 2015 or earlier – Eds.].

Bouchet et al.'s (1996) discovery of *Ascaris*, probably of human fecal origin, in a Late Paleolithic French cave makes clear the possibility that human diseases of one sort or another have been with us since the appearance of *Homo* two million years ago.

It is reasonable to suggest that disease and parasitism affecting humans differ from climatic–ecologic 'zone' to 'zone', and that the major change in the Neolithic to farming, irrigation, and animal husbandry changed the incidence and nature of diseases and parasitisms as well as providing possibilities for switching from domesticated animals to humans.

The information on human disease considered below is a sample of what is currently known; it is not a comprehensive account of evidence for ancient disease.

Lieberman (2013, p. 200–202) emphasized that with the advent of agriculture in the Neolithic the incidence and variety of human diseases greatly increased, making epidemics possible, owing to greater population densities in villages and to greater exposure to infected domesticated animals. The increasing prevalence of dental caries in the Neolithic correlates well with the greater use of cereal products.

Thus, we are left with two opposing views of human disease: 1) that new diseases have appeared within the past four to five millennia, as shown by 'objective, specimen-based' evidence, and 2) that the lack of adequate unequivocal diagnostic capabilities inherent in the available materials makes the first view an artifact of the evidence. One would, of course, expect that the increasingly urban living (Black, 1966) of our species would result in major changes in the incidence of certain infectious diseases such as measles, tuberculosis, and cholera. Black (1966), while making his point about the correlation between human population size and incidence, goes on to infer that measles in humans evolved relatively recently, a view based on assumptions that lack support, in view of the failure to follow relatively small human populations long enough to make the point certain.

The other alternative is to take the data concerning disease recognized in other taxa, even invertebrates, and to generalize to humans, for whom the rather poor fossil record precludes really definitive conclusions.

Reinhard and Pucu (2014) pointed out the great disparity between disease loads revealed from the available samples in the New World, as contrasted with the Old World, with differing population densities and other factors.

The Record

There is an extensive literature on the antiquity of disease in humans. For example, see Brothwell and Sandison (1967) for a review, and the many volumes on human paleopathology and the journal of the Paleopathology Association, *Paleopathology Newsletter*. Cockburn (1971) reviewed the known records of infectious diseases in human populations. The problem boils down to trying to understand whether the genetic, deficiency, and organism-caused diseases of humans are of great antiquity – going back essentially to the 'appearance' of our species, or genus – or whether they are of recent origins, with the ancillary question whether their origins have something to do with the increasing levels of urbanization that have characterized us in our movement away from the hunter-gatherer level toward the more social, agricultural, and ultimately highly urban creature of the present. Cockburn (1963) supported the more recent origins hypothesis, though without any really positive evidence of a paleontologic type, and Rothschild et al. (1988) presented an 'example' of the type of evidence plus logic used to arrive at the conclusion that a human disease may be 'new', as contrasted with going back very far in the hominid or higher-primate family tree. Brothwell (*in* Brothwell and Sandison, 1967, Figure 1) reviewed some of the correlates involved. Even for a layman it is clear, while reading the available literature, that the paleopathology of human materials of great antiquity is a very difficult, commonly ambiguous specialty in terms of precise diagnosis (see Ortner and Putscher, 1981, for a critical review, and Price and Molleson, 1974, for a specific example). The intrinsic difficulty of interpreting skeletal pathology is illustrated by Rothschild and Turnbull's 1987 work on a Pleistocene bear's remains that provided evidence of treponemal infection, and Neiberger's (1988) rebuttal of their conclusions, followed by Rothschild's (1988) comments on the latter. Larsen (1997) provided an excellent overview of human disease and trauma evidence capable of being preserved in the archaeological–anthropological record.

Waldron (1994) provided a very readable account emphasizing the problems involved in trying to estimate the prevalence of disease in human skeletal populations, while providing an overall view that is applicable to all organisms.

Finally, Roberts (2013, *in* Stutz and Tarlow, ed.; see also Wood et al., 1992, for an extended discussion) made the important point that paleopathological evidence deals chiefly with skeletal evidence of disease in individuals who *survived* the disease, leaving 'blank' the individuals who died in the same population without having had time to develop any skeletal evidence.

Reinhard et al. (2013) make the important point that evidence for human disease and parasites increased substantially with the Neolithic owing to transmissions from domesticated animals, whereas this was not the case in the New World. The widespread evidence in the New World for various parasites that were also common in the Old World suggests an ultimately African origin from remote *Homo* ancestors.

Sources of Evidence Other Than Skeletal

Information concerning human disease comes from several sources. The first comes from study of skeletal pathology, including that known from allied higher primates. The second comes from the study of evidence preserved in mummies, discussed below, and also from the study of fecal remains, mostly preserved in ancient privies, which provide information concerning intestinal parasites, among other things. Most of this information is on a non-evolutionary time scale, dealing with information from at best only a few thousands of years.

The most comprehensive account of human paleopathology is that of Aufderheide and Rodriguez-Martin (1998), which includes a chapter on dental materials by Langsjoen.

Mitchell (2015) provided a most comprehensive account of human parasites, with emphasis on the Roman period.

Araújo et al. (2008) pointed out that the presence of relatively warm-climate human parasites in the New World, clearly derived from Old World sources, is possible only during a time(s) of climatic amelioration, i.e., interglacial intervals during the later Pleistocene.

Barnes (2005; see also Schultz, 1967) provided a summary that emphasized relationships with allied anthropoids, as well as a certain amount of literature on historical examples derived chiefly from mummies and coprolites. They provided an account of human disease that considered known occurrences of similar diseases in related mammals; for example, among the higher primates the following diseases are known: malaria, schistosomiasis, Ebola virus, smallpox, herpes, HIV/AIDS, yellow fever, and dengue. This finding implies that these diseases arise from a common ancestral primate. Her approach has limitations, as we have relatively limited knowledge concerning diseases present in most wild animals. Also, the problem of disease-causing organisms 'switching', or spilling over, from one organism to another is always a possibility.

Rothschild and Martin (2006a) provided well documented information regarding paleopathology. Pérez (1996) summarized and referred to a mass of more recent work on human paleopathology.

Mummies

Cockburn and Cockburn (1980) edited a volume on mummies that provided a wealth of information on human disease from the historical past, including Egyptian, Chinese, and Peruvian mummies, with the inference that almost all the diseases detected have been with us for a long time, although this is not 'evolutionary time'. In the Cockburns's volume, Sandison (1980) pointed out that the interpretation of soft-tissue abnormalities requires coping with the chemical and physical results of embalming, fungal activity, and post mortem insect attack. Aufderheide (2003) provided another account of various mummies with some emphasis on soft-tissue evidence. Pringle (2001) provided a volume discussing various mummies, ancient to modern, that contains a wealth of useful information.

Feces

Wilke and Hall (1975; see also Fry, 1985) provided an annotated bibliography of human fecal contents, parasitic and otherwise, chiefly from archaeological sites. Evidence about human parasites has been obtained from coprolites (Bryant and Williams-Dean, 1975, for much of the available evidence), but here again the data extend back only a few thousands of years.

Cancer

In their discussion of cancer in the human record, Aufderheide and Rodriguez-Martin (1998, p. 373–374), reviewed various possibilities, concluding that the apparent lower incidence of cancer in archaeological and anthropological materials is more likely attributable to a series of sampling problems. Specifically, they emphasized that the earlier age(s) of death in earlier human populations than is the case today, and even in the latter part of the Nineteenth Century, eliminates the modern fact that most cancer shows up in the elderly, say after about age 55, and that the bone cancers present in younger people are at such a low incidence today that they have small likelihood of showing up in the very small samples of ancient materials available for study.

This view contrasts with that put forward by David and Zimmerman (2010), who preferred to ascribe the lower incidence of reliable evidence for bone cancer as well as soft-tissue evidence from mummies as due to modern environmental factors.

This is clearly a sampling problem that will be increasingly answered in the future as more studies are made using modern techniques. Roberts and Manchester (2005) discussed the limited data available concerning bone cancers.

Monge et al. (2013) made a strong case for the presence of a presumably benign fibrous dysplasia in a 120,000-year-old Neandertal rib from Krapina in Croatia, by far the oldest human evidence of such things, with the implication that 'environmental factors' were not involved. Strouhal (1998) reviewed evidence concerning bony tumors of the jaw. Waldron (2009; see also Brothwell, 2012) discussed bony tumors at some length.

The Siberian Times, 14 October 2014, is cited for the presence of breast cancer in a woman in her twenties, buried in the Altai, 2500 years BP, based on scans done by Drs. A. Letyagin and A. Savelov, Novosibirsk.

Urteaga and Pack (1966) describe evidence for melanoma in several Peruvian Inca mummies dated 2400 BP.

Rheumatic and Arthritic Diseases, Joint Diseases

Dieppe and Rogers (1993) reviewed the skeletal paleopathology of human rheumatic disorders. Bourke (1967) discussed evidence of arthritic disease in ancient human remains.

Waldron (2009; see also Stirland, 1991) provided an extensive account of joint diseases, as well as Paget's Disease, and allied problems. Roches et al. (2002) described Paget's Disease from a 300–350 A.D. case in Normandy.

Infectious Diseases

Chagas Disease

Guhl et al. (1997, 1999) used molecular techniques to identify Chagas Disease in 4000-year-old human mummies from the Atacama Desert. Poinar (2013) identified *Triatoma antiquus* in Dominican amber, a hemipteran genus responsible today for the spread of Chagas Disease owing to the presence in it today of *Trypanosoma cruzi*, the protozoan responsible; *Trypanosoma dominicana* was recognized in the Dominican amber, attesting to the antiquity of the relationship. Aufderheide (*in* Pringle, 2001, p. 62–72) cited 'megacolon', a Chagas Disease–caused condition in ancient mummies from the U.S. Southwest. Aufderheide et al. (2004) used DNA evidence to indicate 9000-year-old Chagas Disease in northern Chile and southern Peru. Lima et al. (2008) identified *T. cruzi* from a Brazilian site dated 4500–7000 years BP.

Smallpox

Baber (1996, p. 80; see also Holwell, 1767) cited inoculation (presumably variolation) for smallpox in India in 1737. Smallpox may have appeared with the Neolithic. Edward Jenner (see Willis, 1997) popularized the practice of vaccination beginning in 1796 with cowpox as the preferred smallpox prevention treatment (variolation had a small percentage of patients, possibly 2–3%, who got real smallpox from variolation and died).

Tuberculosis

Strouhal (1987) described evidence of tuberculosis, including Pott's disease, affecting human vertebrae from an ancient Nubian skeleton and one from Egypt, while reviewing other published occurrences of a few thousand years' antiquity; none appear to have been recognized in pre-Holocene hominid remains. Canci et al. (1996) suggested tuberculous spondylitis in a Neolithic individual from Italy, with the suggestion that animal husbandry might have led to the spread from animals to humans.

Comas et al. (2013), sequencing *Mycobacteria tuberculosis* from various regions and then comparing the results with human genomes from the same regions, made possible the conclusion that human tuberculosis originated approximately 70,000 years BP in Africa and then diversified greatly with the advent of the Neolithic into various geographic strains. This conclusion suggests that relations to domesticated ungulates is a strictly Neolithic phenomenon.

Barnes et al. (2010) found that urbanization for a lengthy period made resistance to both tuberculosis and leprosy greater. Larsen (1997) cited evidence dating to 5000 years BP.

Murphy et al. (2009) used both paleopathological and biomolecular techniques to identify tuberculosis in Iron Age individuals from Tuva. Hershkovitz et al. (2008) recognized *Mycobacterium tuberculosis* in human bone from a Neolithic settlement in the Eastern Mediterranean dated 9250–8160 years BP. Waldron (2009) provided a useful account of the skeletal effects of tuberculosis. Roberts (2012) reviewed the evidence of tuberculosis from the past dating back to the Sixth Millennium BC. Mays and Taylor (2003) described an Iron Age case of tuberculosis from Dorset, supported by osteological and biomolecular data. Taylor et al. (2007) identified *Mycobacterium bovis* from late Iron Age Siberian human remains, including one juvenile with Potts Disease.

Peptic Ulcers and *Helicobacter pylori*

Linz et al. (2007), using molecular data, found that the source region for *Helicopylori* is in East Africa approximately 58,000 years BP, and that its genetic diversity decreases with distance from East Africa. Maixner et al. (2016) demonstrated the presence of *H. pylori* in the Chalcolithic Alpine Iceman.

Typhoid Fever

Papagrigorakis et al. (2006) made a strong case for the presence of typhoid fever in Athens 430–426 BC. Shapiro et al. (2006) and Littman (2009) raised a number of objections to Papagrigorakis' conclusion, but clearly the matter is unsettled.

Typhus

Raoult et al. (2006) recognized evidence of *Rickettsia prowazekii* in tooth pulp from Napoleon's Grand Army with associated lice as the transmitting agent.

Malaria

Various types of malaria infect cold-blooded and warm-blooded vertebrates today (Boucot and Poinar, 2010, Figure 54). One of the primitive types is *Haemoproteus* in birds and reptiles (Bennett and Peirce, 1988; Telford, 1984). A biting midge of the genus *Protoculicoides* (Diptera: Ceratopogonidae) in Burmese amber contained developing stages of an early lineage of malaria, *Paleohaemoproteus burmacis* (Poinar and Telford, 2005); see color plate 21 after page 132 in Boucot and Poinar (2010). The above information makes it possible that human malaria resulted from a 'switch' from one phylogenetic host type to another rather than having an original higher-primate origin.

Miller et al. (1994) employed biochemical methods to demonstrate *Plasmodium falciparum* in Egyptian mummies with ages back to 5200 years BP. Anastasiou and Mitchell (*in* Mitchell, 2015, Table 7.1) cited *Plasmodium falciparum* in Egypt at 2820–2630 BC. Khairat et al. (2013) demonstrated molecular evidence of *Plasmodium falciparum* in Egyptian mummy material dated at 806 BC–124 A.D. Baum and Bar-Gal (2003) made the interesting point that genesis of malaria in humans may be related to the origins of agriculture in West Africa about 10,000 years ago, and they used the prevalence of sickle-cell anemia there as part of the evidence. Angel (1966) provided archaeological evidence suggesting that malaria has been present as early as 6500 BC in marshy areas but not in dry regions. Maat and Baig (1990) described a case of sickle cells preserved in a body from the region of Kuwait, with an age of approximately 2130 years, associated with other bodies whose bones lack sickle cells, which raises the possibility of this as an indicator of malaria, although there are also other possibilities. Tayles (1996) described some 4,000-year-old remains from central Thailand that indicate anemia associated with malaria.

Poinar and Telford (2005) detected malarial parasites in the body cavity and midgut lumen of an earlier Cretaceous Burmese biting midge (Ceratopogonidae).

Toxoplasmosis

Khairat et al. (2013) demonstrated the presence of *Toxoplasma gondii* using molecular evidence from Egyptian mummy material dated as 806 BC–124 A.D.

Leprosy

Inskip et al. (2015) cited evidence of leprosy from the Second Century BC, and a possible case from the Fourth Millennium BC in Hungary. Lynnerup and Boldsen (2012) cited an Indian example from 2000–2500 BC. Skinsnes and Chang (1985) reviewed what is known and what is thought about leprosy in China from approximately 3000 years BP to the present.

Leishmaniasis

There is ample evidence for the presence of trypanosomatid leishmannial parasites (Poinar, 2004a, 2004b) [break in text? – Eds.]. Anastasiou and Mitchell (2015, Table 7.1; see also Zink et al., 2006, for both Egypt and Nubia) cited *Leishmania donovani* from Egypt at 2050–1650 BC.

Syphilis

Rothschild and Rothschild (1995, 1996a, 1996b, 1998a; see also Rothschild 2005 and Rothschild et al., 2000) provided a wealth of data indicating that syphilis evolved from yaws in the New World during pre-Columbian time, and that yaws was formerly cosmopolitan, with bejel, the other treponemal disease, being strictly Old World. The authors employed large populations to determine the skeletal differences between the three diseases, as well as the time when syphilis first appeared in the record. Their data strongly indicate that syphilis was brought from the New World to the Old World following the Columbian voyages. Rothschild et al. (1995) provided skeletal evidence for the presence of yaws in the Middle Pleistocene of East Africa in *Homo erectus.* Waldron (2009) provided a useful account of the characteristics of syphilitic bone, and discussed the possibility that it was not brought to the Old World by the Columbian voyagers.

Lyme Disease

Keller et al. (2012) described molecular evidence for the presence of *Borrelia burgdorferi,* the protozoan responsible for Lyme Disease in the 5300-year-old Ötzi the 'Iceman' mummy from the Alps. This is by far the oldest known occurrence of evidence for Lyme Disease in humans. Poinar (2014e) described Dominican amber with inclusions of spirochete-like cells (from a tick) that resemble those causing borreliosis.

Plague and Trench Fever

Rosen (2007) discussed the well-known symptoms of bubonic plague recorded from its 540 A.D. occurrence in Constantinople and goes on to discuss subsequent, chiefly European occurrences. Harbeck et al. (2013) discussed the *Yersinia pestis* DNA evidence for the Justinian Plague, 541 A.D. Rasmussen et al. (2015) described the molecular evidence for virulent plague originating approximately 3000 years BP, with the non-virulent ancestral form occurring in the Bronze Age. Poinar (2015b) described microorganisms in the rectum of a Dominican-amber flea that might be bacteria of *Yersinia* type. Wagner et al. (2014) used genomic analysis to decipher the source(s) of different outbreaks of the plague.

Tran et al. (2011) reported DNA evidence suggesting that plague generated by *Y. pestis* in colder regions lacking the rodents commonly involved in its transmission might instead have been transmitted by *Bartonella quintana,* a bacterium borne by lice, *B. quintana* itself being the cause of trench fever.

Raoult et al. (2006) described evidence for louse-borne *Bartonella quintana* from Napoleon's Grand Army.

Rabies

Rabies is 'an acute viral infection of the central nervous system', first recognized in the Mesopotamian Eshnunna Code about 4000 BC; for a review, see Tarantola (2017).

Sinusitis

Armentano et al. (1999) described a case of sinusitis from the Chalcolithic/Bronze Age of the Barcelona region.

Gout

Rothschild and Thillaud (1991) discussed the presence of gout in Neandertals, possibly pyrophosphate deposition disease.

Sleeping Sickness

Meyer (2003, Figure 189) illustrated a Late Eocene *Glossina* fly, and Poschmann and Wedmann (2005) depicted an Oligocene example. These examples raise the possibility of a sleeping-sickness vector.

Ergotism

Poinar et al. (2015) described a grass fungus that today produces ergot on an infected grass kernel in Burmese amber from the earlier Cretaceous (Albian). The presence of this plant parasite with its burden of alkaloids opens up the strong possibility that with the advent of farming in the Neolithic, humans were exposed for the first time to ergotism from eating infected grain. Ergotism ('Saint Anthony's Fire') does not leave skeletal evidence, so 'proving' its presence is obviously difficult, the only possibility being the 'amputation' of some terminal digits. This unique evidence raises the possibility that some Neolithic farmers raising rye, a favorite host of the fungus, might have been victims. Dark and Gent (2001, p. 69) discussed Iron Age evidence of ergot in Europe. Aldhouse-Green (2015, p. 59–60) cited ergot in the gut of Grauballe Man.

Rates of Evolution

It is clear that prokaryotic, bacterial, viral, and some protozoan diseases are not conservative; that is, they are subject to rapid mutation rates and evolve very rapidly (Willis, 1996; Frank, 2002)! This situation contrasts radically with the conservative behavior of metazoan disease-causing organisms.

Diarrhea

Stöllner et al. (2003, Figure 19, and p. 151; see also Aspöck et al., 2002.) cited diarrhea from the Early Iron Age salt mines at Dürrnberg-bei-Hallein.

Amoebic Dysentery and Amoebic Liver Abscess

Le Bailly and Bouchet (2006) summarized known cases of *Entamoeba histolytica* from the record, chiefly Old World, with evidence from Cyprus dated at 7500–7000 BC. Gonçalves et al. (2003, Table XIV) cited *Entamoeba* occurrences from 160 A.D. onwards. Le Bailly and Bouchet (2015) reviewed the instances of well-dated *Entamoeba histolytica,* with the oldest from 5000–2000 BC from Greece. Witenberg (1961) cited *Entamoeba histolytica* from 1800-year-old coprolites from

Israel. Goncalves et al. (2004, Table 1) listed various New World and Old World sites where *Entamoeba histolytica* has been identified, with an 8760-years BP Brazilian site as the oldest.

Protozoans

Tankersley et al. (1994) reported on the occurrence of *Giardia* in American Indian paleofeces dated at 2420 BC, plus or minus 90 years. Gonçalves et al. (2003, Table XIV) cited *Giardia* occurrences from the 160 A.D. onwards. Witenberg (1961) cited *Giardia* from 1800-year-old coprolites from Israel.

Lice and Fleas

Buckman (2003) provided excellent illustrations of fleas, bedbugs, mites, and lice, as well as of the disease-carrying ticks involved with humans.

Lice

A wealth of information on lice has been obtained from mummies. Ewing (1926) cited head lice from Peruvian Inca mummies. Bresciani et al. (1983) discussed head lice from some Fifteenth Century Greenland Eskimo mummies (see also Bresciani et al., 1991, Figure 163, for an illustration of a hair with an attached egg case of a head louse). Horne (1979) discussed some Aleut mummies of about the same age with head lice. Cockburn (1971, p. 53) cited an occurrence of head lice from a 4000-year-old Peruvian mummy. Sadler (1990) provided an account of Viking Age Greenland body lice, as well as of several parasites from associated sheep. Mumcuoglu and Zias (1988) described head lice from nit combs derived from the First Century BC and the Eighth Century A.D. in Israel. Zias and Mumcuoglu (1991) identified the oldest known, Pre-Pottery Neolithic head lice from Nahal Hemar Cave, Israel, with associated human hair. Mumcuoglu et al. (2003) described First Century A.D. body remains with associated textiles from Israel. Wen et al. (1987) described head lice and pubic lice from a Loulan female mummy from approximately 3800 years BP from the Taklamakan Desert region. Reinhard et al. (2013) cited the oldest known head-louse nits on human hair at 10,000 years BP in northeastern Brazil. This nuisance has always been with us.

Kittler et al. (2003) used molecular-clock data to suggest that body lice, *Pediculus humanus*, originated 72,000 plus or minus 42,000 years ago. Araújo et al. (2000) briefly described a 10,000-year-old head louse (*P. humanus*) from Brazil.

Busvine (1980) reviewed some of the evidence from the present concerning head lice and pubic lice, including the inference that the two species evolved 'allopatrically' after humans lost their complete fur coat; he commented that pubic hair is morphologically distinct from that on the head and that the former's density is less. However, Reed et al. (2007) suggested that pubic lice originated in humans by switching from a gorilla source significantly later than the initial presence of head lice, head lice having originated at the same time in humans and chimpanzees. Weiss (2009) indicated that human pubic lice and gorilla lice might have a common origin.

Kenward (1999) documented Roman and Medieval pubic lice (*Pthirus pubis* L.) from Britain.

MacKay (1938; see also Marshall, 1931, for an ivory comb) illustrated a comb and razors from Mohenjo-Daro, which were probably used chiefly for hair combing. Rudenko (1970) illustrated and briefly discussed horn haircombs from the Pazyryk Iron Age burials. Tosi (1968) described some wooden combs from Bronze Age excavations at Shehr-i-Sokhta, Iran. Mumcuoglu (2008, Figure 13.4) illustrated several ancient Israeli wooden combs used for removing head lice.

Reed et al. (2004, Figure 1) concluded that human body lice diverged from chimpanzee body lice 5.6 million years ago, and that Modern and Archaic Humans were in direct contact.

Fleas

Stoddart (2015, p. 13) pointed out that fleas require a nest in which their eggs can mature, and that gorillas and chimpanzees do not use the same nests over time, making a new one just about every night, hence, no fleas. Schedl (2000) identified the remains of fleas associated with the Alpine Iceman, 5300 BC. Grzimek (1968, p. 530–535, notes that *Pulex irritans*, the human flea, sucks blood from humans only and is not a disease vector. Panagiotakopulu (2001) noted fleas from Pharaonic Egypt. Sadler (1990) provided an account of Viking Age Greenland human fleas.

Bedbugs

Panagiotakopulu and Buckland (1999) note 1500-year-old bedbugs (*Cimex lectularis*) from Pharaonic Egypt.

Parasitic Diseases

Helminths

Aspöck et al. (1999) reviewed the various helminths found in prehistoric human populations in Central Europe, and Aspöck (2000) thoroughly reviewed the current status of our knowledge concerning paleoparasitology.

Kliks (1990) reviewed occurrences of New World pre-Columbian helminths in feces and cadavers to indicate that many items traveled from the Old World with the dogs of transberingeal immigrants. Harrison et al. (1991) provided an excellent case where skeletal pathology in a Native American from Arizona (200 to 1400 A.D.) could be assigned to coccidioidomycosis, which was confirmed by microscopic preparations that revealed both spherules and endospores of parasitic type belonging to *Coccidioides immitis*. Jouy-Avantin et al. (1999) reviewed a number of occurrences of helminthic parasites associated with sub-Recent and Late Pleistocene human remains.

Faulkner (1991) discussed some helminths, including pinworms, roundworms, and a possible hookworm from millennia-old Tennessee coprolites, as well as *Giardia* cysts identified with the aid of monoclonal antibodies specific for cyst proteins. Reinhard et al. (1987) discussed helminth evidence from Paleoindian coprolites from the Colorado Plateau.

Pike (1967) provided a useful summary of parasite eggs from human materials, chiefly feces.

Giant Kidney Worm

Le Bailly et al. (2003) described the occurrence of Dioctophymidae eggs from a Swiss Neolithic site.

Schistosomiasis

Loebl (1995) discussed evidence of schistosomiasis, bilharzia from Egyptian mummies, and also a possible example of Symmers' fibrosis, one of the symptoms of that disease. Anastasiou et al. (2014) documented a 6000–6500 years BP occurrence of a schistosome egg from a human pelvic region in the Euphrates River Valley of northern Syria. Pringle (2001, p. 76–78) discussed the behavior of the schistosome worm within the human body, with some of the symptoms of schistosomiasis that include red urine, and its presence in an Egyptian mummy at least 4000 years old. Miller et al. (1992) provided evidence of schistosomal infections present in Egyptian mummies, as did Deelder et al. (1990). Despres et al. (1992) used molecular evidence to suggest that the various schistosomes affecting humans arose with the origin of *Homo*. Gonçalves et al. (2003, Table VIII) cited *Schistosoma* occurrences from 3200 BC onwards. Anastasiou and Mitchell (2015, p. 138) cited schistosomiasis evidence from the Middle East at 4500–4000 BC.

Roundworms

Bouchet et al. (1996), in an important paper, described the Late Paleolithic (24,660 to 30,160 years BP) occurrence of *Ascaris*, probably of human fecal origin, in the Grande Grotte d'Arcy-sur-Cure, Yonne Cave, which also features graphic art. This is currently the oldest known evidence of human disease and is supportive of the possibility that many human diseases have been present since the appearance of *Homo* several million years ago.

Stöllner et al. (2003, Figure 20, p. 150) cited roundworms (*Ascaris*) from the Early Iron Age salt mines at Dürrnberg-bei-Hallein. Faulkner (1991) cited roundworm evidence from millennia-old Tennessee coprolites. Aldhouse-Green (2015, p. 57) cited roundworm in the gut of Lindow Man. Evans et al. (1996) described a Late Stone Age coprolite from South Africa that contained eggs of *Trichuris* and *Ascaris*. Rousset et al. (1996) described *Ascaris* from a Second Century A.D. French burial site with an associated skeleton. Gonçalves et al. (2003, Table III) cited *Ascaris* occurrences, chiefly European, from the Late Paleolithic onwards.

Taenia Tapeworms

Hoberg et al. (2001) concluded that *Taenia* tapeworms were acquired by humans, *Homo*, very early on in Africa from eating bovids, as a result of either hunting or scavenging bovid remains (the host cycle includes hyaenids, canids, and felids as the definitive host and bovids as the intermediate host).

Le Bailly et al. (2005) described the tapeworm *Taenia* from European Neolithic sites, the oldest from approximately 3600 BC. Gonçalves et al. (2003, Table X) cited *Taenia* occurrences, chiefly European, from 3200 years BP onward.

Gonçalves et al. (2003, Table XV) cited occurrences of *Echinococcus granulosus*, the dog tapeworm, from 538 BC – 70 A.D. onward.

Stöllner et al. (2003, Figure 20, p. 150, *Taenia*) cited beef or pork tapeworm from the Early Iron Age salt mines at Dürrnberg-bei-Hallein.

Le Bailly et al. (2010) reviewed the Egyptian occurrences of *Taenia*, including both *T. saginata*, the beef form, and *T. solium*, the pork form, with the oldest taeniasis occurrence from c.1200 BC.

David (1997, p. 1762) cited hydatid cysts caused by the dog tapeworm *Echinococcus granulosus* in an Egyptian mummy (age not provided) at Manchester University.

Gonçalves et al. (2003, Table X) cited *Taenia* occurrences from 3200 years BP onward.

Fish Tapeworm

Reinhard and Urban (2003) discussed diphyllobothriasis from Peruvian mummies dated at 2700–2850 BC. Mitchell (2015) suggested that the widespread evidence of Roman Era *Diphyllobothrium* might well correlate with the Roman taste for fish sauce made from uncooked fish (see Haley, 1990, for an account of a Roman fish-sauce trader). Gonçalves et al. (2003, Table XI) cited *Diphyllobothrium* (Table XI) from 4000–10,000 years BP onward. Le Bailly et al. (2005) described some fish tapeworm eggs (*Diphyllobothrium*) from European Neolithic sites. Le Bailly and Bouchet (2013) compiled known records of New World and Old World *Diphyllobothrium*, back to 8000 BC in Peru and to 7500–7600 BC in the Pre-Pottery Neolithic on Cyprus.

Whipworms

Dark (2004) described evidence of *Trichuris* (whipworm) from early Seventh Millennium BP deposits with human hosts implicated. Jones (1986) discussed the presence of whipworms in Lindown Man, dating from about the time of the Roman occupation of Britain. Stöllner et al. (2003, Figure 20, p. 150) cited whipworm from the Early Iron Age salt mines at Dürrnberg-bei-Hallein. Aspöck et al. (2000) discussed *Trichuris* from the Alpine Iceman. Aldhouse-Green (2015, p. 57) cited whipworm in the gut of Lindow Man. Evans et al. (1996) described a Late Stone Age coprolite from South Africa that contained eggs of *Trichuris* and *Ascaris*. Rousset et al. (1996) described *Trichuris* from a Second Century A.D. French burial site with an associated skeleton. Gonçalves et al. (2003, Table IV) cited *Trichuris* occurrences, chiefly European, from 7000–10,000 years BP onward. Bouchet et al. (1995) described *Trichuris* eggs from a Neolithic site (Chalain, Jura) dated at 3080–2950 BC.

Hookworms

Ferreira et al. (1987) identified hookworm eggs, ancylostomid, from Piaui, Brazil in 7230 years BP plus or minus 80 years, from human coprolites.

Lancet Fluke and Sheep Liver Fluke

Stöllner et al. (2003, Figure 20, p. 150) cited lancet fluke (*Dicrocoelium*) and sheep liver fluke (*Fasciola*) from the Early Iron Age salt mines at Dürrnberg-bei-Hallein. Gonçalves et al. (2003, Table VII) cited *Fasciola* occurrences from 3600 BC onwards, and (Table IX) *Dicrocoelium* from 3384–3370 BC. Searcey et al. (2013) identified *Dicrocoelium dendriticum* in the liver of a Roman-period Dutch bog body. Bouchet et al. (1995) described *Fasciola* eggs from a Neolithic site (Chalain, Jura) dated at 3080–2950 BC. Dittmar and Teegen (2003) described human-associated *Fasciola hepatica* from a 4500-year-old German site.

Guinea Worm

David (1997, p. 1761) cited a calcified nodule with the remains of *Dracunculus medinensis* in a Manchester University Egyptian mummy (age not provided). Gonçalves et al. (2003, Table XV) cited Egyptian *Dracunculus* occurrences from 1450 BC onwards.

Pinworm

Horne (2002) discussed the presence of pinworm eggs from Egyptian fecal samples dated 30 BC–395 A.D. Pinworms (Zias et al., 2006) were also recognized at Qumran in Israel of about the same age. Gonçalves et al. (2003, Table V), cited *Enterobius vermicularis* occurrences, chiefly New World. Numerous New World occurrences of pinworm are known.

Echinostomiasis

Echinostomiasis is caused by a worm that involves aquatic mollusks and was found in a mummified Brazilian dated at 600–1200 years BP (Sianto et al., 2005).

Anthracosis

Hodder (ed., 2007) discussed many examples of carbon deposits in the rib-cage region of adult females from the Pre-Pottery Neolithic B of Çatalhöyük, probably representing anthracosis induced from cooking ovens.

Lead Poisoning

Nriagu (1983) devoted a volume to a discussion of lead poisoning from Classical Antiquity, with application to the present. It is clear that lead poisoning is strictly a Copper Age and younger problem made possible by the mining and smelting of metalliferous materials high in lead. Mercury poisoning has a similar history. In both cases certain occupations and localities are more likely to suffer. Budd et al. (2004) provided a very useful account of lead levels in England from about 5500 years BP to the Sixteenth Century A.D., showing extremely low lead levels in the Neolithic followed by greatly higher lead levels in Roman and younger times. Lead-dust pollution in Swiss glaciers is directly related to English mining taxes paid on lead and silver, as well as to various historical events (Gibbons, 2020).

Osteomyelitis

Hershkovitz et al. (1993) described a male skeleton with lesions most likely caused by osteomyelitis from the Epipaleolithic of Ohala II in Israel. Waldron (2009) provided a useful account of the various organisms responsible for the infections causing osteomyelitis.

Rickets and Scurvy

Roberts and Manchester (2005) devoted space to rickets, caused by vitamin D deficiency, with examples back to the Late Paleolithic, and Waldron (2009) devoted considerable space to its characteristics.

Ferreira (2002) described features of a Fourth to Sixth Century A.D. infant skeleton that were ascribed to scurvy, a vitamin C deficiency disease.

Caries in Humans

D.E. Lieberman (2013, p. 193–194) emphasized the marked increase in the incidence of caries with the onset of agriculture in the Neolithic, attributable to the far starchier diet. Larsen (1997) carefully reviewed occurrences and causes in humans of caries and periodontal disease. Clemen (1956) provided evidence of dental caries in various Pleistocene hominids from South Africa. Tillier et al. (1995) described a Middle Paleolithic example of dental caries that is very convincing and clearly pre-agricultural. Alt et al. (1998; see also Hillson, 1996) provided an elegant summary of caries in hominids (see particularly the Caselitz paper, pp. 203–226). Swindler et al. (1995) concluded that caries, periodontal disease, and severe dental attrition were present in Egyptian remains from the New Kingdom (1550 to 1070 BC); that is, there was little difference from the present. Langsjoen (in Aufderheide and Rodriguez-Martin, 1998) dealt with dental materials from 'antiquity'. While not involving caries, Hershkowitz et al.'s (1997) description of dental calculus in a Miocene ape, Sivapithecus, with scanning-electron-microscope (SEM) evidence of bacterial presence suggests that higher-primate dental problems have been with us for a long time. Although not involving caries in humans, the account by Sala Burgos et al. (2007) of a Miocene artiodactyl from Spain with a carious tooth is of interest. Ducrocq et al. (1995), while discussing a Late Eocene dental anomaly in an anthrocotherid, also provide a summary of other mammalian tooth anomalies from the fossil record, including some primates. Anthony (2007, Figure 16.12) pointed out that caries in humans is most prevalent in bread-eating populations, i.e., is a product of agriculture and its consequences. But Hershkovitz and Gopher (2008) indicated that Late Mesolithic Natufians had as high an incidence of caries as did early agricultural Neolithic humans, i.e., that Natufian hunter-gatherers and farmers probably had similar cereal- and pulse-based diets that led to higher incidence of caries. Cohen (2008, p. 487) pointed out that women tend to have higher rates of caries than men owing to the effects of pregnancy and, also, that caries is more prevalent in agricultural populations. Roberts (2013, in Stutz and Tarlow, eds., Figure 6.3) diagrammed tooth loss, tooth abscess, and caries in Britain from the Neolithic to the present. Caries, likely associated with lead poisoning (Bartsiokas and Day, 1993), was suggested as the cause of the severe dental caries associated with 'Rhodesian Man', with an age estimated as 125,000–300,000 years BP.

Roberts and Manchester (2005) devoted an extended chapter to caries in humans and also to other dental conditions. Lukacs (2012) discussed various dental problems, past and present, in some detail.

Coppa et al. (2006) described the use of flint-tipped tools to drill Neolithic (7500–9000 years BP) teeth from Mehrgarh in Baluchistan, with the assumption that the skills developed in bead drilling were adequate to the task; use of a flint-tipped bow drill is suggested. Some type of filling might have been employed but is not preserved. Bennike and Fredebo (1986) described a Danish example of a tooth drilled with a stone drill, presumably 'powered' by a bow drill, with an age of 4000–5000 years BP. Rudgley (1999, p. 136–137) discussed past evidence of dentistry.

Fractures

Roberts and Manchester (2005) devoted considerable attention to the various types of bone fractures, their morphologies and their causes, including those involved with violence. Waldron (2009) also devoted considerable space to the consideration of fractures due to various causes.

Joint Disease

Roberts and Manchester (2005) discussed the many types of joint disease affecting various joints, and their potential causes.

Developmental Disorders

Barnes (2012) provided an extensive account of skeletal developmental disorders that could be usefully applied to human remains. Brothwell and Powers (1968) discussed numerous ancient developmental malformations. Murphy (1999) described polydactyly in a Zambian 1100–1500 A.D. burial, and noted that polydactyly is far more frequent in African and African-American populations than in Caucasians. Murphy (2000) discussed a number of developmental defects and evidences of disability from Iron Age remains in southern Siberia, including hypoplastic mandibles, congenital scoliosis, possible meningocele and hydrocephalus, developmental dysplasia and congenital dislocation of the hip, slipped femoral capital epiphysis, clubfoot, neurofibromatosis, hemifacial macrosomia, and frontometaphyseal dysplasia. Frayer et al. (1987) described an Italian Late Paleolithic dwarf. Tillier et al. (2001) described a Middle Paleolithic case of hydrocephalus from Israel. Brothwell (1967b) considered a number of congenital disorders from the past, including clubfoot from the British Neolithic, hydrocephaly from the German Neolithic, Bronze Age British scoliosis, dwarfism from pre-dynastic Egypt, and French Neolithic hip dysplasia.

Infanticide and Uxoricide

Although not involving disease in the strict sense, the problem of infanticide as a possible higher anthropoid trait in which step-parents are very much more heavily involved than genetic parents is significant (Daly and Wilson, 1988a, 1988b). Also related is Wilson and Daly's (1996) finding of a relationship between uxoricide (wife murder) and age of the victim; uxoricide is very much higher among younger women of reproductive age.

Roberts and Manchester (2005) discuss infanticide in the past with emphasis on the various factors involved. Scrimshaw (1984) discussed infanticide and the many 'reasons' for its existence, drawing on information from the present.

Boyd and Silk (2003) provided a useful discussion of infanticide in non-human primates but warn against applying the same 'model' to *Homo*.

Trepanation and Skull Deformation

Lillie (1998; see also Weisgerber, 1999, p. 333–337) discussed evidence of trepanation in a human, dated 7600–8,000 years BP. Vallois (1971) described a trepanned hydrocephalic infant's skull from the French Magdalenian, and provided a useful bibliography of the practice. Richards and Anton (1991) describe the morphology of hydrocephalics in some detail, including their cranial and post-cranial features. Crubézy et al. (2001) described examples of trepanation known in Europe, including the Mesolithic of the Ukraine and the Mediterranean, with the oldest instances Epipaleolithic and Mesolithic, approximately 10,000 years BP, and with relatively sophisticated technique suggested in the Neolithic of this region. Kunter (1970) summarized various occurrences, chiefly European Neolithic and younger, from the older literature. Rudgley (1999, p. 126–136) surveyed the record of past trepanation. Roberts and Manchester (2005) discussed evidence of trepanation from the archaeological record.

Meiklejohn et al. (1992) discussed examples of artificial cranial deformation from the earlier Neolithic of the Near East; its significance is unclear.

Cannabis and Opium

Rudgley (1994) provided a compendium of presently known hallucinogens, including various mushrooms, peyote, belladonna, henbane, cannabis, harmel, LSD, mescaline, harmaline, and rare Amazonian plants, inebriants including alcohol and various organic solvents and volatile chemicals, hypnotics, including mandrake, kava, tranquilizers and narcotics including opium, and stimulants, including tea, coffee, cocoa, coca, cola, qat, pituri, betel, tobacco, cocaine, and amphetamines.

Zias (1995) reviewed positive evidence, and evidence from older 'literature', for the use of *Cannabis* during antiquity.

Miller (1991) provided a useful site-by-site summary of many of the Near East localities with evidences of plant cultivation, taxon by taxon, for the Neolithic and later intervals, and Kroll (1991) did the same for southeastern Europe, Küster (1991) for central Europe south of the Danube, Knörzer (1991) for Germany north of the Danube, Wasylikowa et al. (1991) for East-Central Europe, Hopf (1991) for South and Southwestern Europe, Bekels (*in* Miller, 1991) for Western Europe, Greig (1991) for Britain, and Jensen (1991) for the Nordic countries; these papers provide a very comprehensive view of Neolithic and younger crop presences, even including *Cannabis* from a few regions. Russo et al. (2008) discussed the identification of *Cannabis* from a 2700 years BP site at Turpan in Xinjiang. Brunner (1973) reviewed the 'Classic' evidence for the use of hemp and its products, with the earliest use of marijuana as a drug coming from Herodotus in the mid Fifth Century BC, referring to Scythian use.

Sherratt (1991; see also Zohary and Hopf, 1993) reviewed what is known about the use of cannabis and opium in the past, with the suggestion that they were fairly widespread by the Bronze Age or earlier. Merrillees (1962) reported Bronze Age opium trade in the Levant and suggested its use based on the morphology of certain containers.

The Siberian Times, 14 October 2014, cited *Cannabis* in a pouch buried with a female who suffered from breast cancer, with the implication that the plant was used to ease pain; the burial is 2500 years BP from the Altai. Rudenko (1970, pp. 62, 284–285) described *Cannabis* seed and equipment of the type briefly cited by Herodotus from the Iron Age Pazyryk burials.

Kritikos and Papadaki (1967) summarized a mass of information from archaeological and Classical sources plus cuneiform tablets, suggesting that opium and its use were well known in the eastern Mediterranean region at least in the Third Millennium BC, and possibly earlier in the Bronze Age in Mesopotamia. Karageorghis (1976) described what he interpreted as an opium pipe of the Twelfth Century BC from Cyprus. On the other hand, Krikorian (1975) went to some lengths in making the case that the opium poppy, including seeds, has not been reliably shown to have been present early on in the Ancient Near East. This is clearly an area where further work is needed.

Africa (1961) reviewed the evidence that the Second Century A.D. Roman Emperor Marcus Aurelius took opium in the form of 'theriac', a mixture of a large number of plant and animal products (see also Griffin, 2004, for a discussion of the 'contents' of theriac).

Medicine and Medicinal Plants

Johns (1990, p. 251; see also pp. 258, 267) commented that 'the pharmacological use of plants is a fundamental characteristic of our species'. Chaves and Reinhard (2006) discussed a variety of plants from Piauí, Brazil, that might have been used for various medicinal purposes.

Wadley et al. (2011) discussed use of anti-mosquito and larvicidal plants, 77,000 years BP, from South Africa.

Young (2011; see also Johns, 1990, p. 269) reviewed the medicinal properties of clays in some detail, and cited their use in antiquity. They are known to be useful in treating some intestinal problems, including diarrhea, and they tend to adsorb various compounds, some of which may be poisonous. Clark (2001, pp. 659–662, 297–298) discussed some Acheulian clay lumps from the Kalambo Falls an the Tanzania–Zambia border, and reviewed the use of kaolinitic clays for dietary purposes in minimizing the effects of toxic plant substances and for diarrhea; he did not conclude that the Kalambo Falls material was used by humans although associated with artifacts, although he did not entirely dismiss the possibility.

Hardy et al. (2012) discussed dental-calculus evidence for the use of yarrow and camomile by Neandertals.

Pyramarn (1989) cites the presence of *Croton* seeds at a Late Stone Age Thai site; croton oil is a strong purgative.

Crane (1999, p. 508) cited ancient usage of honey for medical purposes as early as 2100–2000 BC.

Care of the Handicapped

Summary

Hublin (2009) reviewed much of the evidence concerning the survival of severely handicapped individuals from 500,000 years BP to the present, including cranial problems, edentulous conditions, limb problems, and the like in terms of 'compassion'. His summary provided further evidence concerning the care of handicapped individuals from the hominin past.

Ackernecht (1978) provided perspectives on medical anthropology in primitive societies. Alt et al. (1997) dealt with evidence of Stone Age cranial surgery.

Early Paleolithic

Lordkipanidze et al. (2006) described a Georgian hominin skull with an almost totally edentulous jaw, the inference being that this adult individual may well have been assisted by its fellows in obtaining suitable food as far back as 1.77 million years ago; the specimen is assigned to *Homo erectus*.

Middle Paleolithic

Crubézy amd Trinkaus (1992) described hyperostotic disease (DISH) in a Middle Paleolithic Neandertal from Iraq; this is chiefly a disease of the elderly and would presumably have partially handicapped the individual.

Trinkaus and Zimmerman (1982) suggested that the injuries sustained by several Mousterian Neandertals would have required support from others. Solecki (1971) described a Mousterian Neandertal skeleton from Shanidar, in the Zagros, Iraq, that showed symptoms of severe handicap and survival.

Trinkaus and Zimmerman (1982) described evidence of pre-mortem trauma present in Shanidar Mousterian Neandertals, with the inference that much of this trauma would have been sufficiently disabling to have necessitated care from co-occurring members of the group.

Buquet-Marcon et al. (2007) referred to several examples of Neandertal amputations from Shanidar (Trinkhaus and Zimmerman, 1982) and Krapina (Kricun et al., 1999).

Late Paleolithic

Frayer et al. (1987) described a Late Paleolithic Italian burial of an adult dwarf, with the obvious conclusion that this individual had received care throughout life, followed by a proper burial. Hershkovitz et al. (1993) described an Epipaleolithic male skeleton from Ohalo II, in Israel, that indicated evidence of needing community support owing to osteomyelitis.

Neolithic to Present

Dickel and Doran (1989) described a case of *spina bifida* from Florida of approximately 7000 years BP, and cited a Moroccan example from 10,000–12,000 years BP.

Buquet-Marcon et al. (2007) described evidence from the French Neolithic, from approximately 7000 years BP, for a successful arm amputation with evidence that the patient survived, and was edentulous, i.e., handicapped and taken good care of.

Oxenham et al. (2009; see also Tilley and Oxenham, 2011) discussed a Neolithic Vietnamese burial of a severely handicapped young adult, with a 'series of skeletal abnormalities', who must have required continual assistance over its lifespan.

Hershkovitz et al. (1993) inferred the need for aid from his fellows from an Ohalo II (23,000 years BP) individual with skeletal anomalies indicating a handicapped condition.

Orscheidt and Haidle (2006, p. 162; see also Orschiedt et al., 2003) discussed the condition of an Early Neolithic German at Herxheim who had suffered from blows to the skull that had healed, but with the need for care from others indicated.

Murphy (2000) discussed a number of Siberian Iron Age cases of developmental defects that would have involved disability and the need for assistance (see **Developmental Disabilties,** above).

Communication

Language

Anthony (2007) reviewed the area of linguistics and makes it very clear that most languages of the distant past are lost in the mists of time. He considered that the Indo-European languages originated from a prototypic Indo-European about 4000–5000 years BP.

Lal et al. (2001) discussed the FOXP2 gene as uniquely human and responsible for human speech capability.

McNeill (ed., 2000) edited a volume devoted to the place of gesture as a part of language or even as a substitute for language in some situations, emphasizing the importance of gesture in human relations.

P. Lieberman (2011) documented the difficulty of ascertaining when true language capability arose in *Homo*, concluding that this is an area of uncertainty at present, with the possibility that earlier *Homo* did possess a fully developed language capability.

P. Lieberman (2006) concluded that fully modern language capability appeared with *H. sapiens* but that earlier *Homo* species did have a language capability. P. Lieberman (1991, 2013; see also Lieberman and Crelin, 1971, P. Lieberman 2013b, P. Lieberman, 2012; P. Lieberman, 2009; Lieberman and McCarthy, 2007) made a case based on detailed anatomical data that fully human language became possible in its *Homo sapiens* form only with the appearance of *H. sapiens* and not with *H. erectus* or *H. neanderthalensis*. Maricic et al. (2012) discussed the molecular basis for the difference between *H. sapiens* and *H. neanderthalensis*. Lieberman and McCarthy (1999) reviewed the ontogeny of cranial base angulation in humans and chimpanzees in relation to language capability, with the implication that only *H. sapiens* had a fully modern capability. P. Lieberman (2013) discussed the ontogeny of language ability in modern humans as well as the accompanying anatomical factors and their relationships to older hominins and other hominids. Lieberman and McCarthy (2007) indicate that fully modern language capability evolved only at about 50,000 years BP, i.e., that earlier *Homo sapiens* lacked a fully developed language capability, which also implies that it is only with the Aurignacian that fully modern language, as well as artistic capabilities, appear on the scene, although that Acheulian Venus poses a problem here.

Another approach to the antiquity of human language is afforded by the presence on Timor of Early Paleolithic stone tools of the type commonly ascribed to *H. erectus* (Antón and Swisher, 2004), with a minimum age of about 1 million years BP. It is very unlikely that a seafaring, colonizing trip to Timor by *H. erectus* could have been carried out in the absence of language. This trip, of course, would have involved seagoing transport of males and females with enough water to ensure survival. Whether such a colonizing voyage would have been undertaken without some prior knowledge provided by previous visits to Timor, intentional or unintentional, by some simple craft such as a paddled dugout or raft is, of course, speculative. It is unlikely that the 'limited' linguistic capabilities of a hominid at the chimpanzee level would have been adequate for such a colonizing voyage. Similarly, the presence on Flores of stone tools with an age of approximately 0.73 million years (van den Bergh et al., 1996; see also Morwood et al., 1998, for further age confirmation), are indicative of the presence of *H. erectus* and the ability to cross a marine barrier.

MacLarnon and Hewitt (1999) pointed out that the expanded thoracic vertebrate canal essential for the fine control of human speech is missing from *Australopithecus* and the great apes – a key item.

Saul (2002) argued that language was intentionally invented within a single generation during an attempt to counter ('beat') death, a matter whose inevitability has not always been self-evident.

Writing

Diamond (2012) pointed out that, before the invention of writing, grandparents and elders provided the 'tribal' memory incorporated in the many oral traditions.

Scarre (2005, ed., Figures 5, 12) cited Aegean writing at 1450 BC, Hittite at 1450 BC, Egyptian at 3000 BC, Mesopotamian Cuneiform at 3100 BC, Indus Valley at 2500 BC, and Chinese at 1200 BC. Chang (1986, p. 295–303; see also Hessler, 2006, on oracle bones) cited Second Millennium BC Shang Dynasty writing at 1700–1100 BC. Keightley (2005) reviewed the Shang Dynasty evidence preserved on scapulae and turtle shells in terms of their potential relationship to writing. Evans (1964, p. 612–646) dealt with Minoan scripts, whose Minoan B was finally understood in 1952.

Postgate (1992, p. 51–70) provided a very useful description of cuneiform writing.

Schmandt-Besserat (1978) proposed that Mesopotamian writing is derived from a system of small clay 'tokens' that were first used at the beginnings of the Neolithic c. 11,000 years BP.

Marshall (1931) discussed the Indus Valley 'Script' present at Bronze Age Mohenjo-Daro at about 3000 BC, a set of symbols whose relationship to true writing is uncertain.

D'Errico (1994, 1995) discussed the potential significance of the systematic markings on the Paleolithic La Marche Antler and the Tossal de la Roca bone. He concluded that these most likely represent notations, a kind of artificial memory system rather than the more fanciful suggestions of Marshack (1972) that lunar counts, calendars, etc. are involved.

Paper

Bell (1983) summarized information concerning 'pre-paper' types: Egyptian papyrus at about 3000 BC (see Tallet, 2012, p. 152, Third Millennium BC), and rice paper at 1634 A.D.

Art

Summary

Beginning with the Aurignacian and its Asian equivalent on Sulawesi, it is clear that *Homo sapiens* has had the ability to produce naturalistic graphic and plastic art that compares favorably with modern examples, and that the parallel examples featuring very exaggerated graphic and plastic forms from stick figures to caricature-like plastic art can be compared with modern abstract art, with the caveat that we have no real understanding of what thought might have been involved with the caricature-like art any more than we can understand what lies behind modern abstract art. What might our distant descendants think of Brancusi's or Picasso's art? Delporte (1984) extensively reviewed European, chiefly Upper Paleolithic, art with the various interpretations of its causal significance, with no conclusions about its ultimate and perhaps unknowable significance.

Art covers various categories, including graphic arts, plastic arts, decorative art, music, and dance, among other forms. Art is a uniquely human character, with no evidence for it in other animals. [Given the opportunity, some elephants will draw lines – Eds.] The motivations for art in the broad sense include such things as pure decoration ('art for art's sake'), art that promotes group morale (such things as group identity, war, and various threats to the group), religious art, art aimed at promoting male–female togetherness that may end in reproduction, and art aimed at achieving success in farming and the hunt.

Currently available information, see below, indicates the presence of both realistic plastic and graphic art by the Aurignacian, and equivalent-age graphic art from Sulawesi [and Borneo], fully formed, but this first appearance may be a sampling artifact since [it is unlikely] that humans 'suddenly' discovered the ability and impetus to make graphic and plastic art featuring the human and animal forms only in the Aurignacian. Mania and Mania (1988), while discussing some 'deliberate engravings on bone artifacts of *Homo erectus*' in the German Lower Paleolithic, also discussed the enigmatic appearance of incised lines on bone artifacts in the pre-Aurignacian, including the Mousterian, [also referring] to the 'modern-type' realistic art of the Aurignacian; there does not appear to be any satisfactory answer yet to this enigma. However, the recognition of incised pattern on freshwater mussel shells from Trinil, Java, precludes any interpretation other than their intentional making by *Homo erectus*. Also at stake here is the enigma of whether graphic art appeared independently in Western Europe and Indonesia or was 'imported' from one to the other or from Africa. In view of the relatively intensive work on European sites as contrasted with the minimal work done to date in Asia, it is far too early to solve these questions.

Graphic Arts

Consideration of the behavioral significance of graphic art can begin with the very naturalistic cave art of the earlier Upper Paleolithic, discussed below. Does this first appearance in the record of naturalistic art truly represent the beginning of graphic art produced by *Homo*, or is it an artifact of the record?

There is currently no known naturalistic art made by humans during the first million years or so of *Homo*'s existence in the fossil record. Current knowledge of *H. erectus* suggests their existence in a warm, lower-latitude set of environments that did not emphasize caves. It is reasonable to assume that any *H. erectus* 'Leonardo' two million years ago would have been limited to drawing or painting on animal skins or bark, substances that would have had essentially no chance of being preserved in the record. Paints for artistic productions would have been readily available in the form of black, red, orange, and yellow ochers, with animal fats and vegetable oils available to be mixed with the colorants.

Blue and purple pigments, except for finely ground lapis lazuli, are 'artificial' items with such things as Egyptian Blue, Han Blue, and Han Purple being involved with Bronze Age for the Egyptians, and probably First Millennium BC for the Han items (see Berke, 2000, for a detailed discussion). Tyrian purple appears to be another First to Second Millennium BC item (McGovern and Michel, 1985).

Tattoos are preserved only in mummies. This restricts the preservation of tattoos in the record to no more than the 5000-years BP 'iceman' cited below under 'Tattoos'. The human proclivity for tattooing is thus precluded from leaving an extended record back in time.

Skeletal evidence of *H. sapiens* appears in Ethiopia about 200,000 years BP (McDougall et al., 2005), suggesting that the later Upper Paleolithic appearance of naturalistic art is not related to an evolutionary event affecting modern humans. Upper Paleolithic graphic cave-art has been found in Northern Spain, southern France, Romania, Sulawesi, and Borneo. One needs to consider here the environmental significance of caves with this preserved art. Virtually all caves are the products of the solution of carbonate rocks – limestone and dolomite – by water percolating from sources above the cave site. This accounts as well for the production of depositional cave features such as stalactites, stalagmites, and dripstone. The preservation of cave-wall art depends on subsequent dryness, aridity, that prevents further wall solution. This special situation is uncommon for very lengthy time intervals, which explains in large part the extreme rarity of graphic art within caves.

Next one needs to consider the behavioral significance of petroglyphs. Petroglyphs include those 'painted' on the rock surface and those that were pecked or 'engraved'. Petroglyphs are largely restricted to arid regions or to rock underhangs in very localized dry environments. Few appear to be more than 40,000 years old. They are evanescent features subject to erosion and rock weathering during humid intervals and are likely to provide little idea of the initiation of human graphic art.

D'Errico et al. (2003) considered the questions involved with the potential significance of 'engravings' from the past on bone, blocks of ocher, and other materials. Marshack (1972) took some of them to be lunar counts.

Brooks and Wakankar (1976) described a large number of Indian 'Stone Age' rock paintings, ascribing ages from the Mesolithic to present, with a few possibly Lower Paleolithic examples, with evidence for dating less clear than one might wish.

Lower Paleolithic

Joordens et al. (2014) made very clear that the geometric engravings on a freshwater mussel shell from Trinil, Java, in the same bed as the remains of *Homo erectus* with an age of a good half million years BP, leave no doubt about their intentional making, i.e., these are not some type of random scratchings. Their comment that 'The manufacture of geometric engravings is generally interpreted as indicative of modern cognition and behavior' is telling. One can consider the possibility for accidental engraving, while powdering a piece of ocher, as in the South African examples, but the Javanese specimen cannot be attributed to any type of 'accident'.

Valladas et al. (1992) provided radiometric dates for some European prehistoric cave paintings that are between 1 and 2 millennia BP.

Mania and Mania (1988) described some Middle Pleistocene (300,000–350,000 years BP) engravings on bone that suggest intentionally made graphic art with associated remains of *Homo erectus*, and occurring at an occupation site featuring remains of dwellings and other features.

Middle Paleolithic

Henshilwood et al. (2002) described several clearly engraved pieces of ocher from Blombos Cave, South Africa, and d'Errico et al. (2001) discussed an engraved bone fragment from this same c. 70,000-year-old Middle Stone Age site, with the implication that the engraving was intentional.

Upper Paleolithic

Bahn (2007) provided a brief survey of Upper Paleolithic European cave art that dealt with most of the known localities, chiefly Magdalenian, but also Solutrian and earlier. Leroi-Gourhan (1967, 1968) provided comprehensive accounts, with very fine illustrative materials, covering Aurignacian through Magdalenian art, graphic and plastic, with emphasis on Magdalenian cave art plus an account of the problems of dating the individual items using 'stylistic' approaches that have since been thoroughly supplanted by radiometric dates. Valladas et al. (1992) used radiocarbon dating to provide a 14,000 years BP date for the Altamira cave paintings. The oldest presently known European graphic art is the very naturalistic animal art preserved in the Aurignacian Chauvet Cave, southeastern France, dated at approximately 32,000 years BP (Chauvet et al., 1996 [since revised – Eds.]). Clottes and Courtin (1996) described the Magdalenian cave art of a very naturalistic type from the subterranean Cosquer Cave in the south of France.

Pike et al. (2012) provided a set of uranium-series ages extending back to the Early Aurignacian for a group of 11 caves from Northern Spain, with one minimum age extending back 40,800 years.

Sandars (1985) discussed and illustrated some of the many examples of Magdalenian graphic art, chiefly from the walls of caves.

Ghemis et al. (2011) described some of the cave art from the Aurignacian/Gravettian Coliboaia Cave, located in Transylvania. These animal depictions are similar to the well-known graphic materials of about the same age from southern France and northern Spain. They make it clear that this early Upper Paleolithic type of art was widespread in Europe, which raises the question about whether its absence in southern Asia is more a reflection of lack of sampling than anything else.

Aubert et al. (2014) described in some detail the approximately 40,000-year-old graphic art from caves on Sulawesi, the oldest known from Asia. Marchant and Mott (2016) provided excellent illustrations of the Sulawesi paintings.

Several painted rock slabs with animals depicted from the Late Stone Age of Southwest Africa, the first from Africa with realistic art, were described by Wendt (1974, 1976).

Kumar et al. (1988) described several [Upper?] Paleolithic occurrences of engraved ostrich eggshells from India.

Natufian

Rosenberg and Redding (2000, *in* Kuijt, ed.) cited and briefly mentioned ornamented Natufian stone bowls from Anatolia.

Neolithic

We have, unfortunately, very little in the way of later graphic arts until the advent of decorated pottery in the Neolithic and later. Hodder (2006, Plate 1) illustrated a very naturalistic wall fresco of a running figure from the pre-pottery Neolithic of Çatalhöyük

Mellaart's (1967) treatment of Çatal Hüyük provides examples of wall art, painted on plastered building-walls, that include some relatively sophisticated geometric designs, relatively 'primitive' stick figures, and animal and bird figures, as well as some 'primitive' plastic art featuring both human and animal subjects.

Orrelle and Gopher (2000, *in* Kuijt, ed.) discussed the linear decorations present on Levantine ceramics.

Barham and Mitchell (2008, Chapter 9) reviewed the widespread Mid-Holocene and younger African rock-art occurrences and their potential significance, with some examples of very naturalistic animal art. Deacon and Deacon (1999) provided additional information about South African rock art. Beaumont and Vogel (1989) provided radiometric dates of 4000 years BP for the South African rock art, i.e., relatively young as these things go. Lewis-Williams (2002) provided his very personal interpretations of African rock art, which many will question.

Plastic Arts

Plastic arts cover such things as primitive ceramic figurines of animals and humans as well as the Aurignacian through Magdalenian cave art and bas reliefs. Also worth considering here are the plastered skulls described from the Near Eastern Pre-Ceramic Neogene (see here under **Burials, etc.**), as well as the beautiful later Bronze Age and younger art of the 'classical' Mediterranean.

One wonders if it will ever be possible to learn what were in the minds of the sculptors thousands of years BP. These relatively realistic sculptures raise the question of just what might be the 'reason' for the presence in the Neolithic, and even the Acheulian, of many anatomically very crude representations of the human figure.

D'Errico and Villa (1997) examined a suite, from Europe, of grooved bones and bones with holes, interpreted as the work of older humans, and concluded that most could just as easily be interpreted as the results of carnivores, i.e., that one must be careful in accepting such items as evidence of human activities.

Acheulian

Goren-Inbar (1985) described an Acheulian Venus figure in volcanic scoria from the Golan Heights. It is well dated radiometrically (Goren-Inbar and Peltz, 1995), pushing human plastic-art capability back to between 233,000 and 800,000 years BP, which are the ages of the two basalt flows that overlie and underlie the Acheulian materials!

Middle Paleolithic

Stepanchuk (1993) described Middle Paleolithic Crimean bone material with engraved lines, referred to similar material of this age from Europe, and speculated about its significance. Henshilwood et al. (2002) described some intentionally engraved African Middle Stone Age pieces of ocher indicating 'intent' and an essentially modern technology.

Rodriguez-Vidal et al. (2014) described engraved cave art made by Neandertals at Gorham's Cave, Gibraltar. This is the oldest known intentionally made art by Neandertals, at least 39,000 years BP, and represents cognition well before the Aurignacian cave art of modern humans.

Henshilwood et al. (2002) described South African evidence of 'engraved' ocher from approximately 77,000 years BP. D'Errico et al. (2001) described a Middle Stone Age bone fragment with parallel engraves from Blombos Cave.

Upper Paleolithic

Trying to understand the 'significance' of the human and animal figurines of the Upper Paleolithic, Aurignacian and younger, as well as those of the Neolithic is a most difficult business. Talalay (1993) considered many possibilities, including such things as religious and cultic, magic, dolls and toys, and fertility.

McDermott (1996) reviewed the known occurrences of the Gravettian Eurasian Venus figures known from France to Siberia. Abramova (1967) reviewed Upper Paleolithic art in the USSR, including materials from Siberia, with many Venus-figure examples, as well as other items. They appear to be the earliest known, geographically widespread, well-documented examples of anatomically accurate plastic art featuring the human form, although having some very exaggerated anatomical features. Their significance has been a subject of intense interest and debate ever since the first discoveries in the late Nineteenth Century, and has ranged over a wide spectrum of possibilities with no end in sight.

Hahn (1986) made the point that the Upper Paleolithic ivory and stone figurines display realistic representations of the various animals (mammoths, felids, bison, etc.) but very exaggerated human female 'Venus' representations with exaggerated breasts and buttocks, and with the 'mysterious 'felid-headed' human figurine leaving one guessing about significance. We clearly have little understanding of the significance of these figurines.

Soffer (1985) cited and illustrated various Upper Paleolithic incised bone materials from the Russian Plains, and also Venuses. Marshack (1979) illustrated a number of Russian Plain Upper Paleolithic incised bone fragments, some reasonably complex, and attempted to interpret their 'meanings'; it is difficult to decide whether these incised patterns are mere doodling or whether they have any other meanings.

Bahn (1989) noted the age of a Late Pleistocene female figure from Galgenberg, near Krems, Austria, dated at about 30,000 years, and discussed a still earlier, more controversial item from Israel dated at 230,000 to 800,000 years. Bednarik (1989) described the Aurignacian Galgenberg female figurine in some detail and also reviewed the evidence for dating it.

Conard (2009) described a well-dated human figurine from a basal Aurignacian German cave site dated as far back as at least 35,000 years BP; it displays the exaggerated female secondary sexual characteristics long known from significantly younger figurines such as the famous Venus of Willendorf. The behavioral significance of these figurines is unknown, although they may have had something to do with fertility rites (this latest example has a ring instead of a head, which would have enabled the owner to wear it around the neck on a thong).

Delporte (1993) has considered Paleolithic figurines in some detail, Gravettian in particular, with emphasis the similarities and differences from various parts of Eurasia.

White et al. (2012) described and dated an earlier Aurignacian vulvar representation and discussed similar materials from European locales.

Soffer (1985) discussed carved artifacts of female figures of Aurignacian type and other items scribed on ivory and bone from the Central Russian Plain.

Sandars (1985) described the widespread European evidence for plastic art in the Magdalenian.

Natufian

Bar-Yosef and Valla (eds., 1991) discussed various Natufian bone and limestone carvings. Noy (1991) described some relatively crude plastic art from Nahal Oren, Israel, in bone and stone.

Neolithic

A Neolithic Turkish female figurine, shown by Hodder (2006, Figure 94), compares favorably with the Gravettian items referred to here, showing that our species has had the capability of providing realistic sculptured examples of the human form from at least the Gravettian and basal Aurignacian forward.

Mania and Mania (1988) describe last-interglacial, later Lower Paleolithic bone engravings associated with the remains of *Homo erectus;* accompanying discussions provide a good sample of current thinking about data extending back for several hundred thousand years, associated with hearths, dwelling structures, and tools.

Rosenberg and Redding (2000, *in* Kuijt, ed.) cited and briefly mentioned ornamented Natufian stone pestles from a sedentary hunter-gatherer site. Schmidt (1995) illustrated several Gobekli Tepe figures of Early Neolithic age. Bar-Yosef and Valla (eds., 1991) provided descriptions of Natufian carvings using bone and stone from the Levant.

Rollefson (2000, *in* Kuijt, ed.) briefly described some crude LPPNB (Late Pre-Pottery Neolithic B) statues made of lime plaster from the Levant. Voigt (2000, *in* Kuijt, ed.) discussed numerous Pre-Pottery Neolithic B animal and human figurines, and figured many items. Rollefson et al. (1992) briefly discussed and figured crude figures from the Pre-Pottery Neolithic B at Ain Ghazal, Jordan. Schmandt-Besserat (1998) described a number of Pre-Pottery Neolithic B figurines from Ain Ghazal, composed of lime plaster over a base of reed bundles bound with twine; they are very crude, in contrast to the contemporary plastered skulls, not to mention the 'classic' Graeco-Roman cultural tradition of the later Bronze and Iron Ages, which continues to the present; their function is uncertain. Hauptmann (1997) briefly cited a number of Pre-Pottery Neolithic B figurines from Nevali Çori, Turkey.

Rollefson (1986) described some Pre-Pottery Neolithic B statuettes of animals and humans from Ain Ghazal, Jordan, with a relatively realistic aspect.

Garfinkel and Miller (2002) illustrated and discussed a number of Yarmukian fired-clay figurines, one with cowrie 'eyes', from Israel.

Bronze Age

Vat (1940) illustrated several very well done naturalistic Bronze Age stone figures from Harappa.

Ornaments

Summary

The use of ornaments is a preeminent human characteristic shared by no other primate. The following lists some items of human ornamentation.

Tattoos

Tattoos are human ornamentation of an obvious sort, but owing to the vagaries of preservation their record is clearly minimal.

Van der Velden et al. (1995) and Sjøvold et al. (1995) described and discussed the tattoos on the Alpine 'Iceman', and cited tattoos from other ancient occurrences. Deter-Wolf (2013) reviewed the known occurrences of actual tattoos, with the oldest known being an approximately 8000 years BP Peruvian mummy, and cites a French Magdalenian tattooing tool-kit as the oldest known positive evidence. He goes on to suggest that Middle Stone Age South African Blombos Cave artifacts might include parts of a tattooing tool kit. The most elegant tattoos recovered thus far are from some Altai burials (*Siberian Times Reporter*, 14 August 2012; drawn by Elena Shumakova, Institute of Archaeology and Ethnography, Siberian Branch of Russian Academy of Science; see also Rudenko, 1953) with various mythical animals and the like, dated approximately 2500 years BP.

Decorative arts

Jewelry

Newman (2007) listed some of the many functions of jewelry from past and present: personal adornment, expressions of love, affection, and commitment, good-luck charms and talismans with magical powers, portable storage of wealth and hedges against inflation, religious symbols, artistic expressions of beauty, ideal gifts and remembrances for special occasions, useful purpose such as belt buckles, buttons, pins, cuff links, watches, signets, medical ID bracelets, rings with secret compartments for perfumes, messages, and poisons, healing aids, political statements...the list is almost endless.

The truly fabulous jewelry known from the Near East and Egyptian later Bronze Age belongs here. Mention could also be made of much of the later Bronze Age gold work from many of the Steppe graves. Marshall (1931) dealt extensively with the Bronze Age jewelry from Mohenjo-Daro.

Radovčič et al. (2015) described some eagle claws from the Krapina Neandertal site that were modified for some kind of jewelry assemblage.

The Victoria and Albert Museum, London, has a chronologically displayed collection of jewelry in Europe from earliest times.

Beads

Beads and pendants are the overall most abundant forms of human ornamentation and have the best record owing to the resistance of their various compositions to weathering and other destructive natural forces.

Newman (2007) listed some of the functions of beads: protective charms and religious objects, social markers of ethnic and individual identity, including age, marital status, rank or position, currency and indicators of wealth, forms of ornamentation, and means of communication.

Beads are a category that can be considered under the decorative arts, although the adornment of the human body with bead necklaces and bracelets may also indicate their use in attracting the opposite sex. Vanhaeren (2005) considered the presence and possible functions of beads and summarized their African and Eurasian occurrences through time. The presence of beads from an early time is evidence for the presence of string or cords made from various materials as well as for the ability to make a knot to secure the necklace. Moorey (1994, p. 106–110) discussed bead-drilling techniques in some detail; these techniques, of course, could have been applied to any need for drilling a hole into various substances for various purposes. Wright (1982; see also Gorelick and Gwinnett, 1987) described the use of the bow drill for drilling beads and other materials. Tosi (1968) described Bronze Age evidence from Shahr-I Sokta, Iran, of bead making involving lapis lazuli, carnelian, and rock-crystal beads and pendants together with the tools used in drilling them. Postgate (1992, Figure 11.3) illustrated Early Mesopotamian lapis lazuli bead making. Moorey (1994, p. 56–58) described Bronze Age drilling equipment usable for making bowls, variants of the bow drill used in bead making (p. 103–110). Moorey (1994, p. 79–103) discussed the various stones and minerals available in Mesopotamia, especially discussing lapis lazuli for bead making and other purposes.

D'Errico et al. (2003) considered the potential significance of beads from the past. Dubin (1987) provides a comprehensive illustrated 'History of beads from 30,000 BC to the Present'.

Acheulian

Bednarik (1997) described some Acheulian ostrich-eggshell beads from Libya.

Middle Stone Age

Henshilwood et al. (2004) described bored snail shells used as beads from the South African Middle Stone age; they are approximately 75,000 years old and are among the currently oldest evidence of a typically 'modern' human behavior (i.e., ornaments). Jacobs et al. (2008) discussed Middle Stone Age South African beads.

Bar-Yosef Mayer et al. (2009) described a group of Middle Paleolithic Israelian *Glycymeris* shells (80,000–100,000 years ago) with natural perforations that had been modified, many of which were red-ocher stained, i.e., good evidence of human use and probable stringing. Vanhaeren et al. (2006) described some Middle Paleolithic shell beads from Israel and Algeria.

Vanhaeren et al. (2006) provided evidence suggesting that *Nassarius*-shell beads made by humans, from Israel and Algeria, can be dated as far back as 100,000 to 135,000 years BP! Bouzouggar et al. (2007; see also Kuhn and Stiner, 2007; Henshilwood, 2007, Figure 11.4; d'Errico and Vanhaeren, 2007) thoroughly reviewed what is known about shell beads from the record, while describing an 82,000-year-old Moroccan occurrence with some beads containing remnants of red ocher, concluding that this facet of human behavior is very ancient.

Vanhaeren et al. (2013) reviewed the evidence of Middle Stone Age Blombos Cave, South Africa, of *Nassarius*-shell beads and provided information, both experimental and observational, about their 'stringing' and use, with comments about shell-bead occurrences elsewhere in North and South Africa.

Péquart et al. (1937), while describing the Late Mesolithic burials at Téviec, an island off the south coast of Brittany, described shell beads and other ornaments accompanying several of the twenty-odd graves.

Zilhão et al. (2010) described bored *Pecten* shells, with associated ocher, from the Mousterian of Iberia, i.e., with Neandertals. Soressi and d'Errico (2007) briefly discussed and illustrated bored material, including animal teeth that might have been used as beads, from the Mousterian.

Paleolithic

D'Errico et al. (2003, Figure 1) illustrated Chatelperronian pierced ornaments, pre-Aurignacian. White (1992) dealt extensively with Aurignacian beads made from various materials; the shells are large ones selected from a nearby intertidal source. See also d'Errico et al. (2005), who reviewed bead evidence from various other localities. White (1993) extensively described Aurignacian Eurasian mammoth-ivory beads and their mode of production and uses with burials. White (1989b, 1993) described Aurignacian beads and pendants and their mode of manufacture.

Derevianko and Shunkov (2004) illustrated early Paleolithic beads made from stone, ostrich eggshell, and mammoth ivory, and cylinder beads from the Altai. Derevianko and Rybin (2003; see table on their p. 43) discussed early Paleolithic beads from the Russian Altai.

Soffer (1985) illustrated and discussed shell and amber beads from Russian Plain Upper Paleolithic sites.

Einwögerer et al. (2006) discussed and illustrated some Upper Paleolithic ivory beads from Lower Austria associated with an infant burial. D'Errico et al. (2012) described ostrich eggshell and snail beads of approximately 40,000 years BP from South Africa. Kuhn et al. (2001) described earliest Paleolithic shell beads from the Levant.

White (1989b), while discussing basket-shaped beads from southern French Aurignacian deposits, mentions that they are distinct in form from contemporary southern German beads. Stiner (1999) described numerous Italian Riviera marine shell beads of Aurignacian and Gravettian age with gastropods dominating and with some *Dentalium* as well, whereas associated bivalves were mostly used for food.

Bader and Lavrushin (eds., 1998) described one of the most elegant ancient bead occurrences known from the Gravettian at a site near Vladimir, east of Moscow. At the site, Sungir, several burials occur that feature thousands of drilled mammoth-ivory beads, as well as some fox-tooth beads that presumably were used to ornament outer clothing in a cold-climate area subject to permafrost, including head and body coverings as well as bracelets; underclothes were also noted. The burials also feature the use of much red ocher.

Ambrose (1998b) described some early Late Stone Age ostrich-eggshell beads from East Africa.

Kumar et al. (1988) described various occurrences of ostrich eggshell beads from the Upper Paleolithic of India, one of which was associated with a burial. Morse (1993) described shell beads from a Western Australian site dated at 30,000 years BP.

Natufian

Belfer-Cohen and Bar-Yosef (*in* Kuijt, ed., 2000) cited Natufian bone beads. Belfer-Cohen (1991) described a number of Natufian bone beads and pendants from Hayonim Cave, as well as bone beads and *Dentalium* beads, many from burials, together with various additional artifacts. Byrd and Monahan (1995) tabulated Natufian burial data that included many burials with *Dentalium* and various bone beads.

Bar-Yosef and Valla (eds., 1991) provided a mass of information about Natufian beads, pendants, and necklaces, including materials such as *Dentalium* and various gastropod-shell beads, some used in burials, from the Levant, as well as beads made from the phalanges of various mammals, stone, turquoise, variscite, malachite, and fossil echinoid spines.

Neolithic

Kingery et al. (1988, p. 233–234) described Pre-Pottery Neolithic beads from Nahal Hemar Cave; these beads sparkle bright green owing to crushed dioptase added on top of the lime-plaster bead coating; the source of the relatively rare mineral dioptase is unknown, although some Copper Age and Bronze Age copper deposits from Jordan and Sinai, Timna in particular, have yielded dioptase from their oxidized zones. Kuijt (1994) cited Pre-Pottery Neolithic A beads. Wright and Garrard (2003) discussed Neolithic stone beads and associated items from Jordan.

Borić (2009) and Kienlin (2012) described Middle Neolitic azurite and malachite beads from Serbia.

Mellaart (1967) illustrated some elegant early Neolithic bead necklaces from Çatal Hüyük. Özdoğan and Özdoğan (1999) cited malachite and copper beads from the Pre-Pottery Neolithic A of Çayönü Tepesi, Turkey.

Bogucki and Grygiel (1983) described some Polish earlier Neolithic beads made from freshwater mussel shells.

Solecki (1969) described a malachite bead from the 'Protoneolithic' of Shanidar Cave, eastern Anatolia, approximately 10,650 years BP, that presumably was derived from a copper deposit.

Coppa et al. (2006; see also Possehl, 1996) cited Lower Neolithic beads from Mehrgarh, Baluchistan, including bone, steatite, shell, calcite, turquoise, lapis lazuli, and carnelian. Kenoyer and Vidale (1992) discussed the drills used to bore this type of bead.

Bronze Age

Donkin (1998) cited various Near East pearl necklaces. Wooley (1934) described a variety of beads made from lapis lazuli, gold, carnelian, silver, glazed frit, agate, steatite, hematite, marble, crystal, and glass paste from the Sumerian Bronze Age at Ur. Vat (1940) provided an extensive account of various materials used in the beads from Harappa. Marshall (1931) dealt with Bronze Age Mohenjo-Daro beads that included lapis lazuli, amazonstone, turquoise, agate, amethyst, steatite, carnelian, gold, silver, and sodalite. Higham (1996) cited the many various materials used for beads in the Bronze Age burials of Southeast Asia, including malachite, jade, turquoise, carnelian, and agate. Inizan (1993) described the manufacture of carnelian beads from Indian material and their occurrences in Third Millennium BC sites such as Ur, as well as at Harappa. Mackay (1937) described the occurrence of Harappan Age bead-manufacturing at Chanhu-daro, Sind, with information about drills and the actual raw materials, plus suggestions about trade in the beads. Possehl (1981; see also Arkell, 1936) described the beadmaking process used at Cambay, in Gujarat, a process probably similar to that used elsewhere as at Chanhu-daro. Kenoyer (1997) adds considerable new information about the beadmaking process at Harappa for both carnelian and steatite. Nguyen (1990) cited nephrite jade beads and other jade ornaments (bracelets and rings) from the Bronze Age of Trang Kenh, Vietnam. Nguyen (1998) discussed the Upper Neolithic nephrite jade articles found at various sites in Vietnam and provides details about their manufacture. Herbaut and Querré (2004) discussed the Neolithic variscite beads occurring in grave mounds of the Carnac area, Brittany.

Iron Age

Rudenko (1970) discussed torques from the Iron Age Pazyryk burials, and pyrite cubes drilled on their edges to act as beads.

Pendants

Pendants include items that may well have been used in necklaces as well as single ornaments.

Middle Paleolithic

Peresani et al. (2013) described the apical portion of a Pliocene–Miocene gastropod from a Mousterian site in Northern Italy (the nearest locality for the gastropod being approximately 100 kilometers away), bearing traces of pigment, that possibly served as a pendant, with the implication about personal adornment.

Upper Paleolithic

Hahn (1972) summarized information about Central and Eastern European pendants, chiefly Aurignacian. White (1989a) pointed out that Aurignacian beads and pendants occur both in burials and outside of them in some quantity, i.e., used not only for burials. Belfer-Cohen and Bar-Yosef (1981) briefly described Aurignacian mammal-tooth pendants from Hayonim Cave.

Sinitsyn (2003) discussed beads and pendants made from shells, bone, and the teeth of polar foxes from Kostenki in Ukraine, an Upper Paleolithic site.

Bailey (1999) cited pierced deer canines and pierced marine shells from the Paleolithic of Klithi in Epirus of about 16,000 years BP.

Bader (1964, 1970) described a Upper Paleolithic burial site from near Vladimir (Sungir), northeastern Russia, that contained an adult male, boys, and others, together with grave goods with indications of pendants and beads sown onto clothing, mammoth-tusk lances, plus other items, and a lot of red ocher, among other things.

Derevianko and Shunkov (2004, Figures 9–14) illustrated various animal-teeth and stone pendants and bone awls from the early Upper Paleolithic of Denisova Cave in the Russian Altai. Derevianko and Rybin (2003; see table on their p. 43) discussed early Upper Paleolithic pendants from the Russian Altai, some with ocher.

Neolithic

Herbaut and Querré (1996) described variscite pendants from the Neolithic of the Carnac area, Brittany.

Bronze Age

Marshall (1931) dealt with Bronze Age Pendants from Mohenjo-Daro.

Bracelets and Headdresses

Derevkianko et al. (2008) described fragments of a 'chloritolite' bracelet from the Upper Paleolithic Denisova Cave. Chloritolite, a rock, is composed of minerals of chlorite family.

Soffer (1985, p. 83) illustrated a Russian Plain Upper Paleolithic mammoth-ivory bracelet.

Vat (1940) discussed bracelets from Bronze Age Harappa. Marshall (1931) dealt with Bronze Age bracelets and rings from Mohenjo-Daro.

Higham (1996) cited various bronze, 'stone', and shell bracelets from the Bronze Age of Southeast Asia.

Garrod and Bate (1937, p. 17–19, Plates VI, VII) briefly discussed and illustrated *Dentalium* headdresses of Natufian Age from the Mount Carmel area, Israel; these items are possibly unique. Wooley (1934) described the beautiful gold Bronze-Age headdresses recovered at Ur.

Earrings

Yang et al. (2007) described the 8000-year-old nephrite jade slit rings used as earrings from the Xinglongwa Culture of Inner Mongolia, and similar earrings from elsewhere in East Asia. They commonly occur as pairs close to the skulls of humans, whereas jade pendants from the graves occur more in the neck region. The slits were made by using either string-sawing or blade-sawing, presumably with quartz sand as the abrasive, and the opening in the earring was made by either quartz-charged stone drills of varying size or by 'pecking'. 'Pecking' was commonly used in the coarse shaping of the jade objects, with scars from this type of percussion sometimes remaining on the objects. Also present are 'stone tubes' (Yang et al., 2007, pp. 9, 12, 188–191), presumably used as beads and jade pendants. Nguyen (1998) cited various nephrite earrings from the Upper Neolithic of Vietnam.

Higham (1996) cited a number of Southeast Asia Bronze Age slit-ring earrings, including many made from 'stone' and jade. Lien (1991) provided brief descriptions and an introduction to the later Neolithic split-ring nephrite earrings of the Peinan-site excavations on Taiwan, known for its numerous slate coffins.

Mascetti and Triossi (1999) surveyed Western earrings, commonly of gold, the oldest appearing about 5000 years BP in the Levant, considerably later than the 8000-year-old East Asian examples cited above. The archaeological evidence seems to suggest that truly elaborate earrings appeared much later in the West than in the East. Vat (1940; see also Marshall, 1931, for nose rings from Mohenjo-Daro) discussed various earrings and some nose rings from Bronze Age Harappa.

Pearls and Mother of Pearl

Donkin (1998) provided a very extensive account of pearls, both modern and ancient, as well as reviewing older literature concerning their origins and distribution. Donkin provided a comprehensive review of the modern distribution of the marine pearl-generating Bivalvia and also of the modern freshwater taxa. Donkin cited an Egyptian blister-pearl occurrence dated at about 1550 BC, and mother-of-pearl objects as far back as the Fourth Millennium BC from Mesopotamia, Arabia, and Iran; he cited an Omani occurrence dating to 4000–3500 BC. Bronze Age mother-of-pearl occurrences from the Middle East were cited.

Pearls and mother of pearl were used in Sumer and Babylonia for decorative items. Wooley (1934) cited a number of occurrences of cockle shells, some use of mother-of-pearl, and also ostrich shell at Ur in Sumer. Bowen (1951) provided an account of the modern Persian Gulf pearl fisheries that provides insight into ancient practices.

The sizes and quantities of pearls from freshwater mussels impressed early European visitors to eastern North America.

Quartz Crystals

D'Errico et al. (1989; see also Moncel et al., 2012) discussed some quartz crystals from an Indian Lower Acheulian site involving *H. erectus*, and their possible use as tools. Soffer (1985, p. 436) cited Russian Plain Upper Paleolithic quartz-crystal occurrences; quartz crystal has been found at many other sites of various ages. Although not as ornaments, the occurrence of quartz crystal in the making of tools was reported from a Middle Stone Age South African site (Delagnes et al., 2006). Beaumont and Vogel (2006) cited quartz crystals from the Middle Stone Age of South Africa. Wadley (1997) cited the use of quartz crystals in South African Neolithic to modern burials. Domínguez-Bella (2012) cited various Spanish localities where quartz crystals had been found at archaeologic sites.

Various Minerals

Moncel et al., (2012, Table 3) tabulated Late Paleolithic European use of various mineral species, including pyrite, galena, fluorite, serpentine, diverse varieties of quartz, and fossil wood. They also cited some Upper Paleolithic European steatite occurrences, including a bead occurrence.

Amber and Lignite plus Jet

Moncel et al. (2012, Tables 4, 5) tabulated various European Upper Paleolithic amber occurrences from hominid sites, some modified into beads. Habu (2004) cited Japanese Jomon amber.

Mirrors

Enoch (2006; see also Mellaart, 1967, p. 158, XII) reviewed the history of mirrors, the earliest known from Çatal Höyük near Konya in Turkey, i.e., Neolithic, made from obsidian and slightly convex. Jarrige (1988) briefly discussed and illustrated several Bronze Age bronze mirrors from Mehrgarh. MacKay (1938) discussed bronze mirrors from Mohenjo-Daro.

Feathers

Peresani et al. (2011) made a good case, based on the cutmarks on bird-wing bones, large to small, that feathers of different colors were used in an Italian Mousterian cave site. Further evidence for the use of feathers might be provided if more attention were given to bird-wing bones from other sites. Bouchud (1953) described a French Magdalenian site that yielded numerous bones of a corvid, interpreted as used as a source of feathers for arrows.

Fossils

A significant number of hominid sites have yielded fossils, chiefly seashells, including sea urchins, but also some mammalian items. Moncel et al. (2012, Table 1) cited a number of occurrences, chiefly European, with dates extending back to the Acheulian. Some specimens were modified to serve as beads.

Ocher and Malachite (see also Beads, Pendants, Ocher Mining, and Burials)

Wreschner (1980; see also Wreschner, 1975, and Schmandt-Besserat, 1980) summarized much of the thinking about the function(s) of ocher in the distant past going back to the Acheulian (Wreschner, 1975). Masset (1980, p. 638–639, *in* Wreschner, 1980) made the important point that 'ocher', composed of hematite and limonite in various forms, is an exceptionally resistant mineral substance with high preservation potential, whereas organic plant and animal colorants have very limited possibilities for surviving in the record, i.e., we should not conclude that they were not employed early on.

Harrell and Storemyr (2009) noted that ground galena and malachite were used by the Egyptians as eye shadow and that paint in both temples and tombs employed ground minerals: lapis lazuli, azurite, goethite ocher, orpiment, gypsum and calcite, hematite ocher, realgar, and malachite.

Abbevillian/Acheulian

Schmandt-Besserat (1980) cited an Abbevillian/Acheulian ocher occurrence from near Nice, France, indicating that *H. erectus* used ocher early on.

Middle Stone Age

Henshilwood et al. (2009; see also Henshilwood et al., 2002; Hovers et al., 2003, Qafzeh Cave; Henshilwood, 2007, Figure 11.3) described pieces of 'engraved' Middle Stone Age ocher from Blombos Cave, South Africa; whether the gravings had any significance other than as evidence for the production of red hematiferous powder is uncertain, but the behavior is certainly very human. Wreschner (1980; see also Marshack, 1981; Watts, 2002) speculated about the use(s) of ocher, other than burials, in the Paleolithic and Mesolithic. Barham (2002) discussed some Middle Pleistocene Zambian ochers, including soft lateritic materials ('soft' indicating dispersed hematite and limonite) and specularite (crystalline hematite). Beaumont (1973) discussed southern African ocher mines of various ages. Barham (2002) described Middle

Stone Age occurrences of ocher and occurrences of abundant specularite from the Twin Rivers site in central Zambia, and the possibility that the use of ocher began in the Late Acheulian.

Marean et al. (2007) cited ocher used as a pigment from a South African Middle Pleistocene coastal site.

Henshilwood et al. (2011) described a unique abalone shell with associated red ocher, probably a grindstone, and other materials suggesting that they represent 100,000-year-old South African Middle Stone Age remains of 'paint' that also included 'oily' material derived from marrow and bone fat. Van Peer et al. (2003) identified ocher remains on possible grinding stones from the Early to Middle Stone Age of Sai Island, northern Sudan. Zilhão et al. (2010) described ocher associated with bored marine shells from the Mousterian of Iberia, i.e., with Neandertals. Soressi and d'Errico (2007) cited ocher as well as manganese oxide from Neandertal sites, both presumably used as pigment.

Roebroeks et al. (2012) described a Dutch red ocher, hematite occurrence at an early Middle Paleolithic early Neandertal site dated at 200,000–250,000 years BP.

Cârciumaru et al. (2002) described a Middle Paleolithic, Carpathian, Romanian cave occurrence of stalagmite segments modified (by hollowing-out upper surfaces) to act as ocher containers: a unique occurrence. Cârciumaru et al. (2009) described these containers in more detail, showing that they are essentially small, shallowly hollowed-out, and dish-like, with evidence of having been made by scraping out the central part of the stalagmite segment, with traces of the scrape marks remaining; ocher is associated with these Mousterian artifacts.

Upper Paleolithic

Einwögerer et al. (2006) discussed the use of red ocher in a Upper Paleolithic infant burial from Lower Austria. Kraybill (1977) pointed out that traces of ocher on grinding stones of one kind or another appear very early, as in the Paleolithic. Schmandt-Besserat (1980) reviewed occurrences and uses of ocher, chiefly from European Upper Paleolithic and younger sites, also providing information about the technologies used in its preparation and uses.

Natufian

Rosenberg and Redding (2000, *in* Kuijt, ed.) cited and briefly mentioned a Natufian occurrence of 'copper ore', presumably used as a pigment (probably a reference to malachite and/or azurite).

Neolithic

Koukouli-Chrysanthaki et al. (1988) discussed Neolithic, Sixth Millennium BC, ocher from Thasos.

Martín-Gil et al.'s (1994) report of a Spanish Neolithic burial with associated powdered cinnabar, though not strictly speaking in the ocher category, is the only report of this kind available to us. Habu (2004) cited Japanese Jomon cinnabar.

Music

There is, of course, no direct evidence of music preserved in the anthropological and archaeological records, but the presence of fragments of musical instruments makes it clear that humans early on were producing music, and presumably were singing, chanting, and music-making with instruments from far back in time.

D'Errico et al. (2003) discussed the problems associated with trying to understand the origins of music.

Zhang et al. (1999) described early Neolithic musical instruments – flutes made from the ulnae of a crane – from China. Zhang and Kuen (2005) described a number of Early Neolithic flutes made from crane wing bones from Central China preserved in burials, about 9000 years BP, one also containing a turtle shell with some pebbles inside, that may be a rattle. These flutes are still playable. MacKay (1938) described pottery rattles from Mohenjo-Daro. Borroff (1971) refers to the sistrum, a rattle-type instrument, that appears during the Bronze Age in Egypt and Southwest Asia, and refers (p. 3–7) without references to various ancient musical instruments.

Zhang et al. (1999) referred to a brief discussion (Slovenian Academy of Sciences, 1997; see also Turk et al., 1995; Lau et al., 1997; Turk et al., 1997) of a Mousterian (43,400–67,000 BC) possible flute fragment from Slovenia, with additional non-referenced citations to Magdalenian and Aurignacian flutes (Marcuse, 1975). D'Errico et al. (1998) reviewed the evidence concerning the Slovenian 'flute' and concluded that the origin of the holes could just as easily reflect carnivore gnawing-activity. Conard et al. (2009) described some German flutes of early Aurignacian age made from bird-wing bones and mammoth ivory. Buisson (1990; see also White, 1986, Figures 135–136) described some Gravettian and Aurignacian flutes from southern France, made from bird-wing bone. Hahn and Münzel (1995) described a German Aurignacian flute made from a swan-wing bone. Renfrew (1972) discussed Bronze Age musical instruments from the eastern Mediterranean.

Wooley (1934) described the elegant Sumerian harps and lyres from the Bronze Age tombs of Ur plus a double pipe in silver and 'long copper cymbals', as well as possible drums and sistra. Borroff (1971, p. 4) figures Assyrian cymbals.

Rudenko (1970) figures and briefly discusses a harp and a drum from the Pazyryk Iron Age burials.

Higham (1996) devoted considerable attention to the Late Bronze Age bronze drums of Southeast Asia, with conclusions about their use being primarily in war rather than as musical instruments. Kempers (1988) provided a comprehensive compilation of known kettledrums from Southeast Asia from Yunnan to South China, Vietnam, Thailand, Cambodia, Malaysia, and Indonesia together with extensive information on their manufacture.

It is easy to suspect that percussion instruments of various types were present in the Paleolithic, but non-preservation, except under very unusual burial conditions, precludes instruments made from wood and animal skin from surviving. Rattles (including the dried seed pods of certain plants as well as rattles made from dried gourds containing pebbles), as well as wooden clappers and panpipes, are possibilities.

Cannibalism

Fernández-Jalvo et al. (1999) reviewed the whole question of cannibalism and described what they concluded was a reliable example from approximately 780,000 years BP. Defleur and White (1999) described evidence of cannibalism that appears convincing from a French Neandertal site.

Bello et al. (2015) described good evidence for Magdalenian cannibalism from Gough's Cave, Somerset, together with a discussion of the evidence for human cannibalism elsewhere. Bello et al. (2011) described the making of three skull cups in the Gough's Cave remains, with their function of the cups uncertain. Andrews and Fernández-Jalvo (2003) made it clear that the evidence for cannibalism at Gough's Cave, where there are human and animal remains with cut marks and other modifications, suggested nutritional cannibalism, although the treatment of one skull did suggest the possibility of ritual cannibalism. Fernández-Jalvo and Andrews (2011) described experimental chewing by modern humans of bones and comparisons with ancient examples from Gough's Cave and elsewhere.

Cáceres et al. (2007) pointed out that the 'function(s) of cannibalism are difficult to ascertain', including endocannibalism (from the same group), exocannibalism (outsiders), ritual cannibalism, survival cannibalism (consumption for food value owing to stress), and magic cannibalism (related to religious beliefs, which may include eating the dead following funerary rituals), i.e., it is not easy to make a specific determination. Cáceres et al. (2007) pointed out in connection with a Spanish Bronze Age occurrence that cannibalism was followed significantly later by burial of the remains.

Boulestin et al. (2009) provided evidence from the terminal Linear Pottery Culture at Herxheim (end of Sixth Millennium BC) that the abundant human remains, infants to older adults, had been cannibalized after being killed, with the 'type' of cannibalism uncertain, but Orschiedt and Haidle (2006) presented extensive evidence that this was a necropolis, a regional center where remains were interred over a short period of time involving ritual purposes. Orschiedt et al. (2003) briefly described and commented on the Herxheim human remains, noting that the skull of one individual revealed antemortem partially healed injuries that were probably incapacitating.

Executions, Human Sacrifice, Banishment, Crime, Suicide

A number of purely human behaviors, absent in the great apes, other primates, and other animals are dealt with here. Most of these behaviors can be recognized only in the Iron Age, with a few exceptions from the Bronze Age.

Bronze Age robbery-murder was discussed briefly by Larsen (2015) in his account of trade, chiefly involving textiles moved by mule train from the upper Tigris to Kanesh in Anatolia. Human sacrifice is invoked by Wooley (1934) while describing the Bronze Age grave at Ur to explain the many females buried with the princess together with the many armed males, all presumably expected to accompany the princess in the afterlife.

Other than the above Bronze Age exceptions, all the available data are from the Iron Age. The death by poison of Socrates is one of the most notable examples. Archaeologists–anthropologists have not described data from the pre–Iron Age suggesting any of the items listed above. This is not surprising in view of the difficulty in assessing such information from older human burials.

Bloom et al. (1995) described a 3600-year-old Israeli adult male whose hand was amputated and the wounded end healed as an adult, raising the possibility that this was a punishment.

Aldhouse-Green (2015) has discussed the many Iron Age to early Roman Era 'bog bodies' from Northern Europe that obviously died violent deaths of one sort or another. Trying to decide in each case whether execution or human sacrifice was involved is difficult and uncertain, but the manners of death are clear and varied.

Murder among contemporary Hadza hunter-gatherers in Tanzania is limited to disputes concerning women, or induced by alcohol.

The recognition of suicide from the distant past appears to be very difficult. The Greeks and Romans under some circumstances approved of suicide, but there does not appear to be any useful information from earlier times.

All of the above suggests that humans may be using some of their 'extra' brain capacity from those extraordinarily large brains for all types of devilment and merryhell, as contrasted with great apes and other animals.

Clothing (see also above: Textiles, Clothing, and Basketry; Beads and Pendants)

Although there is no direct evidence for when clothing was first worn, Kittler et al.'s (2003) determination, using molecular evolution of human lice involved with clothing, suggests an appearance of clothing at least many tens of thousands of years BP: a minimum age of 72,000 years BP plus or minus 42,000 years is suggested. Clothing may have become necessary only after humans left Africa to inhabit cool to cold regions about 100,000 years BP.

Adovasio et al. (1996) discussed Upper Paleolithic fiber technology and interlaced woven finds from Pavlov I., Czech Republic, at about 26,000 years BP.

Kuijt (1994) cited loom weights from the Pre-Pottery Neolithic A. Minozzi et al. (2003) used evidence of non-alimentary, Neolithic Libyan tooth-use caused by vegetal fibers used for such things as rope making, net making, mat making, etc. and cited other non-alimentary tooth use from the record.

Soressi et al. (2013) described several French Mousterian bone tools used for 'smoothing' leather, datable approximately 50,000 years BP.

Needles

Bailey (1999) cited needles and spatulae, useful for making clothing, from the Paleolithic , about 16,000 years BP, from Klithi in Epirus. Adam and Kotjabopoulou (1997) described Upper Paleolithic bone needles from Krithi, northwestern Greece. Stordeur-Yedid (1979; see also White, 1986, Figures 89–90) described various bone needles from the Magdalenian and Solutrean of France, northern Spain, and other European sites. Derevianko and Shunkov (2004) figured bone needles with eyes from the earlier Upper Paleolithic of the Altai. Marshall (1931) dealt with needles from the Bronze Age at Mohenjo-Daro.

Bar-Yosef and Valla (eds., 1991) cited bone needles from various Natufian sites.

Lighting

De Beaune and White (1993; de Beaune, 1987; de Beaune, 1987, translated by Turton, M., 2002) described the relatively simple, but clearly effective, Upper Paleolithic stone lamps consisting of shallow depressions that used a wick of some kind of vegetation and animal fat to produce enough light for uses such as lighting in caves where cave art was being made, as well as for uses outside caves. White (1986, Figures 71–73) illustrated several Upper Paleolithic Magdalenian stone lamps. Evershed et al. (1997) described biochemical evidence for the use of beeswax in Late Minoan Cretan lamps and conical cups. Rudenko (1970) discussed Iron Age Pazyryk burials with an oil lamp.

Stöllner et al. (1999, 2003) cited and illustrated the use of numerous wooden tapers for lighting in the Iron Age Dürrnberg-bei-Hallein Austrian rock-salt mines and also refer to tallow-type lamps used at other Austrian salt mines.

Calendar

Our species from its beginnings must have been keenly aware of the seasons, day and night, sun and moon, and the appearance at night of the heavens, but the absence of preserved records of their thoughts on these topics precludes their consideration, but for this see Saul (2013). Marshack (1985, 1991) provided just about the earliest accounts, Aurignacian in France, at about 32,000 years BP, of what he concluded is a record of phases of the moon covering an extended period of time, engraved on bone; he also considered another bone plaque with boustrophedon marks with calendric significance. If Marshack's interpretations are correct they form a significant body of evidence. Saul (2013) accepts Marshack's conclusions, but d'Errico (1994, 1995) dismisses Marshack's interpretation, preferring to see the marks as an artificial memory system of notation, which seems more reasonable.

Ben-Dov et al. (eds., 2012) reviewed the calendar evidence from the Bronze Age Levant, about 5000 years BP, preserved in various documents, especially cuneiform tablets.

Although having nothing evident to do with calendars, Davis's description of some incised scapulae from the Mousterian of Kebara and from the Aurignacian of the Ha-Yonim caves clearly have to do with some kind of record keeping.

Games

Tach (2013) reported on presumed gaming pieces of Early Bronze Age from southeastern Anatolia. Evans (1964, p. 471–485) illustrated and discussed a Minoan gameboard. Parlett (1999) described in some detail the enormous variety of board games available both today and in the more recent past, which suggests that trying to understand in detail the rules used with remnants of truly ancient games is probably a lost cause. Lhôte (1994) summarized information about games of many types, including board games, with emphasis on those of the Fifteenth and younger centuries A.D., but including Egyptian games from the Bronze Age equivalents. In The Musées de Marseille (1991), various games are discussed from Bronze Age Egypt, including the Game of Senet, which dates back to the Third Millennium BC in Egypt, and some items from Ur as well, through the Roman interval. Wooley (1934) discussed gaming boards and gaming pieces from the Bronze Age Sumerian of Ur. Vat (1940) discussed various game remains and toys from Bronze Age Harappa. MacKay (1938) described gameboards and dice from Mohenjo-Daro. Marshall (1931) provides evidence of games and toys from Mohenjo-Daro. Dales (1968) cited dice from the Bronze Age of the Tigris–Euphrates region and the Indus Valley. Koryakova and Epimakhov (2007, Figure 4.10) illustrated various Late Bronze Age dice from Uralian and western Siberian sites.

Migrations

Antón and Swisher (2004) compiled the widespread evidence for the widespread Old World occurrences of *Homo erectus,* considering the possibility of an early African origin and subsequent dispersal. Evidence from Timor of artifacts indicative of *H. erectus*-type from approximately one million years BP indicates important early dispersal events, as does the presence on Flores of similar evidence from about 0.73 million years (van den Bergh et al., 1996). (See **Movement by Water and 'Boats'**).

Dennell and Roebroeks (2005) examined the evidence for the earliest hominin migration and dispersal out of Africa and point out its fragmentary nature. Benazzi et al. (2015; see also 2011, plus Conard and Bolus, 2015) made it clear that modern humans, *Homo sapiens*, arrived in Europe (Italian evidence) in the pre-Aurignacian Châtelperronian and Uluzzian, Proto-aurignacian, 41,000 years BP.

Vermeersch et al. (1998) concluded that the Middle Paleolithic burial of a modern human at Taramsa Hill, Upper Egypt, dated at 55,000 years BP, is consistent with the out-of-Africa hypothesis of modern humans. Beaumont (1980) cited Sub-Saharan *Homo sapiens* with ages of 90,000 to 110,000 years BP from Southern Africa consistent with early migrations out of Africa. Callaway (2015) discussed the find of human teeth dated approximately 100,000 years BP in Hunan, China, as evidence of a migration from Africa well before *H. sapiens* appeared in Europe.

Deacon (1992) made a case for Southern Africa not having been affected by migration from outside from 125,000 to 10,000 years ago. Ambrose (1998) discussed the potential of a major volcanic eruption on human populations followed by subsequent populations; rather speculative.

Evidence for repeated migrations out of Africa from the Acheulian to the migration of *Homo sapiens* requires review.

Miscellaneous

Copeland et al. (2011) provided new evidence, based on strontium-isotope data from tooth crown enamel of *Australopithecus africanus* and *Paranthropus robustus*, earlier Pleistocene, suggesting that young females grew up in areas separate from where they later were situated. The isotopic 'signature' of plant foods grown on one type of geological substrate differed from those grown on another substrate, in this case from substrates very distinct from the limestone and dolomite cave regions where the fossils occur. Their evidence suggests that females from one area moved when mature into other areas where they were associated with males who had not moved. This type of behavior could well have given rise to 'tribal' affiliations between different family groups. This behavior contrasts, between that of many mammals, ungulates in particular, where the young males leave the family group in order to seek mates elsewhere.

At a minimum, Peters and Schmidt's (2004) description of the unique Pre-Pottery Neolithic Göbekli Tepe site in southeastern Turkey indicates the presence of well-organized communal activity involving various specialists.

Conclusions

My overall conclusions are straightforward. Despite the limited scope of the early evidence dealing with *Homo erectus,* it appears that the basic human behaviors – 'human nature' – have remained stable over the past two million years. There have, of course, been repeated major advances in technology until the relatively exponential 'explosion' of technological progress during the past few millennia, but these are not behavioral changes. For example, the evidence of human infant nutrition and brain growth indicates that the technique of fire-making necessary to provide 'baby food' was available from the beginning. The selectivity of raw materials for tools was shown by *Homo* from the very earliest times at Gona, Ethiopia, where the presence of 'workshops' or 'factories' using materials brought in from elsewhere is strictly human and unlike any ape-like behavior. The quarrying of desirable lithic materials is already recognized in the Acheulian and the late Lower Paleolithic by the associated debris piles, though truly underground works come in significantly later.

When it comes to the violence visited by one human on another, or others, we see major technological advances over time: the earliest use of stone axes, clubs, or rocks to eliminate an enemy, followed by flint-tipped spears plus bows and arrows, followed in turn by swords and chariots, to the ultimate invention of gunpowder and what followed from it up until the advent of the amazing modern weapons of mass destruction.

The same conclusion is obvious when one considers warfare through time; the weapons and tactics have changed, but the basic behaviors remain the same. Consideration of other basic human behaviors permits one to conclude *nihil novum sub sol.* One need only read the daily newspaper to reach this conclusion.

Vigne (2008, p. 186; see also Cauvin, 2000) pointed out that the spread of the 'Neolithic package' occurred over a considerable interval in Europe, reaching northeastern Europe only in the Fourth Millennium BC.

D.E. Lieberman and Bar-Yosef (2005, see their Figure 17.1 for a summary) compiled data that make it clear that morphological changes in time-successive species of *Homo* do not coincide in time with major changes in lithic technologies. *Species change and changes in lithic technologies are decoupled from one another.* It appears from their conclusions that major innovations in lithic technology may have appeared in limited areas, from which they then spread. Such changes were not 'caused' or correlated with the appearance of new taxa with novel behavioral–technological traits.

The initial presence, beginning with the Oldowan of tools and the use of transported raw material, argues for the simultaneous presence of language (see above, and Gibson, 1991). The early presence of humans in Indonesia east of the Wallace Line argues both for the presence of language and for cooperative effort in their transport.

The discovery of undoubted human burials during the Mousterian, of beads beginning in the Acheulian and later, and of well-made tools beginning with the Acheulian, make a good case for the great antiquity of basic human behaviors that persist to the present.

Soddyall (ed., 2006) provided an excellent account of both human origins and human behavior, with emphasis on their antiquity. There is agreement that modern human technology was present in the Neolithic, with accumulating evidence for it in the Upper Paleolithic as well. There is still uncertainty about the Middle Paleolithic, although data from Brown et al. (2012) do indicate good support, as does data from Henshilwood et al. (2004). Higham et al. (2012) provided radiocarbon chronology for the beginnings of the Aurignacian in the German Jura at about 40,000 years BP.

Marshack (1989) reviewed much of the evidence, much of it reliable, for truly human behavior and technology in the Acheulian, Mousterian, and Paleolithic. D'Errico et al. (2012) reviewed elements of human technologies of similar type from San (Bushman) culture from southern Africa as far back at 40,000 years BP. Saul (2013) argued for the continuity of religious belief from an initial 'humanizing' effort to avoid death.

Sandison (1967) reviewed the evidence of human sexual behavior and concluded: 'Human nature appears to have changed remarkably little over the millennia'.

With the Neolithic explosion in technology there was an accompanying explosion in human organizational complexity – government, police, lawyers, etc. – which it would be inappropriate to characterize as progress. Progress...towards what?

Art Boucot's text ends here with the date January 28, 2016 and with notes to himself to 'EXPAND and discuss' the 'Savage Harvest book, and The Bloodlands, T. Snyder, 2010, for the Hitler and Stalin slaughtering'. He also intended to refer to 'organizational complexity – government, police, lawyers, etc., etc.' and to 'overpopulation' and 'pollution, environmental widespread degradation of the land, including soils, the oceans and the atmosphere'. ... And he probably resolved to expand his last comment: 'Progress... towards what? Our aggressive brains have endowed us with aggressive stupidity. Use where appropriate.' [Eds.]

Bibliography

Abramova, Z.A., 1967. Palaeolithic art in the USSR. Arctic Anthropology, vol. IV(2), 180 pp.

Ackernecht, E.H., 1978. Primitive surgery. *In* Health and the human condition: perspectives on medical anthropology, M.H. Logan and E.E. Hunt, Jr., eds. North Scituate, MA, Duxbury Press, 164-180.

Adam, E., and Kotjabopoulou, E., 1997. The organic artefacts from Klithi. *In* Palaeolithic settlement and Quaternary landscapes in northwest Greece. McDonald Institute for Archaeological Research, Cambridge, UK, 245-260.

Adovasio, J.M., 1975. The textile and basketry impressions from Jarmo. Paléorient 3, 223-230.

Adovasio, J.M., et al., 1996. Upper Paleolithic fibre technology: interlaced woven finds from Pavlov I, Czech Republic, c. 26,000 years ago. Antiquity 70, 526-534.

Adovasio, J.M., Hyland, D.C., and Soffer, O., 1997. Textiles and cordage: a preliminary assessment. *In* J. Svoboda, ed., Pavlov I – Northwest; the Upper Palaeolithic burial and its settlement context. Academy of Sciences of the Czech Republic, Institute of Archaeology, Brno 4, 403-424.

Adovasio, J.M., Soffer, O., Hyland, D.C., Illingworth, J.S., Klima, B., and Svoboda, J., 2001. Perishable industries from Dolní Vestonice I: new insights into the nature and origin of the Gravettian. Archaeology, Ethnology and Anthropology of Eurasia 2, 48-65.

Africa, T.W., 1961. The opium addiction of Marcus Aurelius. Journal of the History of Ideas 22(1), 97-102.

Ahlström, G.W., 1978. Wine presses and cup-marks of the Jenin-Megiddo Survey. Bulletin of the American Schools of Oriental Research 231, 19-49.

Aiello, L.C., 1997. Brains and guts in human evolution: the expensive tissue hypothesis. Brazilian Journal of Genetics 20, 9 pp.

Aiello, L.C., and Wells, C.K., 2002. Energetics and the evolution of the genus *Homo*. Annual Review of Anthropology 31, 323-338.

Aiello, L.C., and Wheeler, P., 1995. The expensive-tissue hypothesis. Current Anthropology 36, 199-221.

Aikens, C.M., and N.R. Song, eds., 1992. Pacific Northeast Asia in Prehistory: hunter-fisher-gatherers, farmers and socio-political elites. Washington State University Press.

Aikens, C.M., 1995. First in the World: the Jomon pottery of early Japan. *In* The emergence of pottery: technology and innovation in ancient societies. W.K. Barnett and J.W. Hoopes, eds. Smithsonian Institution Press, Washington, DC, 11-21.

Akazawa, T., 1987. The ecology of the Middle Paleolithic occupation at Douara Cave, Syria. University of Tokyo Museum Bulletin 29, 155-166.

Albarella, U., Dobney, K., and Rowley-Conwy, P., 2006. The domestication of the pig (*Sus scrofa*). *In* Documenting Domestication. University of California Press, 209-227.

Aldhouse-Green, M., 2015. Bog Bodies Uncovered. Thames & Hudson, 223 pp.

Algaze, G., 1995. Fourth Millennium BC trade in Greater Mesopotamia: did it include wine? *In* the Origins and Ancient History of Wine. Gordon & Breach Publishers, 89-96.

Alimov, K., et al., 1998. Prähistorischer Zinnbergbau in Mittelasien. Eurasia Antiqua 4, 137-199.

Alizadeh, A., 1985. A Protoliterate pottery kiln from Chogha Mish. Iran [Iranian Prehistory Project] 23, 39-50.

Allard, P., Bostyn, F., Giligny, F., and Lech, J., eds., 2008. Flint Mining in Prehistoric Europe. British Archaeological Reports International Series 1891, European Association of Archaeologists, 12th Annual Meeting, Cracow, Poland, 163 pp.

Allchin, B., and Allchin, R., 1982. The rise of civilization in India and Pakistan. Cambridge University Press, 379 pp.

Allchin, F.R., 1962. Upon the antiquity and methods of gold mining in Ancient India. Journal of the Economic and Social History of the Orient 5, 195-211.

Allen, J., et al., 1988. Pleistocene dates for the human occupation of New Ireland, northern Melanesia. Nature 331, 707-709.

Alperson-Afil, N., 2008. Continual fire-making by hominins at Gesher Benot Ya'aqov, Israel. Quaternary Science Reviews 27, 1733-1739.

Alperson-Afil, N., et al., 2007. Phantom hearths and the use of fire at Gesher Benot Ya'Aqov, Israel. PaleoAnthropology 2007, 1-15.

Alt, K.W., et al., 1997. Evidence for Stone Age cranial surgery. Nature 387, 360.

Alt, K.W., Rosing, F.W., and Teschler-Nicola, M., eds., 1998. Dental Anthropology: Fundamentals, Limits and Prospects. Springer, New York, 564 pp.

Ambrose, S.H., 1998. Late Pleistocene human population bottlenecks, volcanic winter, and differentiation of modern humans. Journal of Human Evolution 34, 623-651.

Ambrose, S.H., 1998b. Chronology of the Later Stone Age and food production in East Africa. Journal of Archaeological Science 25, 377-392.

An Zhimin, 1989. Prehistoric agriculture in China. In Foraging and Farming: The evolution of plant exploitation, Unwin Hyman, 643-649.

Anastasiou, E., and Mitchell, P.D., 2015. Human intestinal parasites and dysentery in Africa and the Middle East prior to 1500. In P.D. Mitchell, ed., Sanitation, Latrines and Intestinal Parasites in Past Populations. Farnham Ashgate, 121-147.

Anastasiou, E., et al., 2014. Prehistoric schistosomiasis parasite found in the Middle East. The Lancet Infection, 553-554.

András, P., 2008. Staroveký bronz – história a zdroje. Mineralia Slovaka 40, 185-196.

Andrews, P., and Fernández-Jalvo, Y., 2003. Cannibalism in Britain: taphonomy of the Cresswellian (Pleistocene) faunal and human remains from Gough's Cave (Somerset, England). Bulletin of the Natural History Museum (Geology) 58, suppl. 59-81.

Andronic, M., 1989. Cacica – un nou punct neolitic de exploatare a Sării. Materiale Şi Cercetări Arheologice 40, 171-177.

Angel, J.L., 1966. Porotic hyperostosis, anemias, malaria and marshes in the prehistoric Eastern Mediterranean. Science 153, 760-763.

Angelakis, A.N., and Spyridakis, S.V., 1996. The status of water resources in Minoan times: a preliminary study. In NATO ASI Series, vol. 136, 161-191.

Angelakis, A.N., et al., 2005. Urban wastewater and stormwater technologies in ancient Greece. Water Research 39, 210-220.

Angelakis, A.N., et al., 2007. Aqueducts during the Minoan Era. Water Science and Technology: Water Supply 7, 95-101.

Anglim, S., et al., 2002. Fighting techniques of the Ancient World 3000 BC–500 AD. St. Martin's Press, 256 pp.

Anikovich, M.V., et al., 2007. Early Upper Paleolithic in Eastern Europe and implications for the dispersal of modern humans. Science 315, 223-226.

Anthony, D.W., 2007. The horse, the wheel, and language. Princeton University Press, 553 pp.

Anthony, D.W., and Brown, D.R., 1991. The origins of horseback riding. Antiquity 65, 22-38.

Anthony, D.W., and Vinogradov, N., 1995. The birth of the chariot. Archaeology 48, 36-41.

Antón, S.C., and Swisher, C.C., III, 2004. Early dispersal of *Homo* from Africa. Annual Review of Anthropology 33, 271-296.

Aranguren, B., et al., 2007. Grinding flour in Upper Palaeolithic Europe (25,000 years BP). Antiquity 81, 845-855.

Araújo, A., et al., 2000. Ten Thousand Years of Head Lice Infection. Parasitology Today 16(7), 269.

Araújo, A., et al., 2008. Parasites as Probes for Prehistoric Human Migrations? Publications from the Harold W. Manter Laboratory of Parasitology, Lincoln NE, University of Nebraska 49.

Arkell, A.J., 1936. Cambay and the bead trade. Antiquity 10, 292-305.

Armendariz, J., et al., 1994. New evidence of prehistoric arrow wounds in the Iberian Peninsula. International Journal of Osteoarchaeology 4, 215-222.

Armentano, N., Malgosa, A., and Campillo, D., 1999. A case of frontal sinusitis from the Bronze Age site of Can Filuà (Barcelona). International Journal of Osteoarchaeology 9(6), 438-442.

Artzy, M., 1994. Incense, camels and collared rim jars: desert trade routes and maritime outlets in the second millennium. Oxford Journal of Archaeology 13, 121-147.

Asfaw, B., et al., 2002. Remains of *Homo erectus* from Bouri, Middle Awash, Ethiopia. Nature 416, 317-320.

Aspöck, H., 2000. Paläoparasitologie: Zeugen der Vergangenheit. Nova Acta Leopoldina NF 83, Nr. 316, 159-181.

Aspöck, H., et al., 1999. Parasites and parasitic diseases in prehistoric human populations in Central Europe. Helminthologia 36, 139-145.

Aspöck, H., et al., 2000. Parasitological examination of the Iceman. *In* The Iceman and his natural environment, Springer-Verlag, 127-135.

Aspöck, H., et al., 2002. Parasitologische Untersuchungen von im Salz konservierten Exkrementen: Zur Gesundheit der Dürrnberger Bergleute. *In* Dürrnberg und Manching: Wirtschaftsarchäologie im ostkeltischen Raum: Akten des internationalen Kolloquiums in Hallein/Bad Dürrnberg, 123-132.

Atalay, S., 2006. Domesticating clay: the role of clay balls, mini balls and geometric objects in daily life at Çatalhöyük. *In* I. Hodder, ed. Changing materialities at Çatalhöyük: reports from the 1995-99 seasons, McDonald Institute for Archaeological Research, Cambridge, UK, 139-168.

Atwater, A.E., 1979. Biomechanics of overarm throwing movements and of throwing injuries. Exercise and Sport Sciences Reviews 7, 43-85.

Aubert, M., et al., 2014. Pleistocene cave art from Sulawesi, Indonesia. Nature Letters 514, 223-227.

Aubert, M., et al., 2018. Palaeolithic cave art in Borneo. Nature 564, 254-257.

Aufderheide, A.C., 2003. The Scientific Study of Mummies. Cambridge University Press, 608 pp.

Aufderheide, A.C., and Rodríguez-Martin, C., 1998. The Cambridge Encyclopedia of human paleopathology. Cambridge University Press, 478 pp.

Aveling, E.M., and Heron, C., 1998. Identification of birch bark tar at the Mesolithic site of Star Carr. Ancient Biomolecules 2.1, 69, 15 pp.

Baber, Z., 1996. The science of empire: scientific knowledge, civilization and colonial rule in India. State University of New York Press, 298 pp.

Bachechi, L., et al., 1997. An arrow-caused lesion in a Late Upper Palaeolithic human pelvis. Current Anthropology 38, 135-140.

Backwell, L., and d'Errico, F., 2005. The origin of bone tool technology and the identification of early hominid cultural traditions. *In* From Tools to Symbols: from early hominids to modern humans, Wits University Press, 238-275.

Backwell, L.R., and d'Errico, F., 2001. Evidence of termite foraging by Swartkrans early hominids. National Academy of Sciences (USA), Proceedings 98, no. 4, 1358-1363.

Bader, N.O., and Lavrushin, Y.A., 1998. Pozdnepaleoliticheskoe poselenie Sungir (Pogrebeniya i okryzhayushtsaya sreda). Rossiiskaya Akademiya Nauk, Institut Arkheologii, Moscow, Nauchnii MIR, 270 pp.

Bader, O., 1964. The oldest burial? The Illustrated London News, 244, p. 731.

Bader, O., 1970. The boys of Sungir. The Illustrated London News (March 7) 256, p. 24-26.

Badler, V.R., 1996. The archaeological evidence for winemaking, distribution and consumption at Proto-Historic Godin Tepe, Iran. *In* The origins and ancient history of wine, Gordon and Breach Publishers, 45-65.

Bagg, A.M., 2012. Irrigation. *In* A Companion to the Archaeology of the Ancient Near East, Blackwell, 261-277.

Bahn, P.G., 1989. Age and the female form. Nature 342, 345-346.

Bahn, P.G., 2007. Cave Art, a guide to the decorated Ice Age caves of Europe. Frances Lincoln Limited, 224 pp.

Bailey, G., 1999. The Palaeolithic archaeology and palaeogeography of Epirus with particular reference to the investigations of the Klithi rockshelter. The Palaeolithic archaeology of Greece and adjacent areas, Proceedings of the ICOPAG Conference, Ioannina, September 1994, British School at Athens Studies 3, 159-169.

Bakels, C.C., 1991. Western Continental Europe. *In* Progress in Old World palaeoethnobotany, Balkema, 279-298.

Bakker, J.A., Kruk, J., Lanting, A.E., and Milisauskas, S., 1999. The earliest evidence of wheeled vehicles in Europe and the Near East. Antiquity 73, 778-790.

Balter, M., 2014. The killing ground. Science 344, 1080-1083.

Bandy, M., 2008. Global patterns of early village development. *In* The Neolithic Demographic Transition and its consequences, Springer, Science+Business Media B.V., 333-357.

Bar-Adon, P., 1980. The cave of the treasure. The Israel Exploration Society, 243 pp.

Bar-Gal, C.K., et al., 2002. Ancient DNA evidence for the transition from wild to domestic status in Neolithic goats: a case study from the site of Abu Gosh, Israel. Ancient Biomolecules 4, 9-17.

Bar-Oz, G., and Munro, N.D., 2007. Gazelle bone marrow yields and Epipalaeolithic carcass exploitation strategies in the southern Levant. Journal of Archaeological Science 34, 945-956.

Bar-Yosef Mayer, D.E., Vandermeersch, B., and Bar-Yosef, O., 2009. Shells and ochre in Middle Paleolithic Qafzeh Cave, Israel: indications for modern behavior. Journal of Human Evolution 56, 307-314.

Bar-Yosef, O., 1998. The Natufian Culture in the Levant, Threshold to the Origins of Agriculture. Evolutionary Anthropology 6, 159-177.

Bar-Yosef, O., 2004. Eat what is there: hunting and gathering in the world of Neanderthals and their neighbours. International Journal of Osteoarchaeology 14, 333-342.

Bar-Yosef, O., and Meadow, R.H., 1994. The origins of agriculture in the Near East. *In* Last Hunters, First Farmers: new perspectives on the prehistoric transition to agriculture, School of American Research Press, 39-94.

Bar-Yosef, O., and Valla, F.R., eds., 1991. The Natufian culture in the Levant. International Monographs in prehistory, Archaeological series 1, 644 pp.

Bar-Yosef, O., and Vandermeersch, B., 1993. Modern humans in the Levant. Scientific American, 94-100.

Barbassa, J., 2015. Dancing with the Devil in the City of God. Touchstone, 308 pp.

Barber, E.J.W., 1991. Prehistoric textiles. Princeton University Press, 471 pp.

Barber, E.J.W., 2001. The clues in the clothes. Journal of Indo-European studies 38, 1-14.

Barber, E.W., 1994. Women's work: the first 20,000 years. W.W. Norton & Co., New York, 334 pp.

Barber, M., Field, D., and Topping, P., 1999. The Neolithic flint mines of England. English Heritage, 95 pp.

Barber, R.L.N., 1987. The Cyclades in the Bronze Age. University of Iowa Press, 283 pp.

Bard, K.A., 1995. Parenting in primates. *In* The Handbook of Parenting, Biology and Ecology of Parenting, Institute of Education Sciences, vol. 2, 27-58.

Barham, L.S., 1996. Recent research on the Middle Stone Age at Mumbwa Caves, Central Zambia. *In* Aspects of African Archaeology, University of Zimbabwe Press, 191-200.

Barham, L.S., 2000. The Middle Stone Age of Zambia, South Central Africa. CHERUB, the Centre for Human Evolutionary Research, University of Bristol, 303 pp.

Barham, L.S., 2002. Systematic pigment use in the Middle Pleistocene of South Central Africa. Current Anthropology 43, 181-190.

Barham, L.S., 2002b. Backed tools in Middle Pleistocene central Africa and their evolutionary significance. Journal of Human Evolution 43, 585-603.

Barham, L.S., and Mitchell, P., 2008. The First Africans, African Archaeology from the earliest tool makers to most recent foragers. Cambridge University Press, 601 pp.

Barham, L.S., Pinto Llona, A.C., and Stringer, C.B., 2002. Bone tools from Broken Hill (Kabwe) cave, Zambia, and their evolutionary significance. Before Farming 2(3), 1-12.

Barkai, R., and Gopher, A., 2009. Changing the face of the earth: human behavior at Sede Ilan, an extensive Lower–Middle Paleolithic quarry site in Israel. *In* Lithic Materials and Paleolithic Societies, Blackwell Publishing, 174-185.

Barkai, R., and LaPorta, P.C., 2006. Middle Pleistocene landscape of extraction: quarry and workshop complexes in Northern Israel. *In* Axe Age: Acheulian tool-making from quarry to discard, 7-44.

Barkai, R., Gopher, A., and La Porta, P.C., 2002. Palaeolithic landscape extraction: flint surface quarries and workshops at Mt Pua, Israel. Antiquity 76, 672-680.

Barker, G., 2006. The agricultural revolution in prehistory: why did foragers become farmers? Oxford University Press, 598 pp.

Barnes, E., 2005. Diseases and Human Evolution. University of New Mexico Press, Albuquerque, 484 pp.

Barnes, E., 2012. Atlas of Developmental Field Anomalies of the Human Skeleton: A Paleopathology Perspective. Wiley-Blackwell, 210 pp.

Barnes, I., Duda, A., Pybus, O.G., and Thomas, M.G., 2010. Ancient urbanization predicts genetic resistance to tuberculosis. Evolution 65, 842-848.

Barnett, W.K., and Hoopes, J.W., eds., 1995. The emergence of pottery, technology and innovation in ancient societies. Smithsonian Institution Press, 285 pp.

Bartelheim, M., Eckstein, K., Huijsmans, M., Kraub, R., and Pernicka, E., 2003. Chalcolithic metal extraction in Brixlegg, Austria. International Conference Archaeometallury in Europe, 24–26 September 2003, Milan, Italy, Proceedings, vol. 1, 441-447.

Barth, F.E., 1982. Prehistoric saltmining at Hallstatt. Bulletin of the Institute of Archaeology 19, 31-43.

Barton, L., Newsome, S.D., Chen Fa-Hu, Wang Hui, Guilderson, T.P., and Bettinger, R.L., 2009. Agricultural origins and the isotopic identity of domestication in northern China. National Academy of Sciences (USA), Proceedings 106, 5523-5528.

Barton, R.N.E., et al., 1999. Gibraltar Neanderthals and results of recent excavations in Gorham's, Vanguard and Ibex caves. Antiquity 73, 13-23.

Bartsiokas, A., and Day, M.H., 1993. Lead poisoning and dental caries in the Broken Hill hominid. Journal of Human Evolution 24, 243-249.

Bass, G.F., 1967. Cape Gelidonya: A Bronze Age shipwreck. Transactions of the American Philosophical Society, N.S., 57(8), 177 pp.

Bass, G.F., 1986. A Bronze Age shipwreck at Ulu Burun (Kaş): 1984 campaign. American Journal of Archaeology 90, 269-296.

Bass, G.F., ed., 2005. Beneath the Seven Seas. Thames & Hudson, 256 pp.

Baum, J., and Bar-Gal, G.K., 2003. The emergence and co-evolution of human pathogens. In Emerging Pathogens: the Archaeology, Ecology and Evolution of Infectious Disease, C.L. Greenblatt and M. Spigel, eds., Oxford University Press, London, 67-78.

Baumgarten, J., ed., 2005. Dialogue on the Early Neolithic origin of ritual centers. Neo-Lithics 23/05, The Newsletter of Southwest Asian Neolithic Research, 53 pp.

Beaumont, P., et al., eds., 1989. Qanat, kariz and khattara: traditional water systems in the Middle East and North Africa. Middle East and North African Studies Press Limited, 303 pp.

Beaumont, P.B., 1973. The ancient pigment mines of southern Africa. South African Journal of Science 69, 140-146.

Beaumont, P.B., 1980. On the age of Border Cave hominids 1–5. Palaeontologica Africana 23, 21-33.

Beaumont, P.B., and Vogel, J.C., 1989. Patterns in the age and context of rock art in the Northern Cape. South African Archaeological Bulletin 44, 73-81.

Beaumont, P.B., and Vogel, J.C., 2006. On a timescale for the past million years of human history in central South Africa. South African Journal of Science 102, 217-228.

Beaune, S. de: see De Beaune.

Beausang, E., 2000. Childbirth in prehistory: an introduction. European Journal of Archaeology 3, 69-87.

Beck, C.W., and Bouzek, J., eds., 1993. Amber in Archaeology, Proceedings of the Second International Conference on Amber in Archaeology, Liblice, Czech Republic, 1990, 248 pp.

Becker, C.J., 1959. Flint mining in Neolithic Denmark. Antiquity XXXIII, 87-92.

Bednarik, R.G., 1989. The Galgenberg figurine from Krems, Austria. Rock Art Research 6, 118-125.

Bednarik, R.G., 1997. The role of Pleistocene beads in documenting hominid cognition. Rock Art Research 14, 27-41.

Bednarik, R.G., 1999. Maritime navigation in the Lower and Middle Palaeolithic. Comptes Rendus de l'Académie des Sciences de Paris, Sciences de la terre et des planètes 328, 569-563.

Bednarik, R.G., 2002. The maritime dispersal of Pleistocene humans. Migration and Diffusion 3, 6-33.

Bednarik, R.G., 2003. Seafaring in the Pleistocene. Cambridge Archaeological Journal 13, 41-66.

Behre, K.-E., 1991. Zum Brotfund aus dem Ipweger Moor, Ldkr. Wesermarsch. Berichte zur Denkmalpflege in Niedersachsen, Beiheft 11, p. 9.

Beja-Pereira, A., et al., 2003. Gene-culture coevolution between cattle milk protein genes and human lactase genes. Nature Genetics 35, 311-315.

Belfer-Cohen, A., 1988. The Natufian graveyard in Hayonim Cave. Paléorient 14, 297-308.

Belfer-Cohen, A., 1991. Art items from Layer B, Hayonim Cave: A case study of art in a Natufian context. In The Natufian Culture in the Levant. International Monographs in Prehistory, 589-588.

Belfer-Cohen, A., and Bar-Yosef, O., 1981. The Aurignacian at Hayonim Cave. Paléorient 7, 19-42.

Belfer-Cohen, A., and Bar-Yosef, O., 1999. The Levantine Aurignacian: 60 years of research. In Dorothy Garrod and the progress of the Palaeolithic: studies in the prehistoric archaeology of the Near East and Europe, R. Charles and W. Davies, eds., 118-134.

Belfer-Cohen, A., and Hovers, E., 1992. In the eye of the beholder: Mousterian and Natufian burials in the Levant. Current Anthropology 33, 463-471.

Belitzky, S., Goren-Imbar, N., and Werker, E., 1991. A Middle Pleistocene wooden plank with man-made polish. Journal of Human Evolution 20, 349-353.

Bell, L.A, 1983. Papyrus, tapa, amate and rice paper. Liliaceae Press, 146 pp.

Bello, S.M., et al., 2011. Earliest directly-dated Human skull-cups. PLoS ONE 6(2), 12 pp.

Bello, S.M., et al., 2015. Upper Palaeolithic ritualistic cannibalism at Gough's Cave (Somerset, UK): the human remains from head to toe. Journal of Human Evolution 30, 1-20.

Bellomo, R.V., 1994. Methods of determining early hominid behavioral activities associated with the controlled use of fire at FxJj 20 Main, Koobi Fora, Kenya. Journal of Human Evolution 27, 173-195.

Bellwood, P., 2005. First farmers – the origins of agricultural society, Blackwell Publishing, 384 pp.

Bellwood, P., et al., 1992. New dates for Prehistoric Asian rice. Asian Perspectives 31, 161-170.

Ben-Dov et al., eds., 2012. Living the Lunar Calendar. Oxbow Books, Oxford, 387 pp.

Benazzi, S., et al., 2011. Early dispersal of modern humans in Europe and implications for Neanderthal behavior. Nature 479, 525-528.

Benazzi, S., et al., 2014. The makers of the Protoaurignacian and implications for Neandertal extinction. Science Express (April 23), 10 pp.

Bennett, A., 1988. Prehistoric copper smelting in Central Thailand. In Prehistoric Studies, 125-135.

Bennett, A., 1989. The contribution of metallurgical studies to South-East Asian archaeology. World Archaeology 20, 329-351.

Bennett, A., 1990. Prehistoric copper smelting in central Thailand. In Southeast Asian Archaeology 1986. British Archaeological Reports International Series 561, 109-120.

Bennett, G.F., and Peirce, M.A., 1988. Morphological form in the avian haemoproteidae and an annotated checklist of the genus Haemoproteus (Kruse, 1890). Journal of Natural History 22, 1683-1696.

Bennett, M.R., Harris, J.W.K., Richmond, B.G., Braun, D.R., Mbua, E., Kiura, P. Olago, D., Kibunjia, M., Omuombo, C., Behrensmeyer, A.K., Huddart, D., and Gonzalez, S., 2009. Early hominin foot morphology based on 1.5-million-year-old footprints from Ileret, Kenya. Science 323, 1197-1201.

Bennike, P., and Fredebo, L., 1986. Dental Treatment in the Stone Age. *Bulletin of the History of Dentistry* 34, 81–87.

Berge, C., 1984. Multivariate analysis of the pelvis for hominids and other extant primates: implications for the locomotion and systematics of the different species of austalopithecines. Journal of Human Evolution 13, 555-562.

Berger, L.R., et al., 2015. *Homo naledi,* a new species of the genus *Homo* from the Dinaledi Chamber, South Africa. eLIFE, 35 pp.

Berger, T.D., and Trinkhaus, E., 1995. Patterns of trauma among the Neandertals. Journal of Archaeological Science 22, 841-852.

Bergh, G.D., van den: see Van den Bergh.

Bergman, C.A., 1993. The development of the bow in Western Europe: A technological and functional perspective. *In* Hunting and Animal Exploitation in the later Palaeolithic and Mesolithic of Eurasia, Archaeological Papers of the American Anthropological Association 3, 95-105.

Berke, H., 2007. The invention of blue and purple pigments in ancient times. Chemical Society Reviews 36, 15-30.

Berkowitz, M., 1996. World's earliest wine. Archaeology 49, p. 26.

Bermudez de Castro, J.M., Bromage, T.G., and Fernandez Jalvo, Y., 1988. Buccal striations on fossil human anterior teeth: evidence of handedness in the middle and early Upper Pleistocene. Journal of Human Evolution 17, 403-412.

Berna, F., et al., 2012. Microstratigraphic evidence of in situ fire in the Acheulian strata of Wonderwerk Cave, Northern Cape province, South Africa. National Academy of Sciences (USA), Proceedings E1215-1220.

Bettinger, R.L., 2009. Hunter-gatherer foraging: five simple models. Eliot Werner Publications, Inc., Clinton Corners, NY, 111 pp.

Biagi, P., and Cremaschi, M., 1991. The Harappan flint quarries of the Rohri Hills (Sind-Pakistan). Antiquity 65, 97-102.

Bienert, H.-D., 1991. Skull cult in the prehistoric Near East. Journal of Prehistoric Religion 5, 9-23.

Binneman, J., and Beaumont, P., 1992. Use-wear analysis of two Acheulian handaxes from Wonderwerk Cave, Northern Cape. Southern African Field Archaeology 1, 92-97.

Bishop, L.C., et al., 2006. Recent research into Oldowan hominin activities at Kanjera South, Western Kenya. African Archaeological Review 23, 31-40.

Black, F.L., 1966. Measles endemicity in insular populations: critical community size and its evolutionary implications. Journal of Theoretical Biology 11, 207-211.

Boaretto, E., et al., 2009. Specialized flint procurement strategies for hand axes, scrapers and blades in the Late Lower Paleolithic: a ^{10}Be study at Qesem Cave, Israel. Human Evolution 24, 1-12.

Boaz, N.T., and Ciochon, R.L., 2004. Dragon Bone Hill: an Ice-Age saga of *Homo erectus.* Oxford University Press, 232 pp.

Bocquentin, F., and Bar-Yosef, O., 2004. Early Natufian remains: evidence of physical conflict from Mt. Carmel, Israel. Journal of Human Evolution 47, 19-23.

Bocquet-Appel, J.-P., and Bar-Yosef, O., eds., 2008. The Neolithic Demographic Transition and its consequences. Springer, 542 pp.

Boëda, E., Connan, J., and Muhesen, S., 1998. Bitumen as hafting material on Middle Paleolithic artifacts from the El Kowm Basin, Syria. *In* Neandertals and Modern Humans in Western Asia, Plenum Press, 181-204.

Boëda, E., Connan, J., Dessort, D., Muhesen, S., Mercier, N., Valladas, H., and Tisnérat, N., 1996. Bitumen as a hafting material on Middle Palaeolithic artefacts. Nature 380, 336-338.

Boëda, E., Genese, J.M., Griggo, C., Mercier, N., Muhesen, S., Reyss, J.L., Taha, A., and Valladas, H., 1999. A Levallois point embedded in the vertebra of a wild ass (*Equus africanus*): hafting projectiles and Mousterian hunting weapons. Antiquity 73, 394-402.

Bogucki, P., 1986. The Antiquity of Dairying in Temperate Europe. Expedition 28, 51-58.

Bogucki, P., 1988. Forest farmers and stockherders: early agriculture and its consequences in North-Central Europe. Cambridge University Press, 247 pp.

Bogucki, P., 1993. Animal traction and household economies in Neolithic Europe. Antiquity 67, 492-503.

Bogucki, P., and Grygiel, R., 1983. Early farmers of the North European Plain. Scientific American 248, 104-112.

Bogucki, P.J., 1984. Ceramic sieves of the Linear Pottery Culture and their economic implications. Oxford Journal of Archaeology 3, 15-30.

Böhmer, H., and Thompson, J., 1991. The Pazyryk carpet: a technical discussion. Notes in the History of Art 10, 8-15.

Bonifay, E., 1964. La Grotte du Regourdou (Montignac, Dordogne). Stratigraphie et industrie lithique Moustérienne. L'Anthropologie (Paris), tome 68(1-2), 49-64.

Bonneau, D., 1964. La crue du Nil: divinité égyptienne, à travers mille ans d'histoire (332 av. – 641 ap. J.-C.), C. Klincksieck, Paris, 529 pp.

Bonsall, C., ed., 1985 [1989]. The Mesolithic in Europe: Proceedings of the Third International Symposium, Edinburgh, John Donald Publishers Ltd., Edinburgh, 656 pp.

Bonsall, C., et al., 1997. Mesolithic and Early Neolithic in the Iron Gates: a palaeodietary perspective. Journal of European Archaeology 5, 50-92.

Bordes, F., 1961. Mousterian cultures in France. Science 134, 803-810.

Borić, D., 2009. Absolute dating of metallurgical innovations in the Vinča Culture of the Balkans. In Metals and Societies; studies in honour of Barbara S. Ottaway. Verlag Rudolf Habelt, 191-245.

Born, H., 1991. Ein bronzener Prunkhelm der Hallstatzeit. Verlag Axel Guttman, Band I, 103 pp.

Born, H., and Hansen, S., 2001. Helme und Waffen Alteuropas. Band IX, Verlag Philipp von Zabern, 289 pp.

Born, H., and Seidl, U., 1995. Schutzwaffen aus Assyrien und Arartu. Verlag Philipp von Zabern, 195 pp.

Boroffka, N., et al., 2002. Bronze Age tin from central Asia: preliminary notes. In Ancient interactions: east and west in Eurasia, 135-159.

Borroff, E., 1971. Music in Europe and the United States. Prentice Hall, 752 pp.

Boshier, A., and Beaumont, P., 1972. Mining in southern Africa and the emergence of modern man. Optima 22, 2-12.

Bouchet, F., et al., 1996. Paléoparasitologie en contexte pléistocène: premières observations à la Grande Grotte d'Arcy-sur-Cure (Yonne), France. Comptes Rendus de l'Académie des Sciences de Paris – Sciences de la Vie 319, 147-151.

Bouchet, F., Harter, S., and Le Bailly, M., 2003. The State of the Art of Paleoparasitological Research in the Old World. Memoirs of the Oswaldo Cruz Institute, Rio de Janeiro, 98 (Suppl. I), 95-101.

Bouchud, J., 1953. Les Paléolithiques utilisaient-ils les plumes? Bulletin de la Société Préhistorique de France, tome 50, 556-560.

Boucot, A.J., 1990. Evolutionary Paleobiology of Behavior and Coevolution, Elsevier, 725 pp.

Boucot, A.J., and Poinar, G.O., Jr., 2010. Fossil Behavior Compendium. CRC Press, 391 pp.

Boulestin, B., et al., 2009. Mass cannibalism in the Linear Pottery Culture at Herxheim (Palatinate, Germany). Antiquity 83, 968-982.

Bourgarit, D., et al., 2003. Chalcolithic fahlore smelting at Cabrières: reconstruction of smelting processes by archaeometallurgical finds. In Archaeometallurgy in Europe 1, 431-440.

Bourke, J.B., 1967. A review of the palaeopathology of the arthritic diseases. In Diseases in Antiquity, D. Brothwell and A.T. Sandison, eds. Charles C. Thomas Publisher, Springfield, IL, 352-370.

Bouville, V., 1982. Morte violente, les massacres. Histoire et Archeologie 66, 36-41.

Bouzouggar, A., Barton, N., Vanhaeren, M., d'Errico, F., Collcutt, S., Higham, T., Hodge, E., Parfitt, S., Rhodes, E., Schwenninger, J.-L., Stringer, C., Turner, E., Ward, S., Moutmir, A., and Stambouli, A., 2007. 82,000-year-old shell beads from North Africa and implications for the origins of modern human behavior. National Academy of Sciences (USA), Proceedings 104(24), 9964-9969.

Bowen, R.L., Jr., 1951. The pearl fisheries of the Persian Gulf. Middle East Journal 5, 161-180.

Bower, B., 2013. War arose recently, anthropologists contend. Science News, 27 July, 5 pp.

Bown, T.M., et al., 1994. The oldest flagstone road in the world – access to Old Kingdom basalt quarries, northern Fayum Depression, Egypt. Geological Society of America, Rocky Mountain Section, Abstracts 26(6), no. 17480.

Boyd, R., and Silk, J.B., 2003. How humans evolved. W.W. Norton & Co., 3rd edition, 545 pp., plus Appendix and Index.

Boyer, P., 2001. Religion explained: the evolutionary origins of religion. Basic Books, 375 pp.

Bradley, D.G., and Magee, D.A., 2006. Genetics and the origins of domestic cattle. In Documenting Domestication, University of California Press 317-328.

Bradshaw, J.L., 1988. The evolution of human lateral asymmetries: new evidence and second thoughts. Journal of Human Evolution 17, 615-637.

Brain, C.K., 1958. The Transvaal ape-man-bearing cave deposits. Transvaal Museum, Memoir 11, 131 pp.

Brain, C.K., 1988. New information from the Swartkrans Cave of relevance to 'Robust' Australopithecines. In F.E. Grine, ed., Evolutionary history of the Robust Australopithecines. New York, Aldine de Gruyter, 311-324.

Brain, C.K., 1993. The occurrence of burnt bones at Swartkrans and their implications for the control of fire by early hominids. Transvaal Museum, Monograph 8, 229-242.

Brain, C.K., and Shipman, P., 1993. The Swartkrans bone tools. In Swartkrans, a cave's chronicle of early man, Transvaal Museum, Monograph 8(1), 195-215.

Brain, C.K., and Sillen, A., 1988. Evidence from the Swartkrans cave for the earliest use of fire. Nature 336, 464-466.

Bramble, D.M., and Lieberman, D.E., 2004. Endurance running and the evolution of Homo. Nature 432, 345-352.

Braun, D.R., et al., 2008. Oldowan behavior and raw material transport: perspectives from the Kanjera Formation. Journal of Archaeological Science 35, 2329-2345.

Bresciani, J., Dansgaard, W., Fredskild, B., et al., 1991. Living conditions. In The Greenland Mummies, J.P.H. Hansen, J. Meldgaard, and J. Nordqvist, eds., Smithsonian Institution Press, Washington, DC, 51-167.

Bresciani, J., Haarlov, N., Nansen, P., and Moller, G., 1983. Head louse (Pediculus humanus subsp. capitis de Geer) from mummified corpses of Greenlanders, A.D. 1460(±50). Acta Entomologia Fennica 42, 24-27.

Breton, C., et al., 2006. Using multiple types of molecular markers to understand olive phylogeography. *In* Documenting Domestication, University of California Press, 143-152.

Breunig, P., et al. 1996. New Research on the Holocene settlement and environment of the Chad Basin in Nigeria. African Archaeological Review 13(2), 111-145.

Brink, J.S., and Deacon, H.J., 1982. A study of a last interglacial shell midden and bone accumulation at Herolds Bay, Cape Province, South Africa. Palaeoecology of Africa 15, 31-39.

Broodbank, C., 1989. The longboat and society in the Cyclades in the Keros-Syros Culture. American Journal of Archaeology 93, 319-337.

Broodbank, C., 2000. An Island Archaeology of the Early Cyclades. Cambridge University Press, 414 pp.

Broodbank, C., and Strasser, T.F., 1991. Migrant farmers and the Neolithic colonization of Crete. Antiquity 65, 231-245.

Brooks, A.S., 1996. Behavioral perspectives on the origin of modern humans: another look at the African evidence. *In* The Lower and Middle Palaeolithic, International Union of Prehistoric and Protohistoric Sciences, 157-166.

Brooks, A.S., Helgren, D.M., Cramer, J.S., Franklin, A., Hornyak, W., et al., 1995. Dating and context of three Middle Stone Age sites with bone points in the upper Semliki Valley, Zaire. Science 268, 548-553.

Brooks, A.S., Nevell, L., Yellen, J.E., and Hartman, G., 2005. Projectile technologies of the African MSA. *In* Transitions before the transition: evolution and stability in the Middle Paleolithic and Middle Stone Age, Springer, 233-255.

Brooks, R.R.R., and Wakankar, V.S., 1976. Stone Age painting in India. Yale University Press, 116 pp.

Brothwell D., 1967b. Major congenital anomalies of the skeleton: evidence from earlier populations. *In* Diseases in Antiquity, D. Brothwell and A.T. Sandison, eds., Charles C. Thomas: Springfield, IL, pp. 423–444.

Brothwell, D., 1967. The evidence for neoplasms. *In* Diseases in Antiquity, D. Brothwell and A.T. Sandison, eds., Charles C. Thomas Publishers, Springfield, IL, 320-345.

Brothwell, D., 2011. Tumors: Problems of Differential Diagnosis in Paleopathology. *In* A Companion to Paleopathology, A.L. Grauer, ed., Blackwell Publishing, Ch. 23, 420-433.

Brothwell, D., and Sandison, A.T., eds., 1967. Diseases in Antiquity. Charles C. Thomas Publishers, Springfield, IL, 766 pp.

Brothwell, D.R., and Powers, R., 1968. Congenital Malformations of the Skeleton in Earlier Man. *In* The Skeletal Biology of Earlier Human Population, D.R. Brothwell, ed., Pergamon Press, Oxford.

Brown, J., 1970. Note on the division of labor by sex. American Anthropologist 72, 1075-1076.

Brown, K.S., et al., 2009. Fire as an engineering tool of early modern humans. Science 325, 859-862.

Brown, K.S., et al., 2012. An early and enduring advanced technology originating 71,000 years ago in South Africa. Nature Letters 491, 590–593.

Brown, P., Sutikna, T., Norwood, M.J., Soejono, R.P., Jatmiko, E., et al., 2004. A new small-bodied hominin from the Late Pleistocene of Flores, Indonesia. Nature 425, 310-316.

Brunner, T.F., 1973. Marijuana in ancient Greece and Rome? The literary evidence. Bulletin of the History of Medicine 47, 344-355.

Bryant, V.M., and Williams-Dean, G., 1975. The coprolites of man. Scientific American 232(1), 100-109.

Buckley, S., et al., 2014. Dental calculus reveals unique insights into food items, cooking and plant processing in Prehistoric Central Sudan. PLoS ONE 9, issue 7, 10 pp.

Buckman, R., 2003, Human Wildlife. The Johns Hopkins University Press, 203 pp.

Budd, P., Montgomery, J., Evans, J., and Trickett, M., 2004. Human lead exposure in England from approximately 5500 BP to the 16th century AD. Science of the Total Environment 318(1-3), 45-58.

Buisson, D., 1990. Les flûtes paléolithiques d'Isturitz (Pyrénées-Atlantiques). Bulletin de la Société Préhistorique française 87, 420-431.

Bulbeck, D., et al., 2004. Leong Sakapao 1, a second dated Pleistocene site from South Sulawesi, Indonesia. Quaternary Research in Indonesia 18, 111-128.

Bunn, H.T., 2007. Meat made us human. In Evolution of the human diet. Oxford University Press, 191-211.

Buquet-Marcon, C., et al., 2007. The oldest amputation on a Neolithic human skeleton in France. Nature Proceedings, 14 pp.

Burford, A., 1960. Heavy transport in Classical Antiquity. The Economic History Review, XIII(1), 1-18.

Burke, J.G., 1986. Cosmic Debris. University of California Press, 445 pp.

Burney, C., 1972. Urartian irrigation works. Anatolian Studies 22, 179-186.

Busvine, J.R., 1980. The evolution and mutual adaptation of insects, microorganisms and man. In Changing Disease Patterns and Human Behavior, Academic Press, San Diego, CA, 55-68.

Butzer, K.W., 1976. Early Hydraulic Civilization in Egypt. University of Chicago Press, 134 pp.

Byrd, B.F., and Monahan, C.M., 1995. Death, mortuary ritual and Natufian social structure. Journal of Anthropological Archaeology 14, 251-287.

Cáceres, I., et al., 2007. Evidence for Bronze Age Cannibalism in El Mirador Cave (Sierra de Atapuerca, Burgos, Spain). American Journal of Physical Anthropology 133, 899-917.

Callaway, E., 2015. Teeth from China reveal early human trek out of Africa. Nature News, 4 pp.

Campana, D.V., 1989. Natufian and Protoneolithic bone tools: the manufacture and use of bone implements in the Zagros and the Levant. British Archaeological Reports International Series 494, 156 pp.

Campbell, S., and Green, A., eds., 1995. The archaeology of death in the ancient Near East. Oxbow Monograph 51, 297 pp.

Campillo, D., et al., 1993. A mortal wound caused by a flint arrowhead in individual MF-18 of the Neolithic Period exhumed at Sant Quirze del Valles. International Journal of Osteoarchaeology 3, 145-150.

Canci, A., Minozzi, S., and Borgognini Tarli, S.M., 1996. New evidence of tuberculous spondylitis from Neolithic Liguria (Italy). International Journal of Osteoarchaeology 6, 497-501.

Cârciumaru, M., et al., 2002. The Cioarei-Borosteni Cave (Carpathian Mountains, Romania): Middle Palaeolithic finds and technological analysis of the lithic assemblages. Antiquity 76, 681-690.

Cârciumaru, M., et al., 2009. L'ocre et les recipients pour ocre de la grotte Cioarei, village Boroteni, commune Petiani, dép. Gorj, Roumanie. Annales d'Université Targoviste, Section d'Archéologie et d'Histoire, tome XI, no. 1, 7-19.

Carman, J., and Harding, A., eds., 1999. Ancient Warfare: Archaeological perspectives. Sutton Publishing, Stroud, Gloucestershire, UK, 279 pp.

Carmody, R.N., Weintraub, G.S., and Wrangham, R.W., 2011. Energetic consequences of thermal and nonthermal food processing. National Academy of Sciences (USA), Proceedings 19199-19203.

Carter, T., Poupeau, G., Bressy, C., and Pearce, N.J.G., 2006. From chemistry to consumption: towards a history of obsidian use at Çatalhöyük through a programme of inter-laboratory trace-elemental characterization. In I. Hodder, ed., Changing materialities at Çatalhöyük; reports from the 1995-99 seasons, McDonald Institute for Archaeological Research, Cambridge, UK, 285-305.

Cassoli, P.F., and Tagliacozzo, A., 1997. Butchering and Cooking of Birds in the Palaeolithic Site of Grotta Romanelli (Italy). International Journal of Osteoarchaeology 7, 303-320.

Casson, L., 1994. Ships and seafaring in ancient times. University of Texas Press, 160 pp.

Catling, H.W., 1991. Bronze Age trade in the Mediterranean: a view. In Bronze Age trade in the Mediterranean, Conference publication, Paul Åströms Förlag, Jonsered, Göteborg, Studies in Mediterranean Archaeology, XC, 1-13.

Cattelain, P., 1989. Hunting during the Upper Upper Paleolithic: bow, spearthrower, or both? In Projectile Technology, Plenum Press, New York, 213-240.

Cattelain, P., 1999. Un crochet de propulseur solutréen de la grotte de Combe-Saunière 1 (Dordogne). Bulletin de la Société Préhistorique Française 86, 213-216.

Cauvin, J., 1987. L'apparition des premières divinités. Recherche 194, 1472-1480.

Cauvin, J., 2000. The Birth of the Gods and the Origins of Agriculture. Cambridge University Press.

Cauvin, M.-C., et al., 1998. L'obsidienne au Proche et Moyen Orient. British Archaeological Reports International Series 738, 388 pp.

Cerling, T.E., et al., 2011. Diet of Paranthropus boisei in the early Pleistocene of East Africa. National Academy of Sciences (USA), Proceedings 109, 9337-9341.

Chakrabarti, D.K., 1977. Distribution of iron ores and the archaeological evidence of early iron in India. Journal of the Economic and Social History of the Orient 20(II), 166-184.

Chakrabarti, D.K., 1988. The old copper mines of Eastern India. In The beginnings of the use of metals and alloys, MIT Press, 239-244.

Chang, Kwang-chih, 1986. The archaeology of ancient China. Yale University Press, 450 pp.

Chang, T.T., 1989. Domestication and spread of the cultivated rice. In Foraging and Farming: The evolution of plant exploitation. Unwin Hyman, London, 408-417.

Chapman, D.A., and Chapman, S.G., 2013. Reconstructing and testing the Pentwyn pit furnaces. Ancient Arts, 24 pp.

Chapman, J., et al., 2006. The social context of the emergence, development and abandonment of the Varna Cemetery, Bulgaria. European Journal of Archaeology 9, 159-183.

Chapman, R., 1981. The emergence of formal disposal areas and the 'problem' of megalithic tombs in prehistoric Europe. In The archaeology of death, R. Chapman et al. eds., Cambridge University Press, 71-81.

Chapman, R., et al., eds., 1981. The archaeology of death. Cambridge University Press, 159 pp.

Charlier, P., et al., 2012. Toilet hygiene in the classical era. BMJ (British Medical Journal), 5 pp.

Chauvet, J.-M., Deschamps, E.B., and Hillaire, C., 1996. Dawn of art: the Chauvet Cave, the oldest known paintings in the World. Harry N. Abrams, Inc., Publishers, 135 pp.

Chernych, E.N., 1978. Aibunar – a Balkan copper mine of the fourth millennium. Proceedings of the Prehistoric Society 44, 203-217.

Chernykh, E.N., 1992. Ancient metallurgy in the USSR. Cambridge University Press, 335 pp.

Cherry, J.F., 1990. The first colonization of the Mediterranean Islands: a review of recent research. Journal of Mediterranean Archaeology 3, 145-221.

Chirchir, H., et al., 2014. Recent origin of low trabecular bone density in modern humans. National Academy of Sciences (USA), Proceedings, Early Edition, 6 pp.

Churchill, S.E., and Rhodes, J.A., 2009. The evolution of the human capacity for 'killing at a distance': The human fossil evidence for the evolution of Projectile weaponry. In The Evolution of Hominin Diets: Integrating Approaches to the Study of Palaeolithic Subsistence, Springer Science+Business Media B.V., 201-210.

Clare, L., Rohling, E.J., Weniger, B., and Hilpert, J. 2008. Warfare in Late Neolithic\Early Chalcolithic Pisidia, southwestern Turkey. Climate induced social unrest in the late 7th millennium calBC. Documenta Praehistorica 35, 65-92.

Claris, P., and Quartermaine, J., 1989. The Neolithic quarries and axe factory sites of Great Langdale and Scafell Pike: a new field survey. Proceedings of the Prehistoric Society 55, 1-25.

Clark, J.D., and Harris, J.W.K., 1985. Fire and its roles in early hominid lifeways. African Archaeological Review 3, 3-27.

Clark, J.D., 1992. The earlier Stone Age/Lower Palaeolithic in North Africa and the Sahara. In New light on the Northeast African past: current prehistoric research: contributions to a symposium, Cologne 1990, 17-37.

Clark, J.D., 2001. Kalambo Falls prehistoric site. In The earlier cultures: Middle and earlier Stone Age, Cambridge University Press, vol. 3, 701 pp.

Clark, J.D., 2001b. Variability in primary and secondary technologies of the later Acheulian in Africa. In A very remote period indeed: papers on the Palaeolithic presented to Derek Roe, Oxbow Books, 1-18.

Clark, J.D., Philips, J.L., and Staley, P.S., 1974. Interpretation of prehistoric technology from ancient Egyptian and other sources. Part 1: ancient Egyptian bows and arrows and their relevance for African prehistory. Paléorient 2, 323-388.

Clark, J.G.D., 1948. Objects of South Scandinavian flint in the northernmost provinces of Norway, Sweden and Finland, Proceedings of the Prehistoric Society 14, 219-232.

Clark, J.G.D., 1952. Prehistoric Europe: The economic basis. Philosophical Library, New York, 349 pp.

Clark, J.G.D., 1963. Neolithic bows from Somerset, England, and the prehistory of archery in north-western Europe. Proceedings of the Prehistoric Society 29, 50-98.

Clemen, A.J., 1956. Caries in the South African ape-man: some examples of undoubted pathological authenticity believed to be 800,000 years old. British Dental Journal 101, 4-7.

Close, A.E., 1995. Few and far between: early ceramics in North Africa. In The emergence of pottery: technology and innovation in ancient societies, Smithsonian Institution Press, 23-37.

Close, A.E., and Wendorf, F., 1990. North Africa at 18,000 BP. In The World at 18,000 BP, Unwin Hyman, 41-57.

Clottes, J., and Courtin, J., 1996. The cave beneath the sea: Paleolithic images at Cosquer. Harry N. Abrams, Inc., 200 pp.

Clutton-Brock, J., 1981. Domesticated Animals from early times. University of Texas Press and the British Museum (Natural History), 208 pp.

Clutton-Brock, J., 1984. Excavations at Grimes Graves, Norfolk 1972–1976. Fasc. 1. British Museum Publications Limited, 47 pp.

Clutton-Brock, J., 1999. A Natural History of Domesticated Mammals. Cambridge University Press, 238 pp.

Cockburn, A., 1963. The Evolution and Eradication of Infectious Diseases. Johns Hopkins University Press, Baltimore, MD., 255 pp.

Cockburn, T.A., 1971. Infectious diseases in ancient populations. Current Anthropology 12, 45-62.

Cockburn, T.A., and Cockburn, E., 1980. Disease and Ancient Cultures. Cambridge University Press, 340 pp.

Cohen, M.N., 2008. Implications of the NDT for world wide health and mortality in prehistory. In The Neolithic Demographic Transition and its consequences, Springer Science+Business Media B.V., 481-500.

Coleman, J.E., 1977. Kephala, vol. I, A Late Neolithic settlement and cemetery. American School of Classical Studies, Princeton, NJ, 73 pp.

Comas, I., et al., 2013. Out-of-Africa migration and Neolithic coexpansion of *Mycobacerium tuberculosis* with modern humans. Nature Genetics 45(10), 1176-1182.

Conard, N.J., 2005. An overview of the patterns of behavioural change in Africa and Eurasia during the Middle and Late Pleistocene. In From Tools to Symbols, from early hominids to modern humans, Wits University Press, 294-332.

Conard, N.J., 2009. A female figurine from the basal Aurignacian of Hohe Fels Cave in southwestern Germany. Nature Letters 459, 248-252.

Conard, N.J., and Bolus, M., 2015. Chronicling modern human's arrival in Europe. Science Express 23; 3 pp.

Conard, N.J., Malina, M., and Münze, S.C., 2009. New flutes document the earliest musical tradition in southwestern Germany. Nature Letters 460, 737-740.

Connan, J., 1999. Use and trade of bitumen in antiquity and prehistory: molecular archaeology reveals secrets of past civilizations. Philosophical Transactions of the Royal Society, London, B, 354, 33-50.

Conophagos, C.E., 1980. Le Laurium Antique. Ekdotike Hellados S.A., 458 pp.

Conroy, G.C., 1990. Primate Evolution. W.W. Norton & Company, New York, 492 pp.

Constantini, L., 1989. Plant exploitation at Grotta dell'Uzzo, Sicily: new evidence for the transition from Mesolithic to Neolithic subsistence in southern Europe. In Foraging and Farming The evolution of plant exploitation, Unwin Hyman, 197-206.

Constantino, P.J. et al., 2010. Tooth chipping can reveal the diet and bite forces of fossil hominins. Biology Letters, Palaeontology, Royal Society, 4 pp.

Cook, G.T., et al. 2001. A freshwater diet derived ^{14}C reservoir effect at the Stone Age sites in the Iron Gates Gorge. Radiocarbon 43, 453-460.

Copeland, S.R., Sponheimer, M., de Ruiter, D., Lee-thorp, J.A., Codron, D., le Roux, P.J., Grimes, V., and Richards, M.P., 2011. Strontium isotope evidence for landscape use by early hominins. Nature Letters 474, 76-79.

Copley, M.S., et al., 2003. Direct chemical evidence for widespread dairying in prehistoric Britain. National Academy of Sciences (USA), Proceedings 100, 1524-1529.

Coppa, A., et al., 2006. Early Neolithic tradition of dentistry. Nature 440, 755-756.

Cordier, G., 1990. Blessures préhistoriques animales et humaines avec armes ou projectiles conservés. Bulletin de la Société Préhistorique Française 87, 462-481.

Coren, S., and Porae, C., 1977. Fifty centuries of right handedness: the historical record. Science 198, 631-632.

Cosgrove, R., 1989. Thirty thousand years of human colonization in Tasmania: new Pleistocene dates. Science 243, 1706-1707.

Costantini, L., 1981. The beginnings of agriculture in the Kachi Plain: the evidence of Mehrgarh. *In* South Asian Archaeology, Cambridge University Press, 29-33.

Costantini, L., and Costantini Biasini, L., 1985. Agriculture in Baluchistan between the 7th and the 3rd Millennium BC. Newsletter of Baluchistan Studies 2, 16-30.

Courville, C.B., 1950. Cranial injuries in prehistoric man, with particular reference to the Neanderthals. Bulletin of the Los Angeles Neurological Society 15, 1-21.

Covert, H.H., and Kay, R.F., 1981. Dental microwear and diet: Implications for determining the feeding behaviors of extinct primates, with a comment on the dietary pattern of *Sivapithecus.* American Journal of Physical Anthropology 55, 331-336.

Crabtree, D.E., 1972. An Introduction to Flintworking. Occasional Papers of the Idaho State University Museum (Pocatello, ID) 28, 98 pp.

Craddock, P.T., 1995. Early metal mining and production. Edinburgh University Press, 363 pp.

Craddock, P.T., 1999. Paradigms of metallurgical innovation in prehistoric Europe. *In* The beginnings of metallurgy, 175-192.

Craddock, P.T., 2001. From hearth to furnace: evidences for the earliest metal smelting technologies in the Eastern Mediterranean. Paléorient 26, 151-165.

Craddock, P.T., ed., 1980. Scientific studies in early mining and extractive metallurgy. British Museum, Occasional Paper 20, 173 pp.

Craig, O.E., et al., 2005. Did the first farmers of central and eastern Europe produce dairy food? Antiquity 79, 882-894.

Crane, E., 1983. The archaeology of beekeeping. Cornell University Press, 360 pp.

Crane, E., 1999. The world history of beekeeping and honey hunting. Routledge, 682 pp.

Crane, E., and Graham, A.J., 1985. Bee hives of the ancient world. Bee World 66, 25-41, 148-170.

Crawford, H., 2004. Sumer and the Sumerians. Cambridge University Press, 252 pp.

Crew, P., and Crew, S., eds., 1990. Early mining in the British Isles: proceedings of the Early Mining workshop at Plas Tan y Bwlch, Snowdonia National Park Study Centre, 17-19 November, 1989. Plas Tan y Bwlc, Occasional Paper, no. 1, 80 pp.

Crompton, R.H., and Pataky, T.C., 2009. Stepping out. Science 323, 1174-1175.

Cronin, J.E., Boaz, N.T., Stringer, C.B., and Rak, Y., 1981. Tempo and mode in hominid evolution. Nature 292, 113-122.

Crouch, D.P., 1993. Water management in ancient Greek cites. Oxford University Press, 380 pp.

Crubézy, E., and Trinkhaus, E., 1992. Shanidar 1: A case of hyperostotic disease (DISH) in the Middle Paleolithic. American Journal of Physical Anthropology 89, 411-420.

Crubézy, É., et al., 2001. The antiquity of cranial surgery in Europe and in the Mediterranean basin. Comptes Rendus de l'Académie des Sciences, Earth and Planetary Sciences 332, 417-423.

Cucchi, T., and Vigne, J.-D., 2006. Origin and diffusion of the house mouse in the Mediterranean. Human Evolution 21, 95-106.

Curtis, R.I., 2001. Ancient Food Technology. Brill, 465 pp.

D'Errico, F., 1994. Notation versus decoration in the Upper Palaeolithic: a case-study from Tossal de la Roca, Alicante, Spain. Journal of Archaeological Science 21, 185-200.

D'Errico, F., 1995. A new model and its implications for the origin of writing: The La Marche Antler revisited. Cambridge Archaeological Journal 5, 163-206.

D'Errico, F., 2003. The invisible frontier. A multiple species model for the origin of behavioral modernity. Evolutionary Anthropology 12, 188-202.

D'Errico, F., and Backwell, L.R., 2003. Possible evidence of bone tool shaping by Swartkrans early hominids. Journal of Archaeological Science 30, 1559-1576.

D'Errico, F., and Vanhaeren, M., 2007. Evolution or revolution? New evidence for the origin of symbolic behaviour in and out of Africa. Rethinking the Human Revolution, Mcdonald Institute Monographs, 275-286.

D'Errico, F., and Villa, P., 1997. Holes and grooves: the contribution of microscopy and taphonomy to the problem of art origins. Journal of Human Evolution 33, 1-31.

D'Errico, F., Backwell, L., Villa, P., Degano, I., Lucejka, J., and Bamford, M.K., 2012. Early evidence of San material culture represented by organic artifacts from Border Cave, South Africa. National Academy of Sciences (USA), Proceedings, 109(48), E3291-3292.

D'Errico, F., et al., 1989. The collection of non-utilitarian objects by *Homo erectus* in India. *In* HOMINIDAE, Proceedings of the 2nd International Congress of Human Paleontology, 237-239.

D'Errico, F., et al., 1998. A Middle Palaeolithic origin of music? Using cave-bear bone accumulations to assess the Divje Babe I bone 'flute'. Antiquity 72; 9 pp.

D'Errico, F., et al., 2001. An engraved bone fragment from c. 70,000-year-old Middle Stone Age levels at Blombos Cave, South Africa: implications for the origin of symbolism and language. Antiquity 75, 309-318.

D'Errico, F., et al., 2003. Archaeological evidence for the emergence of language, symbolism, and music – an alternative multidisciplinary perspective. Journal of World Prehistory 17, 1-70.

D'Errico, F., Henshilwood, C., Vanhaeren, M., and van Niekerk, Z.K., 2005. *Nassarius kraussianus* shell beads from Blombos Cave: evidence for symbolic behaviour in the Middle Stone Age. Journal of Human Evolution 48, 3-24.

Dales, G.F., 1968. Of dice and men. Journal of the American Oriental Society 88, 14-23.

Daly, M., and Wilson, M., 1988a. Evolutionary social psychology and family homicide. Science 242, 519-524.

Daly, M., and Wilson, M., 1988b. Homicide. Aldine de Gruytrt, New York, 328 pp.

Dark, P., 2004. New evidence for the antiquity of the intestinal parasite *Trichuris* (whipworm) in Europe. Antiquity 78, 676-681.

Dark, P., and Gent, H., 2001. Pests and diseases of prehistoric crops: a yield 'honeymoon' for early grain crops in Europe? Oxford Journal of Archaeology 20, 59-78.

Darlington, P.J., Jr., 1975. Group selection, altruism, reinforcement, and throwing in human evolution. National Academy of Sciences (USA), Proceedings 72, 3748-3752.

Dart, R.A., 1949. The predatory implemental technique of Australopithecus. American Journal of Physical Anthropology 7, 1-38.

Darvill, T., and Thomas, J., 1996. Neolithic houses in Northwest Europe and beyond. Oxbow Monograph 57, 213 pp.

David, A.R., 1997. Disease in Egyptian mummies: the contribution of new technologies. The Lancet 349(9067), 1760-1763.

David, A.R., and Zimmerman, M.R., 2010. Cancer: an old disease, a new disease or something in between? Nature Reviews Cancer 10, 728-733.

Davis, D.D., 1951. The baculum of the gorilla. Fieldiana, Zoology 31(54), 645-647.

Davis, S., 1974. Incised bones from the Mouserian of Kebara Cave (Mount Carmel) and the Aurignacian of Ha-Yonim Cave (Western Gallilee), Israel. Paléorient 2, 181-182.

Davis, S.J.M., 1987. The Archaeology of Animals. Yale University Press, 224 pp.

De Andalusia, the Honorable Lady Maria (edited by Mark S. Harris), 2008. A short history of equestrian stirrups, 24 pp.

De Beaune, S., 1987. Paleolithic lamps and their specialization: a hypothesis. Current Anthropology 28, 569-577. *Also*: de Beaune, S., 1987, translated by M. Turton, 2002. Paleolithic lamps and their specialization: a hypothesis. Bulletin of Primitive Technology, Spring Issue, no. 23, 60-67.

De Beaune, S., 1993. Nonflint stone tools of the Early Upper Paleolithic. *In* Before Lascaux: the complex record of the Upper Paleolithic, CRC Press, 163-192.

De Beaune, S., 2002. Origine du matériel de mouture: innovation et continuité du Paléolithique au Néolithique. Actes du colloque international, La Ferté-sous-Jouarre, Éditions Ibis Press, 17-30.

De Beaune, S., and White, R., 1993. Ice Age lamps. Scientific American, March, 108-113.

De Heinzelin, J., et al., 1999. Environment and behavior of 2.5-million-year-old Bouri hominids. Science 284, 625-629.

De Jesus, P.S., 1980. The development of prehistoric mining and metallurgy in Anatolia, Part 1, British Archaeological Reports International Series 74(i) 206 pp.; 74(ii) 207-495 pp.

De Langhe et al., 2006. Phytolith evidence for the early presence of domesticated banana (*Musa*) in Africa. *In* Documenting Domestication, University of California Press, 68-81.

De Lumley, H., 1969. Une cabane de chasseurs acheuléens vieille de 130 000 ans dans une grotte de Nice. *In* Revue Sciences Progrès Découverte: Les Conférences du Palais de la Découverte, Paris, 119-131.

De Lumley, H., 1969b. Une cabane de chasseurs Acheuleens dans la Grotte du Lazaret à Nice. Archéologia 28, 26-33.

De Lumley, H., 1979. A Paleolithic camp at Nice. *In* Hunters, farmers and civilizations. San Francisco, W.H. Freeman, Ch. 6, 57-65.

Deacon, H.J., 1992. Southern Africa and modern human origins. Philosophical Transactions of the Royal Society London, B, 337, 177-183.

Deacon, H.J., and Deacon, J., 1999. Human Beginnings in South Africa. Altamira Press, 214 pp.

Deelder, A.M., Miller, R.I., DeJonge, N., and Krüger, F.W., 1990. Detection of schistosome antigen in mummies. The Lancet 335, 724-725.

Defleur, A., 1993. Les Sépultures Moustériennes. Centre national de la recherche scientifique, CNRS Editions, 325 pp.

Defleur, A., and White, T., 1999. Neanderthal cannibalism at Moule-Guercy, Ardeche, France. Science 286, 128-131.

Defleur, A., et al., 1999. Neanderthal cannibalism at Moula-Guercy, Ardèche, France. Science 286, 128-131.

Delagnes, A., and Roche, H., 2005. Late Pliocene hominid knapping skills: the case of Lokalalei 2C, West Turkana, Kenya. Journal of Human Evolution 48, 435-472.

Delagnes, A., et al., 2006. Crystal quartz backed tools from the Howiesons Poort at Sibudu Cave. Southern African Humanities 18, 43-56.

Delmas, A.B., and Casanova, M., 1990. The lapis lazuli sources in the Ancient East. South Asian Archaeology 1987, Part 1, 493-505.

Delporte, H., 1984. L'art mobilier et ses rapports avec la faune Paléolithique. Éditions Univérsitaires, Fribourg, 111-142.

Delporte, H., 1993. Gravettian female figurines: a regional survey. *In* Before Lascaux: the complex record of the Early Upper Paleolithic, CRC Press, 243-257.

Denham, T., et al., 2004. New evidence and revised interpretation of early agriculture in Highland New Guinea. Antiquity 78, 839-857.

Dennell, R., 1997. The world's oldest spears. Nature 385, 767-768.

Dennell, R., 2009. The Palaeolithic settlement of Asia. Cambridge University Press, 548 pp.

Dennell, R., and Roebroecks, W., 2005. An Asian perspective on early human dispersal from Africa. Nature 438, 1099-1104.

Dennell, R.W., et al., 1988. Early tool-making in Asia: two-million-year-old artefacts in Pakistan. Antiquity 62, 98-106.

Derevianko, A.P., and Rybin, E.P., 2003. The earliest representatives of symbolic behavior by Paleolithic humans in the Altai Mountains. Archaeology, Ethnology and Anthropology of Eurasia 3, 27-50.

Derevianko, A.P., and Shunkov, M.V., 2004. Formation of the Upper Paleolithic traditions in the Altai. Archaeology, Ethnology and Anthropology of Eurasia 3, 12-40.

Derevianko, A.P., et al., 2008. A Paleolithic bracelet from Denisova Cave. Archaeology, Ethnology and Anthropology of Eurasia 34, 13-25.

Despres, L., et al., 1992. Molecular evidence linking hominid evolution to recent radiation of Schistosomes (Platyhelminthes: Trematoda). Molecular Phylogenetics and Evolution 1, 295-304.

Deter-Wolf, A., 2013. The material culture and Middle Stone Age origins of ancient tattooing. Zurich Studies in Archaeology 9, 15-25.

Dettwyler, K.A., 1995. A time to wean: the hominid blueprint for the natural age of weaning in modern human populations. In Breastfeeding, Biocultural Perspectives, Aldine de Gruyter, 39-73.

Dewey, K., 2003. Guiding principles for complementary feeding of the breastfed child. Pan American Health Organization/World Health Organization, 37 pp.

Diamond, J., 2012. The World until Yesterday. Viking, 498 pp.

Dickel, D.N., and Doran, G.H., 1989. Severe neural tube syndrome from the Early Archaic of Florida. American Journal of Physical Anthropology 80, 325-334.

Dickson, J.H., 1978. Bronze age mead. Antiquity 52, 108-113.

Dieppe, P., and Rogers, J.M., 1993. Skeletal paleopathology of rheumatic disorders. In Arthritis and Allied Conditions, 12th ed., D.J. McCarty and W.J. Koopmans, eds., Lea and Febiger, Philadelphia, PA, 9-16.

Dill, H.G., et al., 2008. Early Bronze Age mining activities on the Island of Seriphos, Greece. Mediterranean Archaeology (Meditarch) 21, 1-2.

Dineley, M., 2004. Barley, Malt and Ale in the Neolithic. British Archaeological Reports International Series 1213, 84 pp.

Dineley, M., and Dineley, G., 2000. Neolithic ale: barley as a source of malt sugars for fermentation. In Plants in Neolithic Britain and beyond, Oxbow Books, 137-155.

Dittmar, K., and Teegen, W.R., 2013. The presence of Fasciola hepatica (liver-fluke) in humans and cattle from a 4,500 year old archaeological site in the Saale-Unstrut valley, Germany. Memoirs of the Oswaldo Cruz Institute, 98 (Suppl. 1), 141-143.

Dixson, A.F., 1987. Baculum length and copulatory behavior in Primates. American Journal of Primatology 13, 51-60.

Dolukhanov, P.M., 1999. War and peace in prehistoric Eastern Europe. In Ancient Warfare, Sutton Publishing, 73-87.

Domínguez-Bella, S. Cassen, S., Pétrequin, P., Přichystal, A., Martínez, J., Ramos, J. and Medina, N., 2015 [2016]. Aroche (Huelva, Andalucía): a new Neolithic axehead of Alpine jade in the southwest of the Iberian Peninsula. Archaeological and Anthropological Sciences 8(1), 205-222.

Dominguez-Bella, S., 2004. Variscite, a prestige mineral in the Neolithic–Aeneolithic Europe. Raw material sources and possible distribution routes. Slovak Geological Magazine 10, 147-152.

Domínguez-Bella, S., 2012. Archaeomineralogy of prehistoric artifacts and gemstones. Seminarios de la Sociedad Española de Mineralogia 12, 5-28.

Domínguez-Bella, S., and Morata-Céspedes, D., 1995. Application of the mineralogical and petrological techniques to archaeometry. Study of the dolmen de Alberite (Villamartín, Cádiz, Spain), materials. Zephyrus XLVIII, 129-142.

Domínguez-Rodrigo, M., 2002. Hunting and scavenging by early humans: the state of the debate. Journal of World Prehistory 16, 1-54.

Domínguez-Rodrigo, M., and Barba, R., 2006. New estimates of tooth mark and percussion mark frequencies at the FLK Zinj site: the carnivore-hominid-carnivore hypothesis falsified. Journal of Human Evolution 50, 170-194.

Dominguez-Rodrigo, M., Rayne Pickering, T., Semaw, S., and Rogers, M.J., 2005. Cutmarked bones from Pliocene archaeological sites at Gona, Afar, Ethiopia: implications for the function of the world's oldest stone tools. Journal of Human Evolution 48, 109-121.

Dominguez-Rodrigo, M., Serrallonga, J., Juan-Tresserras, J., Alcala, L., and Luque, L., 2001. Woodworking activities by early humans: a plant residue analysis on Acheulian stone tools from Peninj (Tanzania). Journal of Human Evolution 40, 289-299.

Donkin, R.A., 1998. Beyond Price: Pearls and Pearl-Fishing: Origins to the Age of Discovery. American Philosophical Society Memoirs 224, 448 pp.

Du Faqing and Gao Wuxun, 1980. The mining of nonferrous metal mineral in ancient China. Nonferrous Metals 32, 93-97.

Dubin, L.S., 1987. The History of Beads: from 30,000 BC to the Present. Harry N. Abrams, Inc., NY, 364 pp.

Dubreuil, L., 2004. Long-term trends in Natufian subsistence: a use-wear analysis of ground stone tools. Journal of Archaeological Science 31, 1613-1629.

Ducrocq, S., Chaimanee, Y., Suteethorn, V., and Jaeger, J.-J., 1995. Dental anomalies in Upper Eocene Anthracotheriidae: a possible case of inbreeding. Lethaia 28, 355-360.

Dudd, S.N., and Evershed, R.P., 1998. Direct demonstration of milk as an element of archaeological economies. Science 282, 1478-1481.

Dunne, J., et al., 2012. First dairying in green Saharan Africa in the fifth millennium BC. Nature 486, 190-194.

Edwards, P.C., 1991. Wadi Hammeh 27: An Early Natufian Site at Pella, Jordan. In The Natufian Culture in the Levant, International Monographs in Prehistory, Archaeological Series 1, 123-148.

Egg, M., and Spindler, K., 2009. Kleidung und Ausrüstung der Kupferzeitlichen Gletschermumie aus den Ötztaler Alpen. Verlag des Römisch-Germanischen Zentralmuseums, Mainz, 262 pp.

Einwögerer, T., 2000. Die jungpaläolithische Station auf dem Wachtberg in Krems, NÖ: eine Rekonstruktion und wissenschaftliche Darlegung der Grabung von J. Bayer aus dem Jahre 1930. Mitteilungen der Prähistorischen Kommission, Band 34, 203 pp.

Einwögerer, T., Friesinger, H., Händel, M., Neugebauer-Maresch, C., Simon, U., and Teschler-Nicola, M., 2006. Upper Palaeolithic infant burial. Nature 444, 285.

Ellis, L., 1984. The Cucuteni-Tripolye Culture. British Archaeological Reports International Series 217, 221 pp.

Ellmers, D., 1984. The earliest evidence for skinboats in Late-Palaeolithic Europe. *In* Aspects of Maritime Archaeology and Ethnography. Greenwich, National Maritime Museum, 41-56.

Endicott, K.L., 1999. Gender relations in hunter-gatherer societies. *In* The Cambridge Encyclopedia of Hunters and Gatherers, 411-418.

Enoch, J.M., 2006. History of mirrors dating back 8000 years. Optometry and Vision Science 83, 775-781.

Esin, U., 1999. Copper objects from the Pre-Pottery Neolithic site of Aşikli (Kizilkaya Village, Province of Aksaray, Turkey). *In* The beginnings of metallurgy, Der Anschnitt, Beiheft 9, 23-30.

Etxeberria, F., and Vegas, J.I., 1992. Heridas por fleche durante la Prehistoria en la Península Ibírica. Munibe Antropología – Arkeologia, Suppl. 8, 129-136.

Evans, A.C., Markus, M.B., Mason, R.J., and Steel, R., 1996. Late stone-age coprolite reveals evidence of prehistoric parasitism. South African Medical Journal 66(3), 274-275.

Evans, Sir A., 1964. The palace of Minos. Biblo and Tannen, 721 pp.

Evershed, R.P., 1997. Fuel for thought? Beeswax in lamps and conical cups from Late Minoan Greece. Antiquity 71, 979-985.

Evershed, R.P., et al., 1991. Epicuticular wax components preserved in potsherds as chemical indicators of leafy vegetables in ancient diets. Antiquity 65, 540-544.

Evershed, R.P., et al., 2008. Earliest date for milk use in the near East and southeastern Europe linked to cattle herding. Nature Letters 455, 528-531.

Ewing, H.E., 1926. A revision of the American lice of the genus *Pediculus,* together with a consideration of the significance of their geographical and host distribution. Proceedings of the United States National Museum 68(1), 1-30.

Fagan, B., 2011. Elixir, a history of water and humankind. Bloomsbury Press, 384 pp.

Fagan, B.M., and Van Noten, F.L., 1966. Wooden implements from Late Stone Age sites at Gwisho Hot-springs, Lochinvar, Zambia. Proceedings of the Prehistoric Society 32, 246-261.

Fagan, B.M., and Van Noten, F.L., 1971. The hunter-gatherers of Gwisho. Musée Royal de l'Afrique Centrale, Tervuren, Annales, Séries IN-8°, Sciences Humaines 74, 228 pp.

Farbstein, R., Radić, D., Brajković, D., and Miracle, P.T., 2012. First Epigravettian ceramic figurines from Europe (Vela Spila, Croatia). PLoS ONE 7, issue 7, 1-14.

Farmer, M.F., 1994. The origins of weapons systems. Current Anthropology 35, 679-681.

Faulkner, C.T., 1991. Prehistoric diet and parasitic infection in Tennessee: evidence from the analysis of desiccated human paleofeces. American Antiquity 56, 687-700.

Féblot-Augustins, J., 1993. Mobility strategies in the Late Middle Palaeolithic of Central Europe and Western Europe: elements of stability and variability. Journal of Anthropological Archaeology 12, 211-265.

Feldesman, M.R., et al., 1990. Femur/Stature ratio and estimates of stature in Mid- and Late-Pleistocene fossil hominids. American Journal of Physical Anthropology 83, 359-372.

Feng Zhao, ed., 1999. Treasures in Silk. ISAT/Costume Squad Ltd., 359 pp.

Fenwick, V., ed., 1978. The Graveney Boat: a Tenth-Century Find from Kent. British Archaeological Reports British Series 53, 48 pp.

Ferioli, P., and Fiandra, E., 1993. Arslantepe locks and the ŠAMAŠ 'Key'. *In* Between the rivers and over the mountains. Gruppo Editoriale Internazionale-Roma, pp. 269-287.

Fernández-Jalvo, Y., and Andrews, P., 2011. When humans chew bones. Journal of Human Evolution 60, 117-123.

Fernández-Jalvo, Y., Diez, J.C., Cáceres, I., and Rosell, J., 1999. Human cannibalism in the Early Pleistocene of Europe (Gran Dolina, Sierra de Atapuerca, Burgos, Spain). Journal of Human Evolution 37, 591-622.

Fernández, H., et al., 2002. Assessing the origin and diffusion of domestic goats using ancient DNA. *In* The First Steps of Animal Domestication. Oxbow Books, 50-54.

Ferreira, L.F., et al., 1987. Encontro de ovos de ancilostomídeos em coprólitos humanos datados de 7230±80 anos, Piauí, Brasil. Anais da Academia Brasileira de Ciências 59, 280-281.

Ferreira, M.T., 2002. A scurvy case in an infant from Monte da Cegonha (Vidigueira – Portugal). Antropologia Portuguesa 19, 57-63.

Figuier, L., 1996 [reprint of edition of 1876]. Une Histoire de l'Asphalte et du Bitume. Trait pour Trait, la Bibliothèque des Métiers, Barbentane, France.

Flad, R., et al., 2005. Archaeological and chemical evidence for early salt production in China. National Academy of Sciences (USA), Proceedings 102, 12618-12622.

Flad, R.K., 2007. Rethinking the context of production through an archaeological study of ancient salt production in the Sichuan Basin, China. Archaeological Papers of the American Anthropological Association 17, 108-128.

Foley, R., 2001. The evolutionary consequences of increased carnivory in hominids. *In* Meat-eating and Human Evolution, Oxford University Press, Ch. 15, 305-331.

Folk, G.E., Jr., and Semken, H.A., Jr., 1991. The evolution of sweat glands. International Journal of Biometeorology 35, 180-186.

Forbes, R.J., 1936. Bitumen and Petroleum in Antiquity. E.J. Brill, Leiden, 109 pp.

Forbes, R.J., 1964. Studies in ancient technology. E.J. Brill, Leiden, 263 pp.

Forbes, R.J., 1970, A short history of the art of distillation. E.J. Brill, Leiden, 405 pp.

Forrer, R., 1932. Les Chars Cultuels Préhistoriques, Préhistoire, E. Leroux, Paris, tome I, fasc. I, 19-123.

Fox, R.B., 1970. The Tabon Caves: archaeological explorations and excavations on Palawan Island, Philippines. Monograph of the National Museum 1, 197 pp.

Frank, S.A., 2002. Immunology and Evolution of Infectious Disease. Princeton University Press, 348 pp.

Frankel, R., 2003. The Olynthus Mill, its origin and diffusion: typology and distribution. American Journal of Archaeology 107, 1-21.

Frayer, D.W., 1997. Ofnet: evidence for a Mesolithic massacre. *In* Troubled Times: Violence and warfare in the past, Gordon and Breach Publishers, 181-216.

Frayer, D.W., Horton, W.A., Macchiarelli, R., and Mussi, M., 1987. Dwarfism in an adolescent from the Italian late Upper Palaeolithic. Nature 330, 60-62.

Fredericksen, C., et al., 1993. Pamwak Rockshelter: a Pleistocene site on Manus Island, Papua New Guinea. *In* Sahul in Review: Pleistocene archaeology in Australia, New Guinea and Island Melanesia. Australian National University, 144-154.

Frierman, J.D., 1971. Lime burning as the precursor of fired ceramics. Israel Exploration Journal 21, 212-216.

Friis-Hansen, J., 1990. Mesolithic cutting arrows: functional analysis of arrows used in the hunting of large game. Antiquity 64, 494-504.

Frost, G.T., 1980. Tool behavior and the origins of laterality. Journal of Human Evolution 9, 447-459.

Fry, D.P., and Söderberg, P., 2013. Lethal aggression in mobile forager bands and implications for the origins of war. Science 341, 270-273.

Fry, G.F., 1985. Analysis of fecal material. In The analysis of Prehistoric Diets, R.I. Gilbert and J.H. Mielke, eds., Academic Press, San Diego, CA, 127-154.

Fuller, D.Q. and Ling Qin, 2010. Declining oaks, increasing artistry, and cultivating rice: the environmental and social context of the emergence of farming in the Lower Yangtze Region. Environmental Archaeology 15, 139-159.

Fuller, D.Q. et al., 2007. Presumed domestication? Evidence for wild rice cultivation and domestication in the fifth millennium BC of the Lower Yangtze region. Antiquity 81, 316-331.

Gabriel, B., 1987. Palaeoecological evidence from neolithic fireplaces in the Sahara. The African Archaeological Review 5, 93-103.

Gabriel. R.A., and Metz, K.S., 1991. From Sumer to Rome. Greenwood Press, 182 pp.

Gagneux, P., and Varki, A., 2001. Genetic differences between humans and great apes. Molecular Phylogenetics and Evolution 18, 2-13.

Galdikas, B.M.F., and Wood, J.W., 1990. Birth spacing patterns in humans and apes. American Journal of Physical Anthropology 83, 185-191.

Gale, N.H., ed., 1991. Bronze Age trade in the Mediterranean, Jonsered, Göteborg, Conference publication, Paul Åströms Förlag, Studies in Mediterranean Archaeology, vol. XC, 398 pp.

Galili, E., Stanley, D.J., Sharvit, J., and Weinstein-Evron, M., 1997. Evidence for earliest olive-oil production in submerged settlements off the Carmel Coast, Israel. Journal of Archaeological Science 24, 1141-1150.

Garbrecht, G., 1980. The water supply system at Tuşpa (Urartu). World Archaeology II, 306-312.

Garbrecht, G., ed., 1987. Historische Talsperren. Verlag Konrad Wittwer, 464 pp.

Garbrecht, G., ed., 1991. Historische Talsperren 2. Verlag Konrad Wittwer 457 pp.

Garfinkel, Y., 1987. Burnt lime products and social implications in the Pre-Pottery Neolithic B villages of the Near East. Paléorient 13, 69-76.

Garfinkel, Y., and Miller, M.A., 2002. Sha'ar Hagolan, vol. 1, Neolithic art in context. Oxbow Books, 262 pp.

Garfinkel, Y., Dag, D., Horwitz, L.K., Lernau, O., and Mienis, H.K., 2002. Ziqim, A Pottery Neolithic site in the Southern Coastal Plain of Israel, a final report. Journal of the Israel Prehistoric Society 32, 73-145.

Garlan, Y., 1982. Les esclaves en Grèce ancienne. François Maspero, 225 pp.

Garrard, A., Colledge, S., and Martin, L., 1996. The emergence of crop cultivation and caprine herding in the 'Marginal zone' of the southern Levant. In The Origins and Spread of Agriculture and Pastoralism in Eurasia. Smithsonian Institution Press, 204-226.

Garrod, D.A.E., 1955. Palaeolithic spear-throwers. Proceedings of the Prehistoric Society 3, 21-35.

Garrod, D.A.E., and Bate, D.M.A., 1937. The Stone Age of Mount Carmel, vol. I. Clarendon Press, 233 pp.

Gasser, A., 2003. World's oldest wheel found in Slovenia. Republic of Slovenia, Government Communications Office of the Republic of Slovenia.

Gat, A., 2006. War in human civilization. Oxford University Press, 822 pp.

Gattiglia, A., and Rossi, M., 1995. Les céramiques de la mine préhistorique de Saint-Véran (Hautes-Alpes) (1). Bulletin de la Société Préhistorique Française, tome 92-94, 509-518.

Geetha, S. et al., 1996. An ethnic method of milk curdling using plants. Ancient Science of life, XVII, 60-61.

Geller, J., 1993. Bread and beer in fourth-millennium Egypt. Food and Foodways 5, 255-267.

Gervers, M., and Gervers, V., 1974. Felt-making craftsmen of the Anatolian and Iranian Plateaux. Textile Museum Journal 4.1, 14-29.

Ghemis, C., et al., 2011. An exceptional archaeological discovery – the 'art gallery' in Coliboaia Cave, Apuseni Mountains, Romania. Acta Archaeologica Carpathica 46, 5-18.

Gibbons, A., 2020. Lead pollution tracks the rise and fall of medieval kings. Science 368, 19-20.

Gibson, K., 1991. Language and intelligence: evolutionary implications. Man, N.S., 26, 255-264.

Gibson, N.E., Wadley, L., and Williamson, B.S., 2004. Microscopic residues as evidence of hafting on backed tools from the 60,000 to 68,000 year-old Howiesons Poort layers of Rose Cottage Cave, South Africa. Southern African Humanities 16, 1-11.

Gignoux, C.R., Henn, B.M., and Mountain, J.L., 2011. Rapid, global demographic expansion after the origins of agriculture. National Academy of Sciences (USA), Proceedings 108, 6044-6049.

Gilad, Y., et al., 2003. Human specific loss of olfactory receptor genes. National Academy of Sciences (USA), Proceedings 100, no. 6, 3324-3327.

Giles, D.L., and Kuijpers, E.P., 1974. Stratiform copper deposit, Northern Anatolia, Turkey: evidence for Early Bronze I (2800 BC) mining activity. Science 186, 823-825.

Giuffra, E., et al., 2000. The origin of the domestic pig: independent domestication and subsequent introgression. Genetics 154, 1785-1791.

Giuliani, G., et al., 2000. Oxygen isotopes and emerald trade routes since antiquity. Science 287, 631-633.

Giumlia-Mair, A., 2005. Copper and copper alloys in the Southeastern Alps: an overview. Archeometry 47, 275-292.

Glover, I.C., 1981. Leang Burung 2: an Upper Palaeolithic rock shelter in South Sulawesi, Indonesia. Modern Quaternary Research in Southeast Asia 6, 1-38.

Glover, I.C., and Higham, C.F.W., 1996. New evidence for early rice cultivation in South, Southeast and East Asia. In The Origins and Spread of Agriculture and Pastoralism in Eurasia, Smithsonian Institution Press 42, 413-441.

Glumac, P.D., and Gringham, R., 1990. Exploitation of the copper minerals at Selevac. In Selevac: a Neolithic village in Yugoslavia. Monumenta archaeologica, vol. 15. Institute of Archaeology. Los Angeles, University of California, 549-565.

Glumac, P.D., and Todd, J.A., 1991. Eneolithic copper smelting slags from the Middle Danube Basin. Archaeometry 90, 155-164.

Golden, J., et al., 2001. Recent discoveries concerning Chalcolithic metallurgy at Shiqmim, Israel. Journal of Archaeological Science 28, 951-963.

Golitko, M., and Keeley, L.H., 2007. Beating ploughshares back into swords: warfare in the Linearbandkeramik. Antiquity 81, 332-342.

Gonçalves, M.L.C., Araujo, A., and Ferreira, L.F., 2003. Human intestinal parasites in the past: new findings and a review. Memoirs of the Oswaldo Cruz Institute 98 (Suppl. 1), 103-118.

Gonçalves, M.L.C., et al., 2004. Amoebiasis distribution in the past: first steps in using an immunoassay technique, Transactions of the Royal Society of Tropical Medicine and Hygiene 98(2), 88-91.

Gopher, A., and Tsuk, T., 1991. Ancient Gold: Rare Finds from the Nahal Qanah Cave. The Israel Museum, Jerusalem, 36 pages in English, 36 pages in Hebrew.

Gopher, A., et al., 1990. Earliest gold artifacts in the Levant. Current Anthropology 31, 436-443.

Gorelick, L., and Gwinnett, A.J., 1987. A history of drills and drilling. New York State Dental Journal, 53, 35-39.

Goren-Inbar, N., 1985. A figurine from the Acheulian site of Berekhat Ram. Mi'tekufat Ha'even 19, 7-12.

Goren-Inbar, N., Alperson, N., Kisler, M.E., Simchoni, O., Melamed, Y., Ben-Nur, A., and Werker, E., 2004. Evidence of hominin control of fire at Gesher Benot Ya'aqol, Israel. Science 304, 725-727.

Goren-Inbar, N., and Peltz, S., 1995. Additional remarks on the Berekhat Ram figurine. Rock Art Research 12, 131-132.

Goren-Inbar, N., Felbel, C.S., Verosub, K.L., Melamed, Y., Kislev, M.E., Tchrnov, E., and Saragusti, I., 2000. Pleistocene milestones on the Out-of-Africa Corridor at Gesher Benot Ya'aqov, Israel. Science 289, 944-947.

Goren-Inbar, N., Lister, A., Werker, E., and Chech, M., 1994. A butchered elephant skull and associated artifacts from the Acheulian site of Gesher Benot Ya'qov, Israel. Paléorient 29(1), 99-112.

Goren-Inbar, N., Sharon, G., Melamed, Y., and Kislev, M., 2002b. Nuts, nut cracking, and pitted stones at Gesher Benot Ya'aqov, Israel. National Academy of Sciences (USA), Proceedings 99, 2455-2460.

Goren-Inbar, N., Werker, E., and Feibel, C.S., 2002. The Acheulian site of Gesher Benot Ya'aqov, Israel: The wood assemblage, Oxbow Books, Oxford, 120 pp.

Goren, Y., Goring-Morris, A.N., and Segal, I., 2001. The technology of skull modeling in the Pre-Pottery Neolithic B (PPNB): Regional variability, the relation of technology and iconography and their archaeological implications. Journal of Archaeological Science 28, 671-690.

Goring-Morris, A.N., and Belfer-Cohen, A., 2008. A roof over one's head: developments in Near Eastern residential architecture across the Epipalaeolithic–Neolithic Transition. In The Neolithic Demographic Transition and Consequences, 239-286.

Gosden, C., and Robertson, N., 1991. Models for Matenkupkum: interpreting a late Pleistocene site from southern New Ireland, Papua New Guinea. Report of the Lapita Homeland Project. Australian National University, Occasional Papers in Prehistory 20, 20-45.

Gosden, C., et al., 1989. Lapita sites of the Bismarck Archipelago. Antiquity 63, 561-586.

Götherström, A., et al., 2005. Cattle domestication in the Near East was followed by hybridization with aurochs bulls in Europe. Proceedings of the Royal Society, B, 272, 2345-2350.

Gowlett, J.A.J., et al., 1981. Early archaeological sites, hominid remains and traces of fire from Chesowanja, Kenya. Nature 294, 125-129.

Graham, C., ed., 1981. Reproductive biology of the great apes. Academic Press, 456 pp.

Gray, H.F., 1940. Sewerage in Ancient and Mediaeval Times. Sewage Works Journal 12, 939-946.

Greenfield, H.J., 1988. The origins of milk and wool production in the Old World. Current Anthropology 29, 573-593.

Greenwood, P., 1968. Fish remains. In The prehistory of Nubia, Southern Methodist University Press, vol. 1, 100-109.

Greig, J.R.A., 1991. The British Isles. In Progress in Old World palaeoethnobotany. Balkema, 299-334.

Griaznov, M., and Boulgakov, A., 1958. L'art ancient de l'Altaï. Musée de l'Ermitage, 95 pp.

Griffin, J.P., 2004. Venetian treacle and the foundation of medicines regulation. British Journal of Clinical Pharmacology 58, 317-325.

Grine, F.E., and Kay, R.F., 1988. Early hominid diets from quantitative image analysis of dental microwear. Nature 333, 765-768.

Grosman, L., Munro, N.D., and Belfer-Cohen, A., 2008. A 12,000-year-old shaman burial from the southern Levant (Israel). National Academy of Sciences (USA), Proceedings, Early edition, 5 pp.

Groube, L., et al., 1986. A 40,000 year-old human occupation site at Huon Peninsula, Papua, New Guinea. Nature 324, 453-455.

Groves, C.P., 1996. Hovering on the brink: nearly but not quite getting to Australia. Perspectives in Human Biology 2, 81-87.

Grünberg, J.M., 2002. Middle Palaeolithic birch-bark pitch. Antiquity 75, 15-16.

Grzimek, B., 1968. Animal Life Encyclopedia, vol. 2, Insecta. Kindler Verlag.

Guhl, F., Jaramillo, C., Yockteng, R., Vallejo, G.A., and Cárdenas-Arroyo, F., 1997. *Trypanosoma cruzi* DNA in human mummies. The Lancet 349, 1370.

Haaland, R., 1992. Fish, pots and grain: Early and Mid-Holocene adaptations in the Central Sudan. The African Archaeological Review 10, 43-64.

Haaland, R., 1999. The puzzle of the late emergence of domesticated sorghum in the Nile Valley. *In* The prehistory of food: appetites for change, Routledge, 397-418.

Haaland, R., 2007. Porridge and Pot, Bread and Oven: Food ways and symbolism in Africa and the Near East from the Neolithic to the present. Cambridge Archaeological Journal 17, 165-182.

Habu, J., 2004. Ancient Jomon of Japan. Cambridge University Press, 32 pp.

Hahn, J., 1972. Aurignacian signs, pendants and art objects in central and Eastern Europe. World Archaeology 3, 252-266.

Hahn, J., 1986. Kraft und Aggression: Die Botschaft der Eiszeitkunst im Aurignacien Süddeutschlands? Verlag Archaeologica Venatoria, Institut für Urgeschichte der Universität Tübingen, 254 pp.

Hahn, J., and Münzel, S., 1995. Knochenflöten aus dem Aurignacien des Geissenklösterle bei Blaubeuren, Alb-Donau-Kreis. Fundberichte aus Baden-Württemberg 20, 1-12.

Haldane, C., 1993. Direct evidence for organic cargoes in the Late Bronze Age. World Archaeology 24, 348-360.

Haley, E.W., 1990. The Fish Sauce Trader L. Iunius Puteolanus. Zeitschrift für Papyrologie und Epigraphik (ZPE) 80, 72-78.

Hallos, J., 2005. '15 Minutes of Fame': exploring the temporal dimension of Middle Pleistocene lithic technology. Journal of Human Evolution 49, 155-179.

Hancock, J.F., 2012. Plant evolution and the origin of crop species, 3rd Ed. CABI International, 245 pp.

Hansen, J.M., 1991. The palaeoethnobotany of Franchthi Cave. Indiana University Press, Fasc. 7, 280 pp.

Harbeck M. et al. 2013. *Yersinia pestis* DNA from Skeletal Remains from the 6th Century AD Reveals Insights into Justinianic Plague. PLoS Pathogens 9(5): e1003349.

Harding, A., 1999. Warfare: a defining characteristic of Bronze Age Europe? *In* Ancient Warfare, Sutton Publishing, 157-173.

Hardy, B.L., 2010. Climate variability and food plant distribution in Pleistocene Europe: implications for Neanderthal diet and subsistence. Quarterly Science Review 29, 662-679.

Hardy, B.L., Kay, M., Marks, A.E., and Monigal, K., 2001. Stone tool function at the Paleolithic sites of Starosele and Buran Kaya III, Crimea: behavioral implications. National Academy of Sciences (USA), Proceedings 6 pp.

Hardy, K., et al., 2012. Neanderthal medics? Evidence for food, cooking, and medicinal plants entrapped in dental calculus. Naturwissenschaften 99, 617-626.

Hardy, K., et al., 2015. The importance of dietary carbohydrate in human evolution. Quarterly Review of Biology 90, 251-268.

Harmand, S., et al., 2015. 3.3-million-year-old stone tools from Lomekwai 3, West Turkana, Kenya. Nature 521, 310-322.

Harrell, J.A., 2004. Archaeological geology of the World's first emerald mine. Geoscience Canada 31, 69-76.

Harrell, J.A., and Bown, T.M., 1994. An Old Kingdom basalt quarry at Widan el-Faras and the Quarry Road to Lake Moeris in the Faiyum. Annual Meeting of the American Research Center in Egypt, Toronto, Canada, Abstracts & Programme, 39-40.

Harrell, J.A., and Osman, A.F., 2007. Ancient amazonite quarries in the Eastern Desert. Egyptian Archaeology 30, 26-28.

Harrell, J.A., and Sidebotham, S.E., 2004. Wadi Abu Diyeiba: an amethyst quarry in Egypt's Eastern Desert. Minerva 15, 12-14.

Harrell, J.A., and Storemyr, P., 2009. Ancient Egyptian quarries – an illustrated overview. Geological Survey of Norway, Special Publication 12, 7-50.

Harris, D.R., 1989. An evolutionary continuum of people-plant interactions. In Foraging and Farming: the evolution of plant exploitation, Unwin Hyman, 11-26.

Harris, D.R., ed., 1996. The origins and Spread of Agriculture and Pastoralism in Eurasia. Smithsonian Institution Press 42, 594 pp.

Harrison, W.R., Merbs, C.F., and Leathers, C.R., 1991. Evidence of coccidioidmycosis in the skeleton of an ancient Arizona Indian. Journal of Infectious diseases 164, 437-438.

Harrold, F.B., 1980. A comparative analysis of Eurasian Palaeolithic burials. World Archaeology 12, 195-211.

Hartmann, A., 1970. Prähistorische Goldfunde aus Europa. Gebr. Mann Verlag, 129 pp.

Hartwell, R., 1967. A cycle of economic change in Imperial China: coal and iron in Northeast China. Journal of the Economic and Social History of the Orient 10, 102-159.

Harvey, P.H., and Clutton-Brock, T.H., 1985. Life history variation in primates. Evolution 39, 559-581.

Hathaway, G.A., 1958. Dams: their effect on some ancient civilizations. Civil Engineering, American Society of Civil Engineers 26, 58-63.

Hauptmann, A., 1987. The earliest periods of copper metallurgy in Feinan, Jordan. In Archäometallurgie der Alten Welt, Beiträge zum internationalen Symposium Old World Archaeometallury, Heidelberg 1987, 119-135.

Hauptmann, A., and Klein, S., 2009. Bronze Age gold in Southern Georgia. ArcheoSciences 33, 75-82.

Hauptmann, A., et al., 1988. Early copper metallurgy in Oman. In The beginning of the use of metals and alloys, R. Maddin, ed., MIT Press, pp. 34-51.

Hauptmann, H., 1997. Nevali Çori. Archaeology in the Near East 4, 131-134.

Hauptmann, H., 2002. Upper Mesopotamia in its regional context during the Early Neolithic. In Neolithic of Central Anatolia: internal developments and external relations during the 9th–6th millennia Cal BC, 263-274.

Hawkes, K., et al., 1997. Hadza women's time allocation, offspring provisioning, and the evolution of long postmenopausal life spans. Current Anthropology 38, 551-577.

Hawkes, K., et al., 1998. Grandmothering, menopause, and the evolution of human life histories. National Academy of Sciences (USA), Proceedings 95, 1336-1339.

Hays, T.R., 1975. Neolithic settlement patterns in Saharan Africa. South African Archaeological Bulletin 30, 29-33.

Helbaek, H., 1963. Textiles from Catal Huyuk. Archaeology 16, 39-46.

Heldal, T., et al., 2005. The geology and archaeology of the ancient silicified sandstone quarries at Gebel Gulab and Gebel Tingar, Aswan (Egypt). Marmora 1, 11-35.

Henderson, J., 2000. The Science and Archaeology of Materials. Routledge, 334 pp.

Henderson, Z., 1992. The context of some Middle Stone Age hearths at Klasies River Shelter 1B: implications for understanding human behaviour. Southern African Field Archaeology 1, 14-26.

Henry, A.G., Brooks, A.S., and Piperno, D.R., 2011. Microfossils in calculus demonstrate consumption of plants and cooked foods in Neanderthal diets (Shanidar III, Iraq; Spy I and II, Belgium). National Academy of Sciences (USA), Proceedings 108, 486-491.

Henry, A.G., et al., 2012. The diet of *Australopithecus sediba*. Nature 487, 90-93.

Henry, A.G., Hudson, H.F., and Piperno, D.R., 2009. Changes in starch grain morphologies from cooking. Journal of Archaeological Science 36, 915-922.

Henshilwood, C., 2007. Fully symbolic *sapiens* behaviour: innovation in the Middle Stone Age at Blombos Cave, South Africa. Rethinking the human revolution, McDonald Institute Monographs, 123-132.

Henshilwood, C., and Marean, C., 2003. The origin of modern human behavior. Current Anthropology 44, 627-651.

Henshilwood, C.S., d'Errico, E., Yates, R., Jacobs, Z., Tribolo, C., et al., 2002. Emergence of modern human behavior: Middle Stone Age engravings from South Africa. Science 295, 1278-1280.

Henshilwood, C.S., d'Errico, F., and Watts, I., 2009. Engraved ochres from the Middle Stone Age levels of Blombos Cave, South Africa. Journal of Human Evolution 57, 27-47.

Henshilwood, C.S., d'Errico, F., Vanhaeren, M., van Niekerk, K., and Jacobs, Z., 2004. Middle Stone Age shell beads from South Africa. Science 304, p. 404.

Henshilwood, C.S., et al., 2001. An early bone tool industry from the Middle Stone Age at Blombos Cave, South Africa: implications for the origins of modern human behavior, symbolism and language. Journal of Human Evolution 41, 631-678.

Henshilwood, C.S., et al., 2011. A 100,000-year-old ochre processing workshop at Blombos Cave. Science 334, 219-222.

Herbaut, F., and Querré, G., 2004. La parure néolithique en variscite dans le sud de l'Armorique. Bulletin de la Société Française 101, 497-520.

Heron, C., Nemcek, N., Bonfield, K.M., Dixon, D., and Ottaway, B.S., 1994. The chemistry of Neolithic beeswax. Naturwissenschaften 81, 266-269.

Herrmann, G., 1968, Lapis lazuli: the early phase of its trade. British Institute for the Study of Iraq 30, 21-57.

Hershkovitz, I., and Gopher, A., 2008. Demographic, biological and cultural aspects of the Neolithic Revolution: A view from the Southern Levant. *In* The Neolithic Demographic Transition and its consequences, Springer Science+Business Media B.V., 441-479.

Hershkovitz, I., and Zohar, I., 1995. Remedy for an 8500 year-old plastered human skull from Kfar Hahoresh, Israel. Journal of Archaeological Science 22, 779-788.

Hershkovitz, I., et al., 1993. Ohalo II Man – unusual findings in the anterior rib cage and shoulder girdle of a 19,000-year-old specimen. International Journal of Osteoarchaeology 3, 177-188.

Hershkovitz, I., et al., 2008. Detection and molecular characterization of 9000-year-old *Mycobacterium tuberculosis* from a Neolithic settlement in the Eastern Mediterranean. PLoS ONE 3, issue 10, 6.

Hershkovitz, I., Wish-Baratz, S., Goren, Y., Goring-Morris, N., Speirs, M.S., Segal, I., Meirav, O., Sherter, U., and Feldman, H., 1995. High-resolution computed tomography and micro-focus radiography on an eight thousand year old plastered skull: how and why it was modeled. *In* Nature et Culture, Colloque de Liège, Liège, Études et Recherches Archéologiques de l'Université de Liège (ERAUL) 68, 667-681.

Hershkowitz, I.H., Kelley, J., Latimer, B., Rothschild, B.M., Simpson, S., Polak, J., and Rosenberg, M., 1997. Oral bacteria in Miocene *Sivapithecus*. Journal of Human Evolution 33, 507-512.

Hessler, P., 2006. Oracle Bones. HarperCollins Publishers, 491 pp.

Higham, C., 1996. The Bronze Age of Southeast Asia. Cambridge University Press, 381 pp.

Higham, T., Basell, L., Jacobi, R., Wood, R., Ramsey, C.B., and Conard, N.J., 2012. Testing models for the beginnings of the Aurignacian and the advent of figurative art and music: The radiocarbon chronology of Geissenklösterle. Journal of Human Evolution 63, 1-13.

Hill, W.C.O., 1946. Note on the male external genitalia of the chimpanzee. Journal of Zoology, Proceedings of the Zoological Society of London 116, 129-132.

Hillman, G., Madeyska, E., and Hather, J., 1989. Wild plant foods and diet at Late Paleolithic Wadi Kubbaniya: The evidence from charred remains. *In* The Prehistory of Wadi Kubbaniya, Southern Methodist University Press, 162-242.

Hillman, G.C., 1989b. Late Palaeolithic plant foods from Wadi Kubbaniya in Upper Egypt: dietary diversity, infant weaning, and seasonality in a riverine environment. *In* Foraging and Farming: the Evolution of Plant Exploitation. Unwin Hyman, 207-239.

Hillman, G.C., Colledge, S.M., and Harris, D.R., 1989a. Plant-food economy during the Epipalaeolithic period at Tell Abu Hureya, Syria: dietary diversity, seasonality, and modes of exploitation. *In* Foraging and Farming: The evolution of plant exploitation. Unwin Hyman, 240-268.

Hillson, S., 1996. Dental Anthropology. Cambridge University Press, 373 pp.

Hoberg, E.P., et al., 2001. Out of Africa: origins of the *Taenia* tapeworms in humans. Proceedings of the Royal Society, London, Biological Sciences, 781-787.

Hodder, I., 2006. Çatalhöyük, the Leopard's Tale. Thames & Hudson, 288 pp.

Hodder, I., 2014. Religion at work in a Neolithic society. Cambridge University Press, 382 pp.

Hodder, I., ed., 2007. Excavating Çatalhöyük, South, North and KOPAL Area reports from the 1995-99 seasons. British Institute at Ankara, BIA Monograph 37, 588 pp.

Hodder, I., ed., 2010. Religion in the emergence of civilization: Çatalhöyük as a case study. Cambridge University Press, 360 pp.

Hodge, A.T., 1992. Roman aqueducts & water supply. Duckworth, 504 pp.

Hoffecker, J.F., Kuz'mina, I.E., Syromyatnikova, E.V., Anikovich, M.V., Sinitsyn, A.A., Popov, V.V., and Holliday, V.T., 2010. Evidence for kill-butchery events of early Upper Paleolithic age at Kostenki, Russia. Journal of Archaeological Science 37, 1073-1089.

Holloway, R.L., 1972. New Australopithecine endocast, SK 1585, from Swartkrans, South Africa. American Journal of Physical Anthropology 37, 173-186.

Holwell, J.Z., 1767. An account of the manner of inoculating for the smallpox in the East Indies. The College of Physicians in London, 151-167.

Holzer, H.F., and Momanzadeh, M., 1971. Ancient copper mines in the Veshnoveh Area, Kuhestan-E- Qom, West-Central Iran. Archaeologia Austriaca XLIX, 1-22.

Hook, D.R., et al., 1991. The early production of copper-alloys in South-East Spain. Archaeometry 90, 65-76.

Hopf, M., 1991. South and Southwest Europe. In Progress in Old World Palaeoethnobotany. Balkema, 241- 277.

Hopkins, K., 1983. Introduction. In Trade in the Ancient Economy. University of California Press, IX- XXV.

Hopper, R.J., 1968. The Laurion Mines: A reconsideration. Annual of the British School at Athens 63, 293-326.

Höppner, B., et al., 2005. Prehistoric copper production in the Inn Valley (Austria), and the earliest copper in Central Europe. Archaeometry 47, 293-315.

Horne, P., 1979. Head lice from an Aleutian mummy. Paleopathology Newsletter 25, 7-8.

Horne, P.D., 2002. First evidence of enterobiasis in ancient Egypt. Journal of Parasitology 88, 1019-1021.

Horwitz, L.K., and Goring-Morris, N., 2004. Animals and ritual during the Levantine PPNB: a case study from the site of Kfar Hahoresh, Israel. Anthropologica 39, 165-178.

Hovers, E., Hani, S., Bar-Yosef, O., and Vandermeersch, B., 2003. An early case of color symbolism. Current Anthropology 44, 491-522.

Hrala, J., Šumberová, R., and Vávra, M., 2000. Velim: a Bronze Age fortified site in Bohemia. Institute of Archaeology, Academy of Sciences of the Czech Republic, 348 pp.

Hublin, J.-J., 2009. The prehistory of compassion. National Academy of Sciences (USA), Proceedings 106, 16, 6429-6430.

Hughes, R.W., ed., 2021. Jade: a gemologist's guide. Lotus Publishing, Bangkok and Boulder, CO.

Hughes, S.S., 1998. Getting to the point: Evolutionary change in prehistoric weaponry. Journal of Archaeological Method and Theory 5, 345-408.

Hundt, H.-J., 1960. Vorgeschichtliche Gewebe aus dem Hallstätter Salzberg. Jahrbuch der Römisch-Germanischen Zentralmuseums Mainz 7, 126-150.

Hundt, H.-J., 1967. Vorgeschichtliche Gewebe aus dem Hallstätter Salzberg. Jahrbuch des Romisch-Germanischen Zentralmuseums Mainz 14, 38-67.

Hundt, H.-J., 1987. Vorgeschichtliche Gewebe aus dem Hallstätter Salzberg. Jahrbuch des Romisch-Germanischen Zentralmuseums Mainz 34, 261-286.

Inizan, M.-L., 1993. At the dawn of trade, cornelian from India to Mesopotamia in the third millennium: the example of Tello. Southeast Asian Archaeology 1991: Proceedings of the eleventh international conference of the association of South Asian archaeologists in Western Europe, pp. 121-134.

Inskip, S.A., 2015. Osteological, biolomecular and geochemical examination of an Early Anglo-Saxon case of Leptromatous leprosy. PLoS ONE, 22 pp.

Jackson, J.S., 1968. Bronze Age copper mining on Mount Gabriel, County Cork, Ireland. Archaeologia Austriaca 43, 92-114.

Jackson, J.S., 1980. Bronze Age copper mining in Counties Cork and Kerry, Ireland. In Scientific Studies in Early Mining and Extractive Metallurgy, 9-30.

Jacobs, Z., Roberts, R.G., Galbraith, R.F., et al., 2008. Ages for the Middle Stone Age of Southern Africa: implications for human behavior and dispersal. Science 322, 733-735.

Jacobsen, T., and Lloyd, S., 1935. Sennacherib's aqueduct at Jerwan. University of Chicago Press, 52 pp.

Jakes, K.A., and Sibley, L.R., 2009. An examination of the phenomenon of textile fabric pseudomorphism. Archaeological Chemistry III, Advances in Chemistry 205, Ch. 20, 403-424.

Jakob-Friesen, K.H., 1956. Eiszeitliche Elefantenjäger in der Lüneberger Heide. Jahrbuch der Römisch-Germanischen Zentralmuseums Mainz 3, 1-22.

James, S.R., 1989. Hominid use of fire in the Lower and Middle Pleistocene. Current Anthropology 30, 1-26.

Janouchevitch, Z.V., and Markevitch, V.I., 1971. Espèces de plantes cultivées des stations primitives au sud-ouest de l'URSS. In VIII Congrès international des sciences préhistoriques et protohistoriques, Belgrade, Les rapports et les communications de la délégation des archéologues de l'URSS, 12 pp.

Jansen, G., 1997. Private toilets at Pompeii: appearance and operation. In Sequence and Space in Pompeii, 121-134.

Jansen, M., 1989. Water supply and sewage disposal at Mohenjo-Daro. World Archaeology 21, 177-192.

Jarrige, J.-F., 1981. Economy and society in the Early Chalcolithic/Bronze Age of Baluchistan: new perspectives from recent excavations at Mehrgarh. In South Asian Archaeology 1979; pp. 93-114.

Jarrige, J.-F., 1988. Le complexe culturel de Mehrgarh (Période VIII) et de Sibri: le 'trésor' de Quetta. In Cités oubliées d'Indus: archéologie du Pakistan, pp. 111-128.

Jensen, H.A., 1991. The Nordic Countries. In Progress in Old World Palaeoethnobotany. Balkema, 335-350.

Jesse, F., 2000. Early Khartoum ceramics in the Wadi Howar (Northwest Sudan). Poznań Archaeological Museum, Studies in African Archaeology 7, 77-87.

Jesus: see De Jesus.

Jochim, M., 2000. The origins of agriculture in south-central Europe. In Europe's First Farmers, 183-196.

Joffe, A.H., 1998. Alcohol and social complexity in ancient Western Asia. Current Anthropology 39, 297-322.

Johns, T., 1990. With bitter herbs they shall eat it. University of Arizona Press, 356 pp.

Jones, A.K.G., 1986. Parasitological investigation on Lindow Man. In Lindow Man: The Body in the Bog, I.M. Steel, J.B. Bourke, and D. Brothwell, eds. Cornell University Press, Ithaca, NY, 136-139.

Jones, G.D.B., 1980. The Roman mines at Riotinto. The Journal of Roman Studies 70, 146-165.

Joordens, J.C.A., et al., 2014. Homo erectus at Trinil on Java used shells for tool production and engraving. Nature Letters, 4 pp.

Jørgensen, L.B., 1992. North European textiles. Aarhus University Press, 285 pp.

Jouy-Avantin, F., Combes, C., de Lumley, H., Miskovsky, J.-C., and Moné, H., 1999. Helminth eggs in animal coprolites from a middle Pleistocene site in Europe. Journal of Parasitology 85, 376-379.

Jovanović, B., 1978. The oldest copper metallurgy in the Balkans. Expedition 21 (Fall), 9-17.

Jovanović, B., 1979. The technology of primary copper mining in South-East Europe. Proceedings of the Prehistoric Society 45, 103-110.

Jovanović, B., 1980. The origins of copper mining in Europe. Scientific American 242, 152-167.

Jovanović, B., 1982. Rudna Glava: Der älteste Kupferbergbau im Zentalbalkan. Das Museum für Bergbau – und Hüttenwesen Besondere Ausgaben Archäologischen Institut Band 17, Beograd, 155 pp.

Jovanović, B., 1988. Early Metallurgy in Yugoslavia. *In* The beginnings of the use of metals and alloys, MIT Press, 69-79.

Jovanović, B., 1995. Continuity of the prehistoric mining in the Central Balkans. *In* Ancient mining and metallurgy in Southeast Europe, 29-35.

Jovanović, B., and Ottaway, B.S., 1976. Copper mining and metallurgy in the Vinča group. Antiquity 50, 104-113.

Kajale, M.D., 1989. Mesolithic exploitation of wild plants in Sri Lanka: archaeobotanical study at the cave site of Beli-Lena. *In* Foraging and Farming: the evolution of plant exploitation, 269-281.

Kaptan, E., 1995. Tin and ancient mining in Turkey. Anatolica 21, 197-203.

Karageorghis, V., 1976. A twelfth-century BC opium pipe from Kition. Antiquity 50, 125-129.

Kardara, C., 1961. Dyeing and weaving works at Isthmia. American Journal of Archaeology 65, 261-266.

Karkanas, P., et al., 2004. The earliest evidence for clay hearths: Aurignacian features in Klisoura Cave I, southern Greece. Antiquity 78, 513-525.

Karmon, N., and Spanier, E., 1987. Archaeological evidence of the purple dye industry from Israel. *In* The royal purple and the biblical blue: Argaman and Tekhelet, 147-158.

Katz, S.J., and Voigt, M.M., 1986. Bread and Beer. Expedition 28(2), 23-34.

Kay, R.F., et al., 1998. The hypoglossal canal and the origin of human vocal behavior. National Academy of Sciences (USA), Proceedings 95, 5417-5419.

Keeley, L., 1996. War Before Civilization: The Myth of the Peaceful Savage. Oxford University Press, 245 pp.

Keeley, L.H., 1977. The functions of Paleolithic flint tools. Scientific American 237, 108-126.

Keeley, L.H., 1997. Frontier warfare in the Early Neolithic. *In* Troubled Times, Gordon and Breach, 303-319.

Keeley, L.H., and Toth, N., 1981. Microwear polishes on early stone tools from Koobi Fora, Kenya. Nature 293, 464-465.

Keeley, L.H., Fontana, M., and Quick, R., 2007. Baffles and bastions: The universal features of fortifications. Journal of Archaeological Research 15, 55-95.

Keightley, D.N., 2005. Marks and labels: early writing in Neolithic and Shang China. *In* Archaeology of Asia, Blackwell, 177-201.

Keith, K., 1998. Spindle whorls, gender, and ethnicity at Late Chalcolithic Hacinebi Tepe. Journal of Field Archaeology 23, 497-515.

Kelany, A., et al., 2009. Granite quarry survey in the Aswan region, Egypt: shedding new light on ancient quarrying. Geological Survey of Norway, Special Publication 12, 87-98.

Keller, A., et al., 2012. New insights into the Tyrolean Iceman's origin and phenotype as inferred by whole-genome sequencing. Nature Communications 3, 698.

Kempers, A.J.B., 1988. The Kettledrums of Southeast Asia. Balkema, 599 pp.

Kennedy, G.E., 2005. From the ape's dilemma to the weanling's dilemma: early weaning and its evolutionary context. Journal of Human Evolution 48, 123-145.

Kenoyer, J.M., 1997. Trade and technology of the Indus Valley: new insights from Harappa, Pakistan. World Archaeology 29, 262-280.

Kenoyer, J.M., and Vidale, M., 1992. A new look at stone drills of the Indus Valley tradition. Materials Issues in Art and Archaeology III, 267, 495-518.

Kenward, H., 1999. Pubic lice (*Pthirus pubis* L.) were present in Roman and Medieval Britain. Antiquity 73, 911-913.

Kenyon, K.M., 1981. Excavations at Jericho. Vol. 3. British School of Archaeology in Jerusalem, 540 pp.

Khairat, R., et. al., 2013. First insights into the metagenome of Egyptian mummies using next-generation sequencing. Journal of Applied Genetics 54(3), 309-325.

Kienlin, T.L., 2012. Working copper in the Chalcolithic: a long-term perspective on the development of metallurgical knowledge in central Europe and the Carpathian basin. *In* Is there a British Chalcolithic?, M.J. Allen et al. eds. Prehistoric Society, 126-143.

Kimbel, W.H., et al., 1996. Late Pliocene *Homo* and Oldowan tools from the Hadar Formation (Kada Hadar Member), Ethiopia. Journal of Human Evolution 31, 549-561.

Kingery, D., et al., 1988. The beginnings of pyrotechnology Part II: production and use of lime and gypsum plaster in the Pre-Pottery Neolithic Near East. Journal of Field Archaeology 15, 219-244.

Kislev, M., Simchoni, O., and Weiss, E., 2002. Reconstruction of the landscape, human economy, and hut use according to seeds and fruit remains from Ohalo II. Hecht Museum, Haifa Issue, 20-23.

Kislev, M.E., and Bar-Yosef, O., 1988. The legumes: the earliest domesticated plants in the Near East? Current Anthropology 29, 175-179.

Kislev, M.E., Hartmann, A., and Bar-Yosef, O., 2006. Early domesticated fig in the Jordan Valley. Science 312, 1372-1374.

Kislev, M.E., Nadel, D., and Carmi, I., 1992. Epipalaeolithic (19,000 BP) cereal and fruit diet at Ohalo II, Sea of Galilee, Israel. Review of Palaeobotany and Palynology 73, 161-166.

Kittler, R., Kayser, M., and Stoneking, M., 2003. Molecular evolution of *Pediculus humanus* and the origin of clothing. Current Biology 13, 1414-1417.

Klees, H., 1998. Sklavenleben im Klassichen Griechenland. Franz Steiner Verlag, Stuttgart, 513 pp.

Klein, R.G., and Cruz-Uribe, K., 1995. Exploitation of large bovids and seals at Middle and later Stone Age sites in South Africa. Journal of Human Evolution 31, 315-334.

Klein, R.G., et al., 2004. The Ysterfontein 1 Middle Stone Age site, South Africa, and early human exploitation of coastal resources. National Academy of Sciences (USA), Proceedings 101, 5708-5715.

Klemm, D., Klemm, R., and Murr, A., 2001. Gold of the Pharaohs – 6000 years of gold mining in Egypt and Nubia. Journal of African Earth Sciences 33(3-4), 643-659.

Klemm, R., and Klemm, D.D., 2008. Stones and Quarries in Ancient Egypt. The British Museum Press, 354 pp.

Kliks, M.M., 1990. Helminths as heirlooms and souvenirs: a review of New World paleoparasitology. Parasitology Today 6, 93-100.

Klima, B., 1956. Coal in the Ice Age: The excavation of a Palaeolithic settlement at Ostrava-Petřkovice in Silesia. Antiquity 30, 98-101.

Klima, B., 1988. A triple burial from the Upper Paleolithic of Dolni Věstonice, Czechoslovakia. Journal of Human Evolution 16, 831-835.

Knapp, A.B., 1991. Spice, drugs, grain and grog: organic goods in Bronze Age East Mediterranean Trade: In Bronze Age trade in the Mediterranean, Conference publication, Paul Åströms Förlag, Jonsered, Göteborg, Studies in Mediterranean Archaeology, vol. XC, 21-68.

Knecht, H., 1993. Early Upper Paleolithic approaches to bone and antler projectile technology. In Hunting and Animal Exploitation in the later Palaeolithic and Mesolithic of Eurasia. Archaeological Papers of the American Anthropological Association 3, 33-47.

Knecht, H., 1993b. Splits and wedges: the techniques and technology of Early Aurignacian antler working. In Before Lascaux, the complex record of the Early Upper Paleolithic, CRC, 137-162.

Knörzer, K.-H., 1991. Deutschland nördlich der Donau. In Progress in Old World palaeoethnobotany, Balkema, 189-206.

Kolen, J., 1999. Hominids without homes: on the nature of Middle Palaeolithic settlement in Europe. In The Middle Palaeolithic occupation of Europe, University of Leiden, 139-175.

Koloski-Ostrow, A.O., 2015. The Archaeology of Sanitation in Roman Italy: Toilets, Sewers, and Water Systems. University of North Carolina Press, 312 pp.

Koryakova, L., and Epimakhov, A.V., 2007. The Urals and Western Siberia in the Bronze and Iron Ages. Cambridge University Press, 383 pp.

Kosambi, D.D., 1963. The beginning of the Iron Age in India. Journal of the Economic and Social History of the Orient 6, 309-318.

Koukouli-Chrysanthaki, C., et al., 1988. Prähistorischer und junger Bergbau auf Eisenpigmente auf Thasos. In Antike Edel- und Buntmetallgewinnung auf Thasos, 241-244.

Kramer, S.N., 1963. The Sumerians. University of Chicago Press, 355 pp.

Kraybill, N., 2011. Pre-agricultural tools for the preparation of foods in the Old World. In Origins of Agriculture, Mouton, 485-521.

Kricun, M., et al., 1999. The Krapina Hominids: A radiographic Atlas of the Skeletal Collection. Croatian Natural History Museum.

Krikorian, A.D., 1975. Were the opium poppy and opium known in the Ancient Near East? Journal of the History of Biology 8, 95-114.

Kristiansen, K., 1999. The emergence of warrior aristocracies in later European prehistory and their long-term history. In Ancient Warfare, Sutton Publishing, 175-189.

Kritikos, P.G., and Papadaki, S.P., 1967. The history of the poppy and of opium and their expansion in antiquity in the eastern Mediterranean area. In UNODC, U.N. Office on Drugs and Crime, 28 pp.

Kroll, H., 1991. Südosteuropa. In Progress in Old World palaeoethnobotany, Balkema, 161-177.

Kuhn, S.L., and Stiner, M.C., 2007. Body ornamentation as information technology: Towards an understanding of the significance of early beads. Rethinking the human revolution. McDonald Institute Monographs, Cambridge, 45-54.

Kuhn, S.L., Stiner, M.C., Reese, D.S., and Güleç, E., 2001. Ornaments of the earliest Upper Paleolithic: new insights from the Levant. National Academy of Sciences (USA), Proceedings 98, 7641-7646.

Kuijt, I., 1994. Pre-Pottery Neolithic: A settlement variability: Evidence for sociopolitical developments in the Southern Levant. Journal of Mediterranean Archaeology 7, 165-192.

Kuijt, I., 1996. Negociating equality through ritual: A consideration of Late Natufian and Prepottery Neolithic: A period mortuary practices. Journal of Archaeological Anthropology 15, 313-336.

Kuijt, I., 2008. Demographic and storage systems during the southern Levantine Neolithic Demographic Transition. In The Neolithic Demographic Transition and its consequences, Springer Science+Business Media B.V., 287-313.

Kuijt, I., ed., 2000. Life in Neolithic Farming Communities, Social Organization, Identity, and Differentiation. Kluwer Academic/Plenum Publishers, 325 pp.

Kumar, G., et al., 1988. Engraved ostrich eggshell objects: new evidence of Upper Palaeolithic art in India. Rock Art Research 5, 43-53.

Kunter, M., 1970. Die Schädeltrepanation in vor- und frühgeschichtlicher Zeit und bei aussereuropäischen Völkern. Bericht der Oberhessischen Gesellschaft für Natur- und Heilkunde zu Giessen, Neue Folge, Naturwissenschaftliche Abteilung, Band 37, 149-159.

Küster, H., 1991. Mitteleuropa südlich der Donau, einschliesslich Alpenraum. *In* Progress in Old World palaeoethnobotany, 179-187.

Kuzmin, Y.V., 2006. Chronology of the earliest pottery in East Asia: progress and pitfalls. Antiquity 80, 362-371.

Kuzmina, E.E., 2001. The first migration wave of Indo-Iranians to the South. The Journal of Indo- European Studies 29, 1-40.

Labriola, L., 2008. First impressions: A preliminary account of mat-impressed pottery in the Prehistoric Aegean. Proceedings of the International Symposium, The Aegean in the Neolithic, Chalcolithic and the Early Bronze Age, Research Center for Maritime Archaeology, Ankara University, Publication no. 1, 309-322.

Laden, G., and Wrangham, R., 2005. The rise of the hominids as an adaptive shift in fallback foods: plant underground storage organs (USOs) and australopith origins. Journal of Human Evolution 49, 482-498.

Lal, C.S.L., et al., 2001. A forkhead-domain gene is mutated in a severe speech and language disorder. Nature 413, 519-523.

Lanting, J.N., and Brindley, A.L., 1996. Irish logboats and their European context. The Journal of Irish Archaeology VII, 85-95.

Larsen, C.S., 1995. Biological changes in human populations with agriculture. Annual Review of Anthropology 24, 185-213.

Larsen, C.S., 1997. Bioarchaeology: interpreting behavior from the human skeleton. Cambridge University Press, 461 pp.

Larsen, M.T., 2015. Ancient Kanesh: a merchant colony in Bronze Age Anatolia. Cambridge University Press, 330 pp.

Larson, G., et al., 2005. Worldwide phylogeography of wild boar reveals multiple centers of pig domestication. Science 307 1618-1621.

Larson, S.G., 2007. Evolutionary transformation of the hominin shoulder. Evolutionary Anthropology 16, 172-187.

Larson, S.G., et al., 2007b. *Homo floresiensis* and the evolution of the hominin shoulder. Journal of Human Evolution 53, 718-731.

Lau, B., et al., 1997. Dating a flautist? Using ESR (electron spin resonance) in the Mousterian cave deposits at Divje Babe I, Slovenia. Geoarchaeology 12, 507-536.

Lawler, A., 2002. Report of oldest boat hints at early trade routes. Science 296, 1791-1792.

Le Bailly, M., and Bouchet, F., 2006. Paléoparasitologie et immunologie. L'exemple d'*Entamoeba histolytica*, ArcheoSciences 30, 129-135.

Le Bailly, M., and Bouchet, F., 2013. Diphyllobothrium in the past: Review and new records, International Journal of Paleopathology 3(3), 182-187.

Le Bailly, M., and Bouchet, F., 2015. A First Attempt to Retrace the History of Dysentery Caused by *Entamoeba histolytica*. *In* P.D. Mitchell, ed., Sanitation, Latrines and Intestinal Parasites in Past Populations, Ashgate, 219-228.

Le Bailly, M., Leuzinger, U., and Bouchet, F., 2003. Dioctophymidae Eggs in Coprolites From Neolithic Site of Arbon–Bleiche 3 (Switzerland). Journal of Parasitology 89(5), 1073-1076.

Le Bailly, M., Leuzinger, U., Schlichterfe, H., and Bouchet, F., 2005. *Diphyllobothrium*: Neolithic parasite? Journal of Parasitology 91, 957-959.

Le Bailly, M., Mouze, S., Da Rocha, G., Heim, J., Lichtenberg, R., Dimand, F., and Bouchet, F., 2010. Identification of Taenia sp. in a Mummy from a Christian Necropolis in El-Deir, Oasis of Kharga, Ancient Egypt. Journal of Parasitology 96(1), 213-215.

Leakey, M.D., 1981. Discoveries at Laetoli in northern Tanzania. Proceedings of the Geologists' Association 92, 81-86.

Leakey, M.D., 1987. Animal prints and trails. *In* M.D. Leakey and J.M. Harris, eds., Laetoli, a Pliocene Site in Northern Tanzania, Oxford Univerity Press 451-489.

Leavesley, M.G., et al., 2002. Buang Merabak: early evidence for human occupation in the Bismarck Archipelago, Papua New Guinea. Australian Archaeology 54(1), 55-57.

Lechtman, H., 1980. The Central Andes: metallurgy without iron. *In* The coming of the age of iron, Yale University Press, 267-334.

Legge, A.J., 1981. The Agricultural Economy. *In* The Grimes Graves, Norfolk: Excavations 1971-72, The Agricultural Economy, vol. 1, Department of the Environment Archaeological Reports, 79-103.

Legge, A.J., and Rowley-Conway, P.A., 1987. Gazelle killing in Stone Age Syria. Scientific American 257, 88-95.

Legge, A.J., and Rowley-Conway, P.A., 2000. The Exploitation of animals. *In* Village on the Euphrates: from foraging to farming at Abu Hureyra. Oxford University Press, 424-471.

Legge, T., 1996. The beginning of caprine domestication in Southwest Asia. *In* The Origins and Spread of Agriculture and Pastoralism in Eurasia. Smithsonian Institution Press, 42, 238-262.

Leidal, R., et al., 1982. Evolution and development of human axillary sweat glands. Program of the Forty-Third Annual Meeting of the Society for Investigative Dermatology, Inc., The Journal of Investigative Dermatology 78, p. 352.

Leisner, V., and Schubart, H., 1966. Die Kupferzeitliche Befestigung von Pedra do Ouro/Portugal. Madrider Mitteilungen 7, 9-60.

Lejju, B.J., et al., 2006. Africa's earliest bananas? Journal of Archaeological Science 33, 102-113.

Lemorini, C., et al., 2014. Old stones' song: use-wear experiments and analysis of the Oldowan quartz and quartzite assemblage from Kanjera South (Kenya). Journal of Human Evolution 72, 10-25.

Leonard, W.R., and Robertson, M.L., 1997. Comparative primate energetics and hominid evolution. American Journal of Physical Anthropology 102, 265-281.

Leonard, W.R., Robertson, M.L., and Snodgrass, J.J., 2007b. Energetic models of human nutritional evolution. *In* Evolution of the Human Diet, Oxford University Press, 344-359.

Leonard, W.R., Snodgrass, J.J., and Robertson, M.L., 2007. Effects of brain evolution on human nutrition and metabolism. Annual Review of Nutrition 27, 311-327.

Lepre, C.J., Roche, H., Kent, D.V., Harmand, S., Quinn, R.L., Brugal, J.-P., Texter, P.-J., Lenoble, A., and Feibel, C.S., 2011. An earlier origin for the Acheulian. Nature 477, 82-85.

Leroi-Gourhan, A., 1967. Préhistoire de l'Art Occidental. Éditions d'Art Lucien Mazenod, 482 pp.

Leroi-Gourhan, A., 1968. The art of prehistoric man in Western Europe. Thames & Hudson, London, 543 pp.

Leroi-Gourhan, A., 1975. The flowers found with Shanidar IV, a Neandertal burial in Iraq. Science 190, 562-564.

Leutenegger, W., 1974. Functional aspects of pelvic morphology in simian primates. Journal of Human Evolution 3, 207-222.

Lev, E., Kislev, M.E., and Bar-Yosef, O., 2005. Mousterian vegetal food in Kebara Cave, Mt. Carmel. Journal of Archaeological Science 32, 475-484.

Levine, M., 1999. The origins of horse husbandry on the Eurasian Steppe. *In* Levine, M., Rassamakin, Y., Kislenko, A., and Tatarintseva, N., 1999. Late prehistoric exploitation of the Eurasian steppe. McDonald Institute Monographs, McDonald Institute for Archaeological Research, Cambridge, 5-58.

Lewis, P.R., and Jones, G.D.B., 1970. Gold-mining in North-West Spain. Journal of Roman Studies 60, 169-185.

Lewis-Williams, J.D., 2002. A cosmos in stone. Altamira Press, 307 pp.

Lewis-Williams, J.D., and Pearce, D., 2005. Inside the Neolithic Mind. Thames & Hudson, 320 pp.

Leyden, J.J., et al., 1981. The microbiology of the human axilla and its relationship to axillary odor. The Journal of Investigative Dermatology 77, 413-416.

Lhôte, J.-M., 1994. Histoire des jeux de société. Flammarion, 672 pp.

Li, Keke, and Xu, Zhifang, 2006. Overview of Dujiangyan irrigation scheme of ancient China with current theory. Irrigation and Drainage 55, 291-298.

Li, Shuicheng, and von Falkenhausen, L., eds., 2006. Salt Archaeology in China, Ancient Salt Production and Landscape Archaeology in the Upper Yangzi Basin: Preliminary Studies, vol. 1, www.sciencep.com, 368 pp.

Li, Shuicheng, and von Falkenhausen, L., eds., 2010. Salt Archaeology in China: Global Comparative Perspectives, vol. 2, www.sciencep.com, 475 pp.

Li, Shuicheng, and von Falkenhausen, L., eds., 2013. Salt Archaeology in China: Ancient Salt Production and Landscape Archaeology in the Upper Yangzi basin: the Site of Zhongba in Perspective, vol. 3, www.sciencep.com, 486 pp.

Lieberman, D.E., 1993. The rise and fall of seasonal mobility among hunter-gatherers. Current Anthropology 34, 599-631.

Lieberman, D.E., 2011. The evolution of the human head. Harvard University Press, 768 pp.

Lieberman, D.E., 2013. The story of the human body. Pantheon Books, New York, 460 pp.

Lieberman, D.E., and Bar-Yosef, O., 2005. Apples and oranges: morphological versus behavioral transitions in the Pleistocene. *In* Interpreting the past: essays on human primate, and mammal evolution in honor of David Pilbeam, Ch. 17, 275-296.

Lieberman, D.E., and McCarthy, R.C., 1999. The ontogeny of cranial base angulation in humans and chimpanzees and its implications for reconstructing pharyngeal dimensions. Journal of Human Evolution 36, 487-517.

Lieberman, D.E., et al., 2004. Effects of food processing on masticatory strain and craniofacial growth in a retrognathic face. Journal of Human Evolution 46, 655-677.

Lieberman, D.E., et al., 2007. The evolution of endurance running and the tyranny of ethnography: A reply to Pickering and Bunn (2007). Journal of Human Evolution 53, 439-442.

Lieberman, P., 1968. Primate vocalizations and human linguistic ability. Journal of the Acoustical Society of America 44, 1574-1584.

Lieberman, P., 1991. Uniquely Human, The evolution of speech, thought, and selfless behavior. Harvard University Press, 210 pp.

Lieberman, P., 2006. Toward an evolutionary biology of language. Harvard University Press, 427 pp.

Lieberman, P., 2009. FOXP2 and human cognition. Cell 137, 800-802.

Lieberman, P., 2012. Vocal tract anatomy and the neural bases of talking. Journal of Phonetics 40, 608-622.

Lieberman, P., 2013. The Unpredictable species. Princeton University Press, 255 pp.

Lieberman, P., 2013b. Synapses, language, and being human. Science 342, 944-945.

Lieberman, P., and Crelin, E.S., 1971. On the speech of Neanderthal Man. Linguistic Inquiry 2, 203-222.

Lieberman, P., and McCarthy, R.C., 2007. Tracking the evolution of language and speech. Expedition 49(2), 15-20.

Lien, Chao-mei, 1991. The Neolithic archaeology of Taiwan and the Peinan excavations. Bulletin of the Indo-Pacific Prehistory Association 11, 339-352.

Lillie, M.C., 1998. Cranial surgery dates back to Mesolithic. Nature 391, 854.

Lima, V.S., et al., 2008. Chagas Disease in Ancient Hunter-Gatherer Population, Brazil. Emerging Infectious Diseases 14(6), 1001–1002.

Linz, B., et al., 2007. An African origin for the intimate association between humans and *Helicobacter pylori.* Nature 445, 915-918.

Littman, R.J., 2009. The Plague of Athens: Epidemiology and Paleopathology. Mount Sinai Journal of Medicine: A Journal of Translational and Personalized Medicine 76(5), 456-467.

Liu Shizong et al., 1993. Antiker Kupferbergbau von Tongling bei Ruichang (Provinz Jiangxi). Der Anschnitt 45, 50-62.

Liverani, M., 2006. Uruk the first city. Equinox, 97 pp.

Lloyd, A.B., 2014. Ancient Egypt. Oxford University Press, 363 pp.

Lockwood, C.A., et al., 1996. Randomization procedures and sexual dimorphism in *Australopithecus afarensis.* Journal of Human Evolution 31, 537-548.

Loebl, W.Y., 1995. A case of Symmers' Fibrosis of the liver during the 18th Dynasty? *In* The Archaeology of death in the ancient Near East, Oxbow Monograph 51, 185-187.

Lombard, M., 2004. Distribution patterns of organic residues on Middle Stone Age points from Sibudu Cave, KwaZulu-Natal, South Africa South African Archaeological Bulletin 59, 37-44.

Lombard, M., 2005. Evidence of hunting and hafting during the Middle Stone Age at Sibudu Cave, KwaZulu–Natal, South Africa: a multianalytical approach. Journal of Human Evolution 48, 279-300.

Lombard, M., 2006. Direct evidence for the use of ochre in the hafting technology of Middle Stone Age tools from Sibudu Cave. Southern African Humanities 18, 57-67.

Lombard, M., 2008. Finding resolution for the Howiesons Poort through the microscope: micro-residue analysis of segments from Sibudu Cave, South Africa. Journal of Archaeological Science 35, 26-41.

Lombard, M., 2011. Quartz-tipped arrows older than 60 ka: further use-trace evidence from Sibudu, KwaZulu–Natal, South Africa. Journal of Archaeological Science 38, 1918-1930.

Lonsdorf, E.V., et al., 2014. Sex differences in wild chimpanzee behavior emerge during infancy. PLoS ONE, 9(6): e99099. 8 pp.

Lordkipanidze, D., et al., 2013. A complete skull from Dmanisi, Georgia, and the evolutionary biology of Early *Homo.* Science 342, 326-331.

Lordkipanidze, D., Vekua, A., Ferring, R., Rightmire, G.P., Zollikofer, P.F., Ponce de Leon, M.S., Agusti, J., Kiladze, G., Mouskhelishvili, A., Nioradze, M., and Tappen, M., 2006. A fourth hominin skull from Dmanisi, Georgia. The Anatomical Record, Part A, 288A, 1146-1157.

Louwe Kooijmans, L.P., 1993. An Early/Middle Bronze Age multiple burial at Wassenaar, the Netherlands. Analecta Praehistorica Leidensia, vol. II, 1-20.

Lovejoy et al., 2009b. The pelvis and femur of *Ardipithecus ramidus*: the emergence of upright walking. Science 326, 71e1-71e6.

Lovejoy, C.O., 1988. Evolution of human walking. Scientific American, November, 118-125.

Lovejoy, C.O., 2005. The natural history of human gait and posture, Part 1. Spine and Pelvis. Gait and Posture 21, 95-112.

Lovejoy, C.O., et al., 2009. Combining prehension and propulsion: the foot of *Ardipithecus ramidus*. Science 326, 72e1-72e8.

Lu, Houyuan, et al., 2005. Millet noodles in Late Neolithic China. Nature Brief Communications 437, 967-968.

Lu, P.J., et al., 2005. The earliest use of corundum and diamond in prehistoric China. Archaeometry 47, 1-12.

Lu, Tracey Lie Dan, 1999. The transition from foraging to farming and the origin of agriculture in China. British Archaeological Reports, International Series 774, 233 pp.

Lucas, A., 1962. Ancient Egyptian Materials and Industries. Fourth edition revised and enlarged by J.R. Harris, Edward Arnold Ltd., London, 523 pp.

Lucas, P.W., Corlett, R.T., and Luke, D.A., 1985. Plio-Pleistocene hominid diets: an approach combining masticatory and ecological analysis. Journal of Human Evolution 14, 187-202.

Lucas, P.W., et al., 2013. Mechanisms and causes of wear in tooth enamel: implications for hominin diets. Journal of the Royal Society, Interface 10, 9 pp.

Ludwig, B.V., and Harris, J.W.K., 1998. Towards a technological reassessment of East African Plio-Pleistocene lithic assemblages. *In* Early human behavior in global context: the rise and diversity of the Lower Paleolithic Period, Routledge, 84-107.

Lycett, S.J., and von Cramon-Taubadel, N., 2008. Acheulian variability and hominin dispersals: a model-bound approach. Journal of Archaeological Science 35, 553-562.

Lynnerup, N., and Boldsen, J., 2011. Leprosy (Hansen's disease). *In* A Companion to Paleopathology, A.L. Grauer, ed., Blackwell, Ch. 25, 458-471.

Maat, G.J.R., and Baig, M.S., 1990. Microscopy electron scanning of fossilized sickle-cells. International Journal of Anthropology 5, 271–275.

MacHugh, D.E., et al., 1997. Microsatellite DNA variation and the evolution, domestication and phylogeography of Taurine and Zebu cattle (*Bos Taurus* and *Bos indicus*). Genetics 146, 1071-1086.

MacKay, E., 1937. Bead making in ancient Sind. Journal of the American Oriental Society 57, 1-15.

MacKay, E.J.H., 1938. Further excavations at Mohenjo-Daro. Government of India Press, vol. I (text), 718 pp.; vol. II (plates).

MacLarnon, A.M., and Hewitt, G.P., 1999. The evolution of human speech: the role of enhanced breathing control. American Journal of Physical Anthropology 109, 341-363.

MacRae, R.J., 1988. Belt, shoulder-bag or basket? An enquiry into handaxe transport and flint sources. Lithics 9, 2-8.

Maddin, R., et al., 1991. Çayönü Tepesi: The earliest archaeological metal artifacts. *In* Découverte du metal. Éditions Picard, Paris, 375-386.

Maeir, A.M., and Garfinkel, Y., 1992. Bone and metal straw-tip beer-strainers from the ancient Near East. Levant, XXIV, 218-223.

Maggi, R., and Pearce, M., 2005. Mid fourth-millennium copper mining in Liguria, northwest Italy: the earliest known copper mines in Western Europe. Antiquity 79, 66-77.

Maggi, R., et al., 1996. The quarrying and workshop site of Valle Lagorara (Liguria – Italy). Accordia Research Papers, The Journal of the Accordia Research Centre 5, 73-96.

Maixner, F., et al., 2016. The 5300-year-old *Helicobacter pyloris* genome of the Iceman. Science 351, 162-165.

Malinowski, R., 1982. Ancient mortars and concretes: aspects of their durability. *In* History of Technology 7, 89-101.

Mania, D., and Mania, U., 1988. Deliberate engravings on bone artifacts of *Homo erectus.* Rock Art Research 5, 91-107.

Mania, D., and Vlćek, E., 1987. *Homo erectus* from Bilzingsleben (GDR) – his culture and his environment. Anthropologie 25(1), 1-45.

Mania, D., et al., 1990. Neumark-Gröbern. Beiträge zur Jagd des mittelpaläolithischen Menschen. Deutscher Verlag der Wissenschaften, Berlin, 319 pp.

Marchal, F., 2000. A new morphometric analysis of the hominid pelvic bone. Journal of Human Evolution 38, 347-365.

Marchant J., and Mott, J., 2016. The awakening. Smithsonian 46(9), 80-95.

Marcuse, S.A., 1975. A Survey of Musical Instruments. Harper & Rowe, New York, 555 pp.

Marean, C.W., et al., 2007. Early human use of marine resources and pigment in South Africa during the Middle Pleistocene. Nature 449, 905-909.

Maricic, T., et al., 2012. A Recent evolutionary change affects a regulatory element in the human FOXP2 gene. Molecular Biology and Evolution 30, 844-852.

Marshack, A., 1972. The Roots of Civilization. McGraw Hill, NY, 413 pp.

Marshack, A., 1979. Upper Paleolithic symbol systems of the Russian Plain: cognitive and comparative analysis. Current Anthropology 20, 271-311.

Marshack, A., 1981. Paleolithic ochre and the early use of color and symbol. Current Anthropology 22, 188-191.

Marshack, A., 1985. Hierarchical evolution of the human capacity: the Paleolithic evidence. Fifty- Fourth James Arthur Lecture on the Evolution of the Human Brain, American Museum of Natural History, 52 pp.

Marshack, A., 1989. Evolution of human capacity: the symbolic evidence. Yearbook of Physical Anthropology 32, 1-34.

Marshack, A., 1991. The Tai plaque and calendrical notation in the Upper Paleolithic. Cambridge Archaeological Journal 1, 25-61.

Marshall, F., 1986. Implications of bone modification in a Neolithic faunal assemblage for the study of early hominid butchery and subsistence practices. Journal of Human Evolution 15, 661-672.

Marshall, Sir John, 1931. Mohenjo-Daro and the Indus Civilization. Arthur Probsthain, vol. I, 364 pp.; vol. II, 365-716; vol. III, plates.

Martín-Gil, J., et al., 1994. Neolítico Uso del cinabrio. Investigación y Ciencia (December) 29-30.

Martin, D.L., and Frayer, D.W., eds., 1997. Troubled times: Violence and Warfare in the Past, Gordon and Breach Publishers, 378 pp.

Martinez, I., et al., 2004. Auditory capacities in Middle Pleistocene humans from the Sierra de Atapuerca in Spain. National Academy of Sciences (USA), Proceedings 101, 9976-9981.

Martinez, N.A., 2002. Le Moulin rotatif manuel au nord-est de la Péninsule ibérique. In Moudre et broyer: l'interprétation fonctionnelle de l'outillage de mouture et de broyage dans la Préhistoire et Antiquité, 183-196.

Marzke, M.W., 1997. Precision grips, hand morphology, and tools. American Journal of Physical Anthropology 102, 91-110.

Marzke, M.W., and Shackley, M.S., 1987. Hominid hand use in the Pliocene and Pleistocene: evidence from experimental archaeology and comparative morphology. Journal of Human Evolution 15, 439-460.

Mascetti, D., and Triossi, A., 1999. The historical background: from antiquity to the 17th century. *In* Earrings from Antiquity to the Present, Thames & Hudson, 9-40.

Mason, R.J., 1993. La Grotte de hearths, Vallée de Makapansgat, Transvaal, Afrique du Sud, 1937-1988. L'Anthropologie (Paris) tome 97, 85-96.

Masset, C., 1980. *In* Wreschner, 1980, Red ochre and human evolution: a case for discussion. Current Anthropology 21, 638-639.

Mateescu, C.N., 1975. Remarks on cattle breeding and agriculture in the Middle and Late Neolithic on the Lower Danube. Dacia, N.S., tome XIX, 13-18.

Mauss, M., 1954. The gift: forms and functions of exchange in Archaic societies. Cohen & West, 130 pp.

Mays, S. and Taylor, G.M., 2003. A first prehistoric case of tuberculosis from Britain. International Journal of Osteoarchaeology 13(4), 189-196.

Mazar, A., et al., 2008. Iron Age beehives at Tel Rehov in the Jordan valley. Antiquity 82, 629-639.

Mazel, A.D., 1992. Early pottery from the eastern part of Southern Africa. South African Archaeological Bulletin 47, 3-7.

Mazza, P.P.A., Martini, F., Sala, B., et al., 2006. A new Palaeolithic discovery: tar-hafted stone tools in a European Mid-Pleistocene bone-bearing bed. Journal of Archaeological Science 33, 1310-1318.

Mbida, C.M., 2004. Yes, there were bananas in Cameroon more than 2000 years ago. InfoMusa 13, 40-42.

Mbida, C.M., et al., 2006. Phytolith evidence for the early presence of domesticated banana (*Musa*) in Africa. *In* Documenting domestication, new genetic and archaeological paradigms. University of California Press, 68-81.

McBrearty, S., 2007. Down with the Revolution. Rethinking the human revolution, McDonald Institute Monographs, 133-151.

McBrearty, S., and Brooks, A.S., 2000. The revolution that wasn't: a new interpretation of the origin of modern human behavior. Journal of Human Evolution 39, 453-563.

McBrearty, S., and Tryon, C., 2006. From Acheulian to Middle Stone Age in the Kapthurin Formation, Kenya. *In* Transitions before the transition: evolution and stability in the Middle Paleolithic and Middle Stone Age, Springer, 237-277.

McDermott, L., 1996. Self-representation in Upper Paleolithic female figurines. Current Anthropology 37, 227-275.

McDougall, I., Brown, F.H., and Fleagle, J.G., 2005. Stratigraphic placement and age of modern humans from Kibish, Ethiopia. Nature 433, 733-736.

McGovern, P.E., et al., 1996. Neolithic resinated wine. Nature 381, 480-481.

McGovern, P.E., et al., eds. 1996. Origins and Ancient History of Wine. Gordon and Breach Publishers, 409 pp.

McGovern, P.E., 2009. Uncorking the Past. University of California Press, 330 pp.

McGovern, P.E., and Michel, R.H., 1985. Royal Purple Dye: tracing the origins of the dye. Analytical Chemistry 57, 1514A-1522A.

McGovern, P.E., and Michel, R.H., 1990. Royal Purple Dye: The chemical reconstruction of the ancient Mediterranean industry. Accounts of Chemical Research 23, 152-158.

McGovern, P.E., et al., 2004. Fermented beverages of pre- and proto-historic China. National Academy of Sciences (USA), Proceedings 101, 17593-17598.

McGovern, P.E., et al., 2005. Chemical identification and cultural implications of a mixed fermented beverage from Late Prehistoric China. Asian Perspectives 44, 249-275.

McGrail, S., 1991. Bronze Age seafaring in the Mediterranean: a view from NW Europe. *In* Bronze Age trade in the Mediterranean, Conference publication, Paul Åströms Förlag, Jonsered, Göteborg, Studies in Mediterranean Archaeology, vol. XC, 83-90.

McGrail, S., 2001. Boats of the World, Oxford University Press, 480 pp.

McGrail, S., 2007. Early sea voyages. The International Journal of Nautical Archaeology 20, 85-93.

McHenry, H.M., 1994. Behavioral ecological implications of early hominid body size. Journal of Human Evolution 27, 77-87.

McHenry, H.M., 1994b. Tempo and mode in human evolution. National Academy of Sciences (USA), Proceedings 91, 6780-6786.

McHenry, H.M., and Coffing, K., 2000. *Australopithecus* to *Homo*: transformations in body and mind. Annual Review of Anthropology 29, 125-146.

McLaren, F.S., 1992. Plums from Douara Cave, Syria: the chemical analysis of charred stone fruits. *In* Res Archaeobotanica: International Workgroup for Palaeoethnobotany: proceedings of the Ninth symposium, Kiel 1992, 195-218.

McNeill, D., ed., 2000. Language and Gesture. Cambridge University Press, 409 pp.

McPherron, S.P., et al., 2010. Evidence for stone-tool-assisted consumption of animal tissue before 3.39 million years age at Dikika, Ethiopia. Nature Letters 466, 857-860.

Meignen, L., Bar-Yosef, O., Goldberg, P., and Weiner, S., 2001. Le feu au Paléolithique Moyen: Recherches sur les structures de combustion et le statut des foyers. L'exemple du Proche-Orient. Paléorient 282, 9-22.

Meiklejohn, C., Agelarakis, A., Akkermans, P.A., Smith, P.E.L., and Solecki, R., 1992. Artificial cranial deformation in the Proto-Neolithic and Neolithic Near East and its possible origin: evidence from four sites. Paléorient 18, 83-97.

Meisler, M.H., and Ting, Chao-Nan, 1993. The remarkable evolutionary history of the human amylase genes. Critical Reviews in Oral Biology and Medicine 4, 503-509.

Mellaart, J., 1963. Excavations at Çatal Hüyük, 1962. Anatolian Studies 13, 43-103.

Mellaart, J., 1964. Excavations at Çatal Hüyük, 1963, 3rd Preliminary Report. Anatolian Studies 14, 39-119.

Mellaart, J., 1967. Çatal Hüyük: A Neolithic town in Anatolia. McGraw-Hill Book Company, New York, 232 pp.

Mercer, R.J., 1999. The origins of warfare in the British Isles. *In* Ancient Warfare, Archaeological Perspectives, Sutton Publishing, 143-156.

Merrick, H.V., Brown, F.H., and Nash, W.P., 1994. Use and movement of obsidian in the Early and Middle Stone Ages of Kenya and Northern Tanzania. Society, Culture and Technology in Africa 11, 29-44.

Merrillees, R.S., 1962. Opium trade in the Bronze Age Levant. Antiquity 36, 287-292.

Meszaros, G., and Vertes, L., 1955. A paint mine from the early upper Palaeolithic age near Lovas (Hungary, County Veszprem). Acta Archaeologica Academiae Scientarum Hungaricae 5, 1-32.

Meyer, H.W., 2003. The Fossils of Florissant. Smithsonian Books, Washington, D.C., 258 pp.

Michel, R.H., et al., 1992. Chemical evidence for ancient beer. Nature 360, p. 24.

Michel, R.H., et al., 1993. The first wine & beer, chemical detection of ancient fermented beverages. Analytical Chemistry 65, 408A-413A.

Miller, G., Tybur, J.M., and Jordan, B.D., 2007. Ovulatory cycle effects on tip earnings by lap dancers: economic evidence for human estrus? Evolution and Human Behavior 28, 375-381.

Miller, J.I., 1969. The spice trade of the Roman Empire. Clarendon Press, 294 pp.

Miller, N.F., 1991. The Near East. *In* Progress in Old World palaeoethnobotany. Balkema, pp. 133-160.

Miller, N.F., 1992. The origins of plant cultivation in the Near East. *In* The origins of agriculture: an international perspective. University of Alabama Press, Tuscaloosa, 39-58.

Miller, R., 1980. Water use in Syria and Palestine from the Neolithic to the Bronze Age. World Archaeology, vol. II, no. 3, 331-341.

Miller, R.L., Armelagos, G.J., Ikraum, S., De Jonge, N., Krijger, F.W., and Deelder, A.M., 1992. Palaeoepidemiology of *Schistosoma* infections in mummies. British Medical Journal (BMJ) 304(6826), 355-356.

Miller, R.L., Ikraum, S., Armelagos, G.J., Walker, R., Harer, W.B., Shiff, C.J., Baggett, D., Carrigan, M., and Marel, S.M., 1994. diagnosis of *Plasmodium falciparum* in mummies using the rapid manual ParaSight™–F test. Transactions of the Royal Society of Tropical Medicine and Hygiene 88, 31-32.

Millett, P., 1983. Maritime loans and the structure of credit in fourth-century Athens. *In* Trade in the Ancient Economy, University of California Press, 36-52.

Milner, G.R., 2005. Nineteenth-Century arrow wounds and perceptions of prehistoric warfare. American Antiquity 70, 144-156.

Milo, R.G. 1998. Evidence for human predation at Klasies River Mouth, South Africa, and its implications for the behaviour of early modern humans. Journal of Archaeological Science 25, 99-133.

Milton, K., 1989. Primate diets and gut morphology: implications for hominid evolution. *In* Food and Evolution: toward a theory of human food habits, Temple University Press, 93-115.

Milton, K., 1999. A hypothesis to explain the role of meat-eating in human evolution. Evolutionary Anthropology 8, 11-21.

Minozzi, S., et al., 2003. Nonalimentary tooth use in Prehistory: An example from Early Holocene in Central Sahara (Uan Muhuggiag, Tadrart Acacus, Libya). American Journal of Physical Anthropology 120, 225-232.

Mishra, S., et al., 1995. Earliest Acheulian industry from Peninsular India. Current Anthropology 36, 847-851.

Misra, V.N., et al., 1977. Bhimbetka: prehistoric man and his art in Central India. Sakal Printing Press, 26 pp.

Mitchell, P.D., 2015. Human parasites in medieval Europe: lifestyle, sanitation and medical treatment. Advances in Parasitology 90, 389-420.

Mitchell, P.D., 2015b. Assessing the impact of sanitation upon health in early human populations from hunter-gatherers to ancient civilizations, using theoretical modelling. *In* Sanitation, Latrines and Intestinal Parasites in Past Populations, P.D. Mitchell, ed., Ashgate, Farnham, UK, pp. 5-17.

Mitchell, P.D., 2017. Human parasites in the Roman world: health consequences of conquering an empire. Parasitology 144, 48-58.

Miyazaki, M., et al., 2003. Mechanism and regulation of body malodor generation (2) – Development of a novel deodorant powder and application to antiperspirants. Journal of the Society of Cosmetological Chemistry 37(3), 202-209.

Moggi-Cecchi, J., and Collard, M., 2002. A fossil stapes from Sterkfontein, South Africa, and the hearing capabilities of early hominids. Journal of Human Evolution 42, 259-265.

Molleson, T., 1981. The archaeology and anthropology of death: what the bones tell us. *In* S.H. Humphries, ed., Mortality and Immortality, Academic Press, London, 15-22.

Molleson, T., 1989. Seed preparation in the Mesolithic: the osteological evidence. Antiquity 63, 356-362.

Molleson, T., 2000. The people of Abu Hureyra. *In* Village on the Euphrates, 302-324.

Molleson, T., et al., 1992. A Neolithic painted skull from Tell Abu Hureyra, Northern Syria. Cambridge Archaeological Journal 2, 230-235.

Molleson, T., et al., 1993. Dietary change and the effects of food preparation on microwear patterns in the Late Neolithic of abu Hureyra, northern Syria. Journal of Human Evolution 24, 455-468.

Momber, G., et al., eds., 2011. Mesolithic occupation at Bouldnor Cliff and the Submerged Prehistoric Landscapes of the Solent. CBA Research Report 164, Council for British Archaeology, 197 pp.

Monah, D., 1991. L'exploitation du sel dans les Carpates Orientales et ses rapports avec le Culture de Cucuteni-Tripolye. *In* Bibliotheca Archaeologica Iassiensis, IV, Le Paléolithique et le Néolithique de la Roumanie en contexte Européen, 387-400.

Monah, D., 2002. L'exploitation préhistorique du sel dans les Carpates orientales. *In* Archéologie du sel: Techniques et sociétés, Internationale Archäologie, ASTK 3, Colloque 12.2, XIVe Congrès Union Internationale des Sciences Préhistoriques et Protohistoriques, Liège, 135-146.

Moncel, M.-H., et al., 2012. Non-utilitarian lithic objects from the European Paleolithic. Archaeology, Ethnology and Anthropology of Eurasia 40, 24-40.

Monge, J., Kricun, M., Radovčić, J., Radovčić, D., Mann, A., and Frayer, D.W., 2013. Fibrous dysplasia in a 120,000+ year old Neandertal from Krapina, Croatia. PLoS ONE 8, issue 6, 1-4.

Monnier, J.-L. et al., 1994. A new regional group of the Lower Palaeolithic in Brittany (France), recently dated by electron spin resonance. Comptes Rendus de l'Academie des Sciences 319, 155-160.

Montagna, W., 1983. The evolution of human skin(?). Journal of Human Evolution 14, 3-22.

Moore, A.M.T., 1995. The inception of potting in Western Asia and its impact on economy and society. *In* The emergence of pottery, Smithsonian Institution Press, 39-53.

Moorey, P.R.S., 1994. Ancient Mesopotamian Materials and Industries. Clarendon Press, Oxford, 414 pp.

Morris, I., 1992. Death-ritual and social structure in Classical Antiquity. Cambridge University Press, 264 pp.

Morris, I., 2014. War! What is it good for? Farrar, Straus and Giroux, 495 pp.

Morris, I., 2015. Foragers, Farmers and Fossil Fuels: How human values evolve. Princeton University Press, 169 pp.

Morse, K., 1993. Shell beads from Mandu Mandu Creek rock-shelter, Cape Range peninsula, Western Australia, dated before 30,000 b.p. Antiquity 67, 877-883.

Morwood, M.J., et al., 1998. Fission-track ages of stone tools and fossils on the east Indonesian island of Flores. Nature 392, 173-176.

Moulherat, C., et al., 2002. First evidence of cotton at Neolithic Mehrgarh, Pakistan: analysis of mineralized fibres from a copper bead. Journal of Archaeological Science 29, 1393-1401.

Mourre, V., Villa, P., and Henshilwood, C.S., 2010. Early use of pressure flaking on lithic artifacts at Blombos Cave, South Africa. Science 330, 659-662.

Movius, H.L., Jr., 1950. A wooden spear of Third Interglacial age from Lower Saxony. Southwestern Journal of Anthropology 6(2), 139-142.

Movius, H.L., Jr., 1953. The Mousterian Cave of Teshik-Tash, southeastern Uzbekistan, central Asia. American School of Prehistoric Research Bulletin 17, 11-71.

Muchlinski, M.N., 2010. A comparative analysis of vibrissa count and infraorbital foramen area in primates and other mammals. Journal of Human Evolution 58, 447-473.

Muckelroy, K., 1981. Middle Bronze Age trade between Britain and Europe: a maritime perspective. Proceedings of the Prehistoric Society 47, 275-297.

Mudzhiri, T.P., and Kvirikadze, M.V., 1979. Polevie issledovaniya drevnikh rudnikov gruzii epokhi bronzi (Gornaya Nauki i Tekhniki, materialy pervego vsesoiiuz koordinatsoveshch, Tbilisi, 15-17, 67-83.

Muhly, J.D., 1986. Prehistoric background leading to the first use of metals in Asia. In Proceedings of symposium on the early metallurgy in Japan and the surrounding area, Bulletin of the Metals Museum, special issue, vol. 11, 21-42.

Mumcuoglu, K.Y., 2008. Human lice: Pediculus and Pthirus. In D. Raoult and M. Drancourt, eds., Paleomicrobiology, Springer, Berlin, Heidelberg, Ch. 13, 215-222.

Mumcuoglu, K.Y., et al., 2003. Body louse remains found in textiles excavated at Masada, Israel. *Journal of Medical Entomology* 40(4), 585-587.

Mumcuoglu, K.Y., and Zias, J., 1988. Head lice, *Pediculus humanus capitis* (Anoplura: Pediculidae) from hair combs excavated in Israel and dated from the first century BC to the eighth century A.D. Journal of Medical Entomology 25, 545-547.

Muñoz, M., and Casadevall, M., 1997. Fish remains from Arbreda Cave (Seriyà, Girona), Northeast Spain and their paleoecological significance. Journal of Quaternary Science 12, 111-115.

Munro, N.D., and Bar-Oz, G., 2005. Gazelle bone fat processing in the Levantine Epipalaeolithic. Journal of Archaeological Science 32, 223-239.

Murphy, E.M., 2000. Mummification and body processing – evidence from the Ironage in southern Siberia. In Davis-Kimball, J., Murphy, E., Koryakova, L., and Yablonsky, L., eds., Kurgans, Ritual Sites, and Settlements: The Eurasian Bronze and Iron Age. British Archaeological Reports International Series 890. Archaeopress, Oxford, pp. 279-292.

Murphy, E.M., Chistov, Y.K., Hopkins, R., Rutland, P., and Taylor, G.M., 2009. Tuberculosis among Iron Age individuals from Tyva [Tuva], South Siberia: palaeopathological and biomolecular findings. Journal of Archaeological Science 36, 2029-2038.

Murphy, K.A., 1999. A prehistoric example of polydactyly from the Iron Age site of Simbusenga, Zambia. American Journal of Physical Anthropology 108, 311-319.

Musées de Marseille, 1991. Jouer dans l'Antiquité, 204 pp.

Musonda, F.B., 1987. The significance of pottery in Zambian Later Stone Age contexts. The African Archaeological Review 5, 147-158.

Nadel, D., ed., 2002. Ohalo II: A 23,000-year-old fisher-hunter-gatherers' camp on the shore of the Sea of Galilee. Reuben and Edith Hecht Museum, University of Haifa, 65 pp.

Nadel, D., et al., 1994. 19,000-year-old twisted fibers from Ohalo II. Current Anthropology 35, 451-458.

Nadel, D., et al., 2011. The Nahal Galim/Nahal Ornit prehistoric flint quarries in Mt. Carmel, Israel. Eurasian Prehistory 8, 51-66.

Nadel, D., Weiss, E., Simchoni, O., Tsatkin, A., Danin, A., and Kislev, M., 2004. Stone Age hut in Israel yields world's oldest evidence of bedding. National Academy of Sciences (USA), Proceedings 101, no. 17, 6821-6826.

Nakazawa, Y., et al., 2009. On stone-boiling technology in the Upper Paleolithic: behavioral implications from an Early Magdalenian hearth in El Mirón Cave, Cantabria, Spain. Journal of Archaeological Science 36, 684-693.

Natapintu, S., 1988. Current research on ancient copper-base metallurgy in Thailand. *In* Prehistoric studies: the stone and metal ages in Thailand, 107-124.

Needham, S., and Evans, J., 1987. Honey and dripping: Neolithic food residues from Runnymede Bridge. Oxford Journal of Archaeology 6, 21-28.

Negash, A., Shackley, M.S., and Alene, M., 2006. Source provenance of obsidian artifacts from the Early Stone Age (ESA) site of Melka Konture, Ethiopia. Journal of Archaeological Science 33, 1647-1650.

Neiberger, E.J., 1988. Syphilis in a Pleistocene bear? Nature 333, 603.

Nelson, K., 2010. Environment, cooking strategies and containers. Journal of Anthropological Archaeology 29, 238-247.

Neudecker, R., 1994. Die Pracht der Latrine. Verlag Dr. Friedrich Pfeil, 175 pp.

Neumann, K., and Hildebrand, E., 2009. Early bananas in Africa: the state of the art. Ethnobotany Research and Applications 17, 353-362.

Newman, R., 2007. Jewelry handbook: how to select, wear & care for jewelry. International Jewelry Publications, 177 pp.

Nezafati, N., et al., 2006. Ancient tin: old question and a new answer. Antiquity 80, no. 308, 4 pp.

Nguyen Kim Dung, 1990. The Lithic Workshop at Tang Kenh. Khao Co Hoc 3, 64-82.

Nguyen Kim Dung, 1998. Nephrite and jadeite manufacturing in prehistoric Vietnam. Khao Co Hoc, 4, 23-40.

Niblett, R., 2001. A Neolithic dugout from a multiperiod site near St Albans, Herts, England. International Journal of Nautical Archaeology 30, 155-195.

Nicholson, P.T., and Shaw, I., 2000. Ancient Egyptian Materials and Technology. Cambridge University Press, 702 pp.

Nikolov, V., and Bacvarov, K., eds., 2012. Salt and gold: the role of salt in prehistoric Europe. Proceedings of the International Symposium (Humboldt-Kolleg) in Provadia, Bulgaria, 372 pp.

Nishida, T., and Turner, L.A., 1996. Food transfer between mother and infant chimpanzees of the Mahale Mountains National Park, Tanzania. International Journal of Primatology 17, 947-968.

Nitta, E., 1996. Iron and salt industries in Isan. Kyoto University Research Information Repository, pp. 43-65.

Noy, T., 1991. The art and decoration of the Natufian of Nahal Oren. *In* The Natufian culture in the Levant, 557-568.

Nriagu, J.O., 1983. Lead and lead poisoning in antiquity. Wiley-Interscience, 437 pp.

O'Brien, W., 1961. Bronze Age copper mining in Britain and Ireland. Shire Publications Ltd., 64 pp.

O'Brien, W., 2015. Prehistoric copper mining in Europe. Oxford University Press 345 pp.

O'Connell, J.F., and Allen, J., 1998. When did humans first arrive in Greater Australia and why is it important to know? Evolutionary Anthropology 6, 132-146.

O'Connell, J.F., and Allen, J., 2004. Dating the colonization of Sahul (Pleistocene Australia–New Guinea): a review of recent research. Journal of Archaeological Science 31, 835-853.

O'Connell, J.F., et al., 1999. Grandmothering and the evolution of *Homo erectus*. Journal of Human Evolution 36, 461-485.

Oakeshott, R.E., 1960, The archaeology of weapons. Frederick A. Praeger, 359 pp.

Oates, D., and Oates, J., 1976. Early irrigation agriculture in Mesopotamia. *In* Problems in economic and social archaeology, Duckworth, 109-152.

Oates, J., 2001. Equid figurines and 'chariot' models. *In* McDonald Institute for Archaeological Research 2, 279-293.

Oates, J., 2003. A note on the early evidence for horse and the riding of equids in Western Asia. *In* Prehistoric Steppe Adaptation and the Horse, McDonald Institute for Archaeological Research, 115-125.

Oates, J., et al., 1977. Seafaring merchants of Ur? Antiquity, vol. LI, 221-234.

Obruchev, D.V., ed., 1967. Fundamentals of Palaeontology, vol. XI. Agnatha, Pisces. Israel Program for Scientific Translation, Jerusalem, 825 pp.

Oded. B. 1979. Mass deportations and deportees in the Neo-Assyrian Empire. Dr. Ludwig Reichert Verlag, 142 pp.

Olsen, S.L., 2006. Early horse domestication on the Eurasian steppe. *In* Documenting Domestication, University of California Press, 245-269.

Oppenheim, A.L., 1970. The Cuneiform. *In* Glass and Glassmaking in Ancient Mesopotamia by A.L. Oppenheim, R.H. Brill, D. Barag, and A. von Saldem, Corning Museum of Glass, Corning, NY, viii + 246 pp.

Opperman, H., 1996. Strathalan Cave B, North-Eastern Cape Province, South Africa: evidence for human behavior 29,000–26,000 years ago. Quaternary International 33, 45-53.

Opperman, H., and Heydenrych, B., 1990. A 22,000 year-old Middle Stone Age camp site with plant food remains from the North-Eastern Cape. South African Archaeological Bulletin 45, 93-99.

Organ, C., et al., 2011. Phylogenetic rate shifts in feeding time during the evolution of *Homo*. National Academy of Sciences (USA), Proceedings 108, 14555-14559.

Oriental Ceramic Society, 1984. Chinese ivories from the Shang to the Qing. British Museum Publications, 200 pp.

Orscheidt, J., 2005. The head burials from Ofnet cave: an example of warlike conflict in the Mesolithic. *In* Warfare, violence and slavery in prehistory. British Archaeological Reports International Series 1374, 67-73.

Orscheidt, J., and Haidle, M.Y., 2006. The LBK enclosure at Herxheim: Theatre of war or ritual centre? References from osteoarchaeological investigations. Journal of Conflict Archaeology 2, 153-167.

Orscheidt, J., et al., 2003. Survival of a multiple skull trauma: the case of an Early Neolithic individual from the LBK Enclosure at Herxheim (Southwest Germany). International Journal of Osteoarchaeology 13, 375-383.

Ortner, D.J., and Putscher, W.G.J., 1981. Identification of pathological conditions in human skeletal remains. Smithsonian Contributions to Anthropology 28, 479 pp.

Osgood, R., Monks, S., and Toms, J., 2000. Bronze Age Warfare. Sutton Publishing, 165 pp.

Oshibkina, S.V., 1985. The material culture of the Veretye-type sites in the region to the east of Lake Onega. *In* The Mesolithic in Europe, John Donald Publishers Ltd, 402-413.

Otterbein, K.F., 2004. How war began. Texas A & M University Press, 292 pp.

Outram, A.K., et al., 2009. The earliest horse harnessing and milking. Science 323, 1332-1335.

Ouzman, S., 1997. Between margin and centre: the archaeology of southern African bored stones. *In* Our gendered past: archaeological studies of gender in southern Africa, Witwatersrand University Press, 71-106.

Owens, E.J., 1983. The Koprologoi at Athens in the Fifth and Fourth Centuries BC. Classical Quarterly 33, 44-50.

Oxenham, M.F., Tilley, L., Matsumura, H., Nguyen, L.C., Nguyen, K.T., Nguyen, K.D., Domett, K., and Huffer, D., 2009. Paralysis and severe disability requiring intensive care in Neolithic Asia. Anthropological Science 11, 107-112.

Özdoğan, M., Özdoğan, A., 1999. Archaeological evidence on the early metallurgy at Çayönü Tepesi. The beginnings of metallurgy, Der Anschnitt, Beiheft, 13-22.

Paddayya, K., et al., 1999. Geoarchaeology of the Acheulian workshop at Isampur, Hunsgi Valley, Karnataka. Man & Environment 24, 167-184.

Paddayya, K., et al., 2002. Recent findings on the Acheulian of the Hunsgi and Baichbal valleys, Karnataka, with special reference to the Isampur excavation and its dating. Current Science 83, 641-647.

Panagiotakopulo, E, and Buckland, P.C., 1999. *Cimex lectularius* L., the common bed bug from Pharaonic Egypt. Antiquity 73, 908-911.

Panagiotakopulo, E., 2001. Fleas from Pharaonic Amarna. Antiquity 75, 499-500.

Panger, M.A., et al., 2002. Older than the Oldowan? Rethinking the emergence of Hominin tool use. Evolutionary Anthropology 11, 235-245.

Papagrigorak, M.J., et al., 2006. DNA examination of ancient dental pulp incriminates typhoid fever as a probable cause of the Plague of Athens. International Journal of Infectious Diseases 10, 206-214. *See* also B. Shapiro et al., 2006.

Park, M.S., et al., 2007. Evolution of the human brain: changing brain size and the fossil record. Neurosurgery 60, 555-562.

Parkington, J., 2001. Milestones: the impact of the systematic exploitation of marine foods on human evolution. *In* Humanity from African naissance to coming millennia, Firenze University Press & Witwatersrand University Press, 327-336.

Parlett, D., 1999. The Oxford history of board games. Oxford University Press, 386 pp.

Parzinger, H., and Boroffka, N., 2003. Das Zinn der Bronzezeit in Mittelasien I. Archäologie in Iran und Turan, Band 5. Verlag Philipp von Zabern, 328 pp.

Pavlides, C., and Gosden, C., 1994. 35,000-year-old sites in the rainforests of West New Britain, Papua New Guinea. Antiquity 68, 604-610.

Peacock, D., and Cutler, L., 2010. A Neolithic Voyage. The International Journal of Nautical Archaeology 39, 116-124.

Peers, C.J., and McBride, A., 1990. Ancient Chinese armies 1500–200 BC. Osprey Publishing, London, 48 pp.

Pellett, P., 1990. Protein requirements in humans. American Journal of Clinical Nutrition 51, 723-737.

Pelto, G.H., Zhang, Y., and Habicht, J.-P., 2010. Premastication: the second arm of infant and young child feeding for health and survival. Maternal and Child Nutrition 6, 4-128.

Pemberton, W., et al., 1988. Canals and bunds, ancient and modern. *In* Bulletin on Sumerian agriculture, 4, 207-221.

Penhallurick, R.D., 1986. Tin in antiquity: its mining and trade throughout the ancient world with particular reference to Cornwall. Institute of Metals, London, 271 pp.

Pennisi, E., 2004. Burials in Cyprus suggest cats were ancient pets. Science 304, 189.

Péquart, M., Péquart, S.-J., Boule, M., and Vallois, H., 1937. Téviec, station-nécropole Mésolithique du Morbihan. Archives du l'Institut de Paléontologie Humaine, Mémoire 18, 227 pp.

Peresani, M., Fiore, I., Gala, M., Romandini, M., and Tagliacozzo, A., 2011. Late Neandertals and the intentional removal of feathers as evidenced from bird bone taphonomy at Fumane Cave, 44 ky B.P., Italy. National Academy of Sciences (USA), Proceedings 108(10), 3888-3893.

Peresani, M., Vanhaeren, M., Quaggiotto, E., Queffelec, A., and d'Errico, F., 2013. An ochered fossil marine shell from the Mousterian of Fumane Cave, Italy. PLoS ONE 8, issue 7, 15 pp.

Pérez, P.J., 1996. Resultados de las investigaciones paleopatológicas en homínidos fosiles. Revista Española de Paleontología, 256-268.

Perrot, J., 1966. Le gisement Natoufien de Mallaha (Eynan), Israël. L'Anthropologie, tome 70, 437-484.

Peterkin, G.L., 1993. Lithic and organic hunting technology in the French Upper Palaeolithic. In Hunting and Animal Exploitation in the later Palaeolithic and Mesolithic of Eurasia, Archaeological Papers of the American Anthropological Association 3, 49-67.

Peters, C.R., 1979. Toward an ecological model of African Plio-Pleistocene hominid adaptations. American Anthropologist, N.S, 81, 261-278.

Peters, J., and Schmidt, K., 2004. Animals of the symbolic world of Pre-Pottery Neolithic Göbekli Tepe, south-eastern Turkey: a preliminary assessment. Anthropozoologica 39, 179-218.

Peters, J., Helmer, D., von den Driesch, A., and Segui, M.S., 1999. Early animal husbandry in the Northern Levant. Paléorient 25, 27-47.

Peters, J., von den Driesch, A., and Helmer, D., 2005. the Upper Euphrates-Tigris basin: cradle of agro-pastoralism? The International Council for Archaeozoology [9th] Conference, Oxbow Books, 96-124.

Petraglia, M., et al., 1999. Isampur, the first Acheulian quarry in India: stone tool manufacture, biface morphology, and behaviors. Journal of Anthropological Research 55, 39-70.

Petraglia, M.D., 2002. The heated and the broken: thermally altered stone, human behavior and archaeological site formation. North American Archaeologist 23, 241-269.

Petrequin, P., Sheridan, A., Cassen, S., Errera, E., Gautheir, E., Klassen, L., Le Maux, N., and Pailler, Y., 2008. Neolithic Alpine axeheads, from the Continent to Great Britain, the Isle of Man and Ireland. Analecta Praehistorica Leidensia 40, 261-279.

Peyrony, D., 1934. La Ferrassie: Moustérien, Périgordien, Aurignacien. Préhistoire 3, 1-92.

Pickering, T.R., and Domínguez-Rodrigo, M., 2006. The acquisition and use of large mammal carcasses by Oldowan hominins in eastern and southern Africa: A selected review and reassessment. In The Oldowan: Case studies into the earliest Stone Age, 113-128.

Pickering, T.R., and Dunn, H.T., 2007. The endurance running hypothesis and hunting and scavenging in savanna-woodlands. Journal of Human Evolution 53, 434-438.

Piggott, S., 1968.The earliest wheeled vehicles and the Caucasian evidence. Proceedings of the Prehistoric Society 34, 266-318.

Piggott, S., 1983. The earliest wheeled transport from the Atlantic Coast to the Caspian Sea. Cornell University Press, 272 pp.

Pigott, V.C. et al., 1997. The archaeology of copper production: excavations in the Khao Wong Prachan Valley, central Thailand. In South-East Asian Archaeology 1992, Proceedings of the Fourth International Conference of the European Association of South-East Asian Archaeologists, 119-157.

Pigott, V.C., 1996. The study of ancient metallurgical technology: a review. Asian Perspectives 35, 89-97.

Pigott, V.C., and Natapintu, S., 1988 [1986]. Archaeological investigations into prehistoric copper production: the Thailand archaeometallurgy project 1984-1986. In The Beginning of the use of metals and alloys, pp. 156-162.

Pigott, V.C., and Weisgerber, G., 1999 [1998]. Mining archaeology in geological context: the prehistoric copper mining complex at Phu Lon, Nong Khai Province, northeast Thailand. Metallurgia Antiqua, Der Anschnitt, Beiheft 8, 135-162.

Pike, A.W., 1967. The recovery of parasite eggs from ancient cesspit and latrine deposits: an approach to the study of early parasite infections. *In* Diseases in Antiquity, D. Brothwell and A.T. Sandison, eds., Charles C. Thomas, Springfield, IL, 184-188.

Pike, A.W.G., et al., 2012. U-Series dating of Paleolithic art in 11 caves in Spain. Science 336, 1409-1413.

Pinhasi, R., et al., 2010. First direct evidence of Chalcolithic footwear from the Near Eastern Highlands. PLoS ONE 5, issue 6, 5 pp.

Pinnock, F., 1988. Observations on the trade of lapis lazuli in the IIIrd Millennium BC. Heidelberger Studien zum Alten Orient, Band 2, Wirtschaft und Gesellschaft von Ebla, 107-110.

Plummer, T., 2004. Flaked stones and old bones: biological and cultural evolution at the dawn of technology. Yearbook of Physical Anthropology 47, 118-164.

Plummer, T., et al., 2001. Late Pliocene Oldowan excavations at Kanjera South, Kenya. Antiquity 75, 809-810.

Poinar, G., Jr., 2013. *Panstrongylus hispaniolae* sp. n. (Hemiptera: Reduviidae: Triatominae), a new fossil triatomine in Dominican amber, with evidence of gut flagellates. Palaeodiversity 6, 1-8.

Poinar, G., Jr., 2015b. a new genus of fleas with associate microorganisms in Dominican amber. Journal of Medical Entomology 52, 1234-1240.

Poinar, G., Jr., et al., 2015. One hundred million year old ergot: psychotropic compounds in the Cretaceous? Palaeodiversity 8, 13-19.

Poinar, G.O., Jr., 2004b. Evidence of vector-borne disease of Early Cretaceous reptiles. Vector-Borne Zoonotic Diseases 4, 281-284.

Poinar, G.O., Jr., 2004a. *Paleoleishmania proterus* n. gen., n. sp., (Trypanosomatidae: Kinetoplastida) from Cretaceous Burmese amber. Protist 155, 305-310.

Poinar, G.O., Jr., 2014e [2015]. Spirochete-like cells in a Dominican amber *Amblyomma* tick (Arachnida: Ixodidae). Historical Biology 27(5), 565-570.

Poinar, G.O., Jr., and Telford, S.R., Jr., 2005. *Paleophaemoproteus burmacis* n. gen., n. sp. (Haemosporida: Plasmodiidae) from an Early Cretaceous Burmese midge (Diptera: Ceratopogonidae). Parasitology 131, 79-84.

Pokutta, D.A., and Howcroft, R., 2013. Children, childhood and food – the diets of subadults in the Únětice Culture of southwestern Poland. British Archaeological Reports International Series, Archaeopress Oxford, pp. 1-14.

Polosmak, N.V., 2000. Pazyryk felt blankets: the Ukok collection. Archaeology, Ethnology and Anthropology of Eurasia 1, 94-100.

Poschmann, M., and Wedmann, S., 2005. Spinnen, Heuschrecken und Wasserwanzen aus dem Westerwald. Fossilien 4, 234-240.

Poss, J.R., 1975. Stones of Destiny: a Story of Man's Quest for Earth's Riches. Michigan Technological University, Houghton, Michigan, 253 pp.

Possehl, G.I., 1981. Cambay beadmaking. Expedition 23, 39-47.

Possehl, G.I., 1996. Mehrgarh. *In* The Oxford Companion to Archaeology, 436-438.

Postgate, J.N., 1992. Early Mesopotamia: Society and economy at the dawn of history. Routledge, 367 pp.

Postgate, J.N., and Powell, M.A., 1984. Bulletin on Sumerian agriculture, vol. I, 152 pp.

Potter, B.A., et al., 2014. New insights into Eastern Beringian mortuary behavior: a terminal Pleistocene double infant burial at Upward Sun River. National Academy of Sciences (USA), Proceedings, Early Edition, 6 pp.

Potts, R., 1998. Environmental hypotheses of hominin evolution. Yearbook of Physical Anthropology 41, 93-136.

Potts, R., 2003. Early human predation. In Predator-Prey Interactions in the Fossil Record, P.H. Kelley, M. Kowalewski, and T.A. Hansen, eds., Kluwer Academic, New York, 359-376.

Potts, R., and Shipman, P., 1981. Cutmarks made by stone tools on bones from Olduvai Gorge, Tanzania. Nature 291, 577-580.

Potts, R., and Sloan, C. 2010. What does it mean to be human? National Geographic Society, 175 pp.

Potts, R., et al., 2004. Small Mid-Pleistocene hominin associated with East African Acheulian technology. Science 305, 75-78.

Powell, J., 2003. Fishing in the Mesolithic and Neolithic – the Cave of Cyclops, Youra. Zooarchaeology in Greece, recent Advances, British School at Athens, Studies 9, 75-84.

Powell, M.A., Jr., 1979. Ancient Mesopotamian Weight Metrology: Methods, Problems and Perspectives. In Studies in Honor of Tom B. Jones, Alter Orient und Altest Testament, vol. 203, M.A. Powell, Jr., and R.H. Sack, eds. Neukirchen: Verlag Butzon & BerckerKevelaer, 71-110.

Pramankij, S., et al., 1997. Modification in the use of Mesolithic tools in case of migration. In South-East Asian Archaeology 1992, 29-43.

Preece, R.C., et al., 2006. Humans in the Hoxnian: habitat, context and fire use at Beeches Pit, West Stow, Suffolk, UK. Journal of Quaternary Science 21, 485-496.

Price, J.L., and Molleson, T.I., 1974. A radiographic examination of the left temporal bone of Kabwe Man, Broken Hill Mine, Zambia. Journal of Archaeological Science 1, 285-289.

Pringle, H., 1997. Ice age communities may be the earliest known net hunters. Science 277, 1203-1204.

Pringle, H., 2001. The Mummy Congress. Hyperion/Theia, 368 pp.

Pulak, C., 1998. The Uluburun shipwreck: an overview. International Journal of Nautical Archaeology 27, 188-224.

Pyramarn, K., 1989. New evidence on plant exploitation and environment during the Hoabinhian (Late Stone Age) from Ban Kao Caves, Thailand. In Foraging and farming: the evolution of plant exploitation, D.R. Harris and G.C. Hillman, eds. Unwin Hyman, 282-291.

Quade, J., Levin, N., Semaw, S., Stout, D., Renne, P., Rogers, M., and Simpson, S., 2004. Paleoenvironments of the earliest stone toolmakers, Gona, Ethiopia. Geological Society of America Bulletin 116, 1529-1544.

Quam, R., et al. Early hominin auditory capacities. Science Advances, 12 pp.

Rabinovich, R., Gaudzinski-Windheuser, S., and Goren-Inbar, N., 2008. Systematic butchering of fallow deer (Dama) at the early middle Pleistocene Acheulian site of Gesher Benot Ya'aqov (Israel). Journal of Human Evolution 54, 134-149.

Radovčič, D., et al., 2015. Evidence for Neanderthal jewelry: modified white-tailed eagle claws at Krapina. PLoS ONE, 14 pp.

Ragir, S., 2000. Diet and food preparation: rethinking early hominid behavior. Evolutionary Anthropology 9, 153-155.

Rak, Y., et al., 1994. A Neandertal infant burial from Amud Cave, Israel. Journal of Human Evolution 26, 313-324.

Raoult, D. et al. 2006. Evidence for louse-transmitted diseases in soldiers of Napoleon's Grand Army in Vilnius. Journal of Infectious Diseases 193(1), 112-120.

Rasmussen, S., et al., 2015. Early divergent strains of *Yersina pestis* in Eurasia 5,000 years ago. Cell 163, 571-582.

Ray, H.P., 1990. Seafaring in the Bay of Bengal in the early centuries AD. Studies in History, N.S., 6, 1.

Reed, D.L., et al., 2004. Genetic analysis of lice supports direct contact between Modern and Archaic Humans. PLoS Biology, vol. 2, issue 11, 1972-1983.

Reed, D.L., Light, J.E., Allen, J.M., et al., 2007. Pair of lice lost or parasites regained: the evolutionary history of anthropoid primate lice. BMC Biology 5, 7.

Reese, D.S., 1996. Cypriot hippo hunters no myth. Journal of Mediterranean Archaeology 9, 107-112.

Reese, D.S., 2005. The Çatalhöyük shells. *In* Papers from the 1995-99 seasons, I. Hodder, ed., McDonald Institute for Archaeological Research, 123-127.

Regert, M., Colinart, S., Degrand, L., and Decavallas, O., 2001. Chemical alteration and use of beeswax through time: accelerated aging tests and analysis of archaeological samples from various environmental contexts. Archaeometry 43, 549-569.

Reich, D., et al., 2011. Denisova admixture and the first modern human dispersals into Southeast Asia and Oceania. American Journal of Human Genetics 89, 516-528.

Reinhard K., and Urban, O., 2003. Diagnosing ancient diphyllobothriasis from Chinchorro mummies. Memoirs of the Oswaldo Cruz Institute 98, 191-193.

Reinhard, K.J., and Pucu de Araújo, E., 2014. Comparative Parasitological Perspectives on Epidemiologic Transitions: The Americas and Europe. *In* Modern Environments and Human Health: Revisiting the Second Epidemiologic Transition, M.K. Zuckerman, ed., John Wiley & Sons, 321-336.

Reinhard, K.J., Helvy, R.H., and Anderson, G.A., 1987. Helminth remains from prehistoric Indian coprolites on the Colorado Plateau. Journal of Parasitology 73, 630-639.

Rendu, W., et al., 2013. Evidence supporting an intentional Neandertal burial at La Chapelle-aux-Saints. National Academy of Sciences (USA), Proceedings, Early Edition, 6 pp.

Renfrew, C., 1972. The Emergence of Civilisation. Methuan & Co., 595 pp.

Renfrew, C., 1986. Varna and the emergence of wealth in prehistoric Europe. *In* The social life of things: commodities in cultural perspective, Cambridge University Press, 141-168.

Rennie, P.J., et al., 1990. The skin microflora and the formation of human axillary odour. International Journal of Cosmetic Science 12, 197-207.

Reno, P.L., et al., 2013. A penile spine/vibrissa enhancer sequence is missing in modern and extinct humans but is retained in multiple primates with penile spines and sensory vibrissae. PLoS ONE 8(12), 7 pp.

Revedin, A., et al., 2010. Thirty thousand-year-old evidence of plant food processing. National Academy of Sciences (USA), Proceedings 107, 5 pp.

Reynolds, T.S., 1983. Stronger than a hundred men: a history of the vertical water wheel. The Johns Hopkins University Press, 454 pp.

Rice, M., 1994. The Archaeology of the Arabian Gulf. Routledge, 369 pp.

Richards, G.D., and Anton, S.C., 1991. Craniofacial configuration and postcranial development of a hydrocephalic child (ca. 2500 BC – 500 A.D.): With a review of cases and comment on diagnostic criteria. American Journal of Physical Anthropology 85, 185-200.

Richards, M.P., and Trinkhaus, E., 2009. Isotopic evidence for the diets of European Neanderthals and early modern humans. National Academy of Sciences (USA), Proceedings, Early Edition, 6 pp.

Riel-Salvatore, J, and Gravel-Miguel, C., 2013, *see* Stutz and Tarlow, eds.

Riel-Salvatore, J., and Clark, G.A., 2001. Middle and Early Upper Paleolithic burials and the use of chronotypology in contemporary Paleolithic research. Current Anthropology 42, 449-479.

Rieser, B., and Schrattenthaler, H., 1998. Urgeschichtlicher kupferbergbau in raum Schwaz-Brixlegg, Tirol. Archaeologia Austriaca 82/83, 135-179.

Rightmire, G.P., 1990. The evolution of *Homo erectus.* Cambridge University Press, 260 pp.

Roach, N.T., Lieberman, D.E., Gill, T.J., IV, Palmer, W.E., and Gill, T.J., III, 2012. The effect of humeral torsion on rotational range of motion in the shoulder and throwing performance. Journal of Anatomy 220, 293-301.

Roach, N.T., Venkadesan, M., Rainbow, M.J., and Lieberman, D.E., 2013. Elastic energy storage in the shoulder and the evolution of high-speed throwing in *Homo.* Nature Letters, 483-486.

Robbins, L.H., 1974. A Late Stone Age fishing settlement in the Lake Rudolf Basin, Kenya. Anthropological Series, vol. 1, no. 2, Publications of the Museum, Michigan State University, 157-216.

Robbins, L.M., 1987. Hominid footprints from site G. *In* Laetoli, a Pliocene Site in Northern Tanzania, M.D. Leakey, and J.M. Harris, eds., Oxford University Press, 497-523.

Robbins, L.M., Murphy, M.L., Stewart, K.M., Campbell, A.C., and Brook, G.A., 1994. Barbed stone points, paleoenvironments, and the antiquity of fish exploitation in the Kalahari Desert, Botswana. Journal of Field Archaeology 21, 257-264.

Robbins, L.R., Murphy, M.L., Brook, G.A., et al., 2000. Archaeology, paleoenvironment, and chronology of the White Paintings Rock Shelter, Tsodilo Hills, northwest Kalahari Desert, Botswana. Journal of Archaeological Science 27, 1086-1113.

Roberts, C., 2011. Re-emerging infections: developments in bioarchaeological contributions to understanding tuberculosis today. *In* A Companion to Paleopathology, A.L. Grauer, ed., Blackwell, Ch.24.

Roberts, C.A., and Manchester, K., 2005. The Archaeology of Disease. Cornell University Press, 338 pp.

Roberts, R.G., et al., 1990. Thermoluminescence dating of a 50,000-year-old human occupation site in northern Australia. Nature 345, 153-156.

Roberts, R.G., et al., 1994. The human colonization of Australia: optical dates of 53,000 and 60,000 years bracket human arrival at Deaf Adder Gorge, Northern Territory. Quaternary Science Reviews 13, 575- 583.

Robinson, M., 1999. Oldest bread in Britain. BBC News Science/Nature, 12 October 1999.

Robinson, M.E., et al., 1999. Re-assessing the logboat from Lurgan Townland, Co. Galway, Ireland. Antiquity 73, 903-908.

Roche, H., 2005. From simple flaking to shaping: stone-knapping evolution among Early Hominins. *In* Stone knapping: the necessary conditions for a uniquely hominin behavior, Ch. 3, McDonald Institute for Archaeological Research, Oxbow Books, 35-52.

Roche, H., Delagnes, A., Brugal, J.-P. et al., 1999. Early hominid stone tool production and technical skill 2.34 myr ago in West Turkana, Kenya. Nature 399, 57-60.

Roches, E., et al., 2002. Microscopic evidence for Paget's disease in two osteoarchaeological samples from early northern France. International Journal of Osteoarchaeology 12(4), 229-234.

Rodriguez-Vidal., et al., 2014. A rock engraving made by Neanderthals in Gibraltar. National Academy of Sciences (USA), Proceedings, Early Edition, 6 pp.

Roe, D., 1992. Investigations into the prehistory of the Central Solomons: some old and some new data from Northwest Guadalcanal. Actes du Colloque Lapita, ORSTOM, Noumea, 91-101.

Roebroeks, W., and Villa, P., 2011. On the earliest evidence for habitual use of fire in Europe. National Academy of Sciences (USA), Proceedings 108(13), 5209-5214.

Roebroeks, W., et al., 2012. Use of red ochre by early Neandertals. National Academy of Sciences (USA), Proceedings 109, 1889-1894.

Rogachev, A.N., 1955. Pogrebenie drevnekamennogo veka na stoyanke Kostenki XIV (Markina Gora). Sovetskaya Etnografiya 1, 29-38.

Rogers, M.J., and Semaw, S., 2009. From nothing to something: the appearance and context of the earliest archaeological record. In Sourcebook of Paleolithic transitions, Springer Science+Business Media, 155-171.

Rollefson, G.O., 1986. Neolithic 'Ain Ghazal (Jordan): ritual and ceremonial. Antiquity 12, 45-52.

Rollefson, G.O., and Schmidt, K., 2005. The early Neolithic origin of ritual centers. Neo-Lithics 2/05, 53 pp.

Rollefson, G.O., Simmons, A.H., and Kafafi, Z., 1992. Neolithic Cultures at 'Ain Ghazal, Jordan. Journal of Field Archaeology 19, 443-470.

Roper, M.K., 1969. A survey of the evidence for intrahuman killing in the Pleistocene. Current Anthropology 10, 427-459.

Rosell, J., et al., 1993. Les Mines prehistòriques de Gavà (Baix Llobregat). Mineralogistes de Catalunya, vol. V, no. 8, pp. 220-239.

Rosen, S.A., 1996. The decline and fall of flint. In Stone tools, Springer, 129-158.

Rosen, W., 2007. Justinian's Flea. Viking, 367 pp.

Rosenberg, K., and Trevathan, W., 1996. Bipedalism and human birth: the obstetrical dilemma revisited. Evolutionary Anthropology 4, 161-168.

Rosenberg, K.R., 1998. Morphological variation in West Asian postcrania: Implications for obstetric and locomotor behavior. In Neandertals and Modern Humans in Western Asia, Plenum Press, NY, 367-379.

Rothenberg, B., and Blanco Freijeiro, A., 1980. Ancient copper mining and smelting at Chinflon (Huelva, SW Spain). In Scientific studies in early mining and extractive metallurgy, 41-62.

Rothschild, B.M., 1988. Existence of syphilis in a Pleistocene bear. Nature 335, p. 595.

Rothschild, B.M., 2005. History of syphilis. Clinical Infectious Diseases 40, 1454-1463.

Rothschild, B.M., and Martin, L.D., 2006a. Skeletal Impact of Disease. New Mexico Museum of Natural History and Science, Albuquerque NM, Bulletin 13, 226 pp.

Rothschild, B.M., and Rothschild, C., 1995. Treponemal disease revisited: skeletal discrimination of yaws, bejel, and venereal syphilis. Clinical Infectious Diseases 20, 1402-1408.

Rothschild, B.M., and Rothschild, C., 1996a. Treponemal disease in the New World. Current Anthropology 37, 555-561.

Rothschild, B.M., and Rothschild, C., 1996b. Treponematoses: origins and 1.5 million years of transition. Human Evolution 11, 225-232.

Rothschild, B.M., and Rothschild, C., 1998a. Skeletal examination-based recognition of treponematoses: a four-continent odyssey of denouement, transition and spread. Bulletin et Mémoires de la Société d'Anthropologie de Paris 10, 29-40.

Rothschild, B.M., and Thillaud, P.L., 1991. Oldest bone disease. Nature 399, p. 288.

Rothschild, B.M., and Turnbull, W., 1987. Treponemal infection in a Pleistocene bear. Nature 329, 61-62.

Rothschild, B.M., Calderon, F.L., Coppa, A., and Rothschild, C., 2000. First European exposure to syphilis: the Dominican Republic in the time of Colombian contact. Clinical Infectious Diseases 31, 936-941.

Rothschild, B.M., Hershkovitz, I., and Rothschild, C., 1995. Origin of yaws in Pleistocene East Africa: *Homo erectus.* Nature 378, 343-344.

Rothschild, B.M., Turner, K.R., and DeLucca, M.A., 1988. Symmetrical erosive peripheral polyarthritis in the Late Archaic Period of Alabama. Nature 241, 1498-1501.

Roux, V., and de Miroschedji, P., 2009. Revisiting the history of the potter's wheel in the Southern Levant. Levant 41, 155-173.

Rowlands, M.J., 1972. Defence: a factor in the organization of settlements. *In* Man, Settlement and Urbanism, pp. 447-462.

Rowlett, R.M., 1999. Did the use of fire for cooking lead to a diet change that resulted in the expansion of brain size in *Homo erectus* from that of *Australopithecus africanus*? Science 284, 741.

Rowlett, R.M., 1999b. Comment on Wrangham et al. Current Anthropology 40(5), 584-585.

Rowlett, R.M., 2000. Fire control by *Homo erectus* in East Africa and Asia. Acta Anthropologica Sinica, Suppl. to vol. 19, 198-208.

Rubinson, K.S., 1990. The textiles from Pazyryk. Expedition 31(1), 49-61.

Rudenko, S.I., 1953. Kultura naseveniya Gornogo Altaiya v Skifskoe Vremya, Izdatel'stvo Akademiya Nauk SSSR, 402 pp.

Rudenko, S.I., 1970, translated by M.W. Thompson, 1970. Frozen Tombs of Siberia. University of California Press, 340 pp.

Rudgley, R., 1994. Essential substances: a cultural history of intoxicants in society. Kodansha International, 195 pp.

Rudgley, R., 1999. The lost civilization of the Stone Age. The Free Press, 310 pp.

Ruff, C.B., 1993. Climatic adaptation and hominid evolution: the thermoregulatory imperative. Evolutionary Anthropology 2, 53-60.

Ruff, C.B., et al., 1993b. Postcranial robusticity in *Homo*. 1. Temporal trends and mechanical interpretation. American Journal of Physical Anthropology 91, 21-53.

Runnels, C.N., 1985. Trade and the demand for millstones in Southern Greece in the Neolithic and the Early Bronze Age. *In* Prehistoric production and exchange: the Aegean and Eastern Mediterranean, 30-43.

Russell, M., 2000. Flint Mines in Neolithic Britain. Tempus Publishing Ltd., 160 pp.

Russell, M.D., 1987. Mortuary practices at the Krapina Neandertal site. American Journal of Physical Anthropology 72, 381-397.

Russo, E.B. 2008. Phytochemical and genetic analyses of ancient cannabis from Central *Asia*. Journal of Experimental Botany 59(15), 4171–4182.

Ryan, K., 2005. Facilitating milk let-down in traditional cattle herding systems: East Africa and beyond. *In* The Zooarchaeology of Milk and Fats, Oxbow Books, 96-106.

Ryder, M.L., 1990. Wool remains from Scythian burials in Siberia. Oxford Journal of Archaeology 9, 313-321.

Ryder, M.L., 1990b. Skin, and wool-textile remains from Hallstatt, Austria. Oxford Journal of Archaeology 9, 37-49.

Ryšánek, J., and Václavů, V., 1989. A distillation apparatus from Spišský Štvrtok. Archeologické rozhledy 41, 196-201.

Sabet, A.M., Tsogoev, V.B., Shibanin, S.P., El Kai, M., and Awad, S., 1976. The placer tin deposits of Abu-Dabbab, Igla, and Nuweibi. Annals of the Geological Survey of Egypt VI, 169-180.

Sablin, M.V., and Khlopachev, G.A., 2002. The earliest Ice Age dogs: evidence from Eliseevichi I. Current Anthropology 43, 795-799.

Sadler, J.P., 1990. Records of ectoparasites on humans and sheep from Viking Age deposits in the former Western Settlement on Greenland. Journal of Medical Entomology 27, 628-631.

Sadr, K., and Sampson, C.G., 2006. Through thick and thin: early pottery in Southern Africa. Journal of African Archaeology 4, 235-252.

Sala Burgos, N., Cuevas González, J., and López Martínez, N., 2007. Estudio paleopatólogico de una hemimandibula de *Tethytragus* (Artiodactyla, Mammalia) del Mioceneo Medio de Somosaguas (Pozuelo de Alarcón, Madrid). Coloquios de Paleontología 57, 7-14.

Salque, M., et al., 2013. Earliest evidence for cheese making in the sixth millennium BC in northern Europe. Nature 493, 522-525.

Sampson, A., 1998. The Neolithic and Mesolithic occupation of the Cave of Cyclops, Youra, Alonnessos, Greece. British School of Athens, 93, 1-22.

Sampson, C.G., 2006. Acheulian quarries at hornfels outcrops in the Upper Karoo region of South Africa. *In* Axe age: Acheulian tool-making from quarry to discard. Equinox, London, 75-107.

Samuel, D., 1996. Archaeology of ancient Egyptian beer. Journal of the American Society of Brewing Chemists 53, 3-12.

Sandars, N.K., 1985. Prehistoric art in Europe. Penguin Books, 508 pp.

Sandison, A.T., 1967. Sexual behavior in ancient Societies. *In* Diseases in Antiquity, D. Brothwell and A.T. Sandison, eds., Charles C. Thomas, Springfield, IL, 734-755.

Sandison, A.T., 1980. Diseases in ancient Egypt. *In* Mummies, Disease and Ancient Cultures, A. Cockburn and E. Cockburn, eds., Cambridge University Press, 29-44.

Santonja, M., and Villa, P., 2006. The Acheulian of Western Europe. *In* The axe age: Acheulian toolmaking. Equinox, London, 429-478.

Sapir-Hen, L., and Ben-Yosef, E., 2013. The introduction of domestic camel to the Southern Levant: evidence from the Aravah Valley. Tel Aviv 40, 277-285.

Saul, J.M., 2002. Was the first language purposefully invented? Mother Tongue: Journal of the Association for the Study of Language in Prehistory, VII, 259-264.

Saul, J.M., 2013. The Tale Told in All Lands. Paris, Les Trois Colonnes, 623 pp.

Saul, J.M., 2019. What the Stork Brought: African click-speakers and the spread of humanity's oldest beliefs. Old Africa Books, Naivasha, Kenya, 109 pp.

Savolainen, P., et al., 2002. Genetic evidence for an East Asian origin of domestic dogs. Science 298, 1610-1613.

Sawyer, S., et al., 2015. Nuclear and mitochondrial DNA sequences from two Denisovan individuals. National Academy of Sciences (USA), Proceedings, Early Edition, 5 pp.

Scarre, C., ed., 2005. The human past. Thomas & Hudson, 2nd edition, 784 pp.

Schedl, W., 2000. Contribution to insect remains from the accompanying equipment of the Iceman. *In* The man in the ice, Harmony Books, vol. 4, 151-155.

Schick, K.D., 1998. A comparative perspective on Paleolithic cultural patterns. *In* Neandertals and Modern Humans in Western Asia, Plenum Press, 449-460.

Schick, K.D., and Toth, N., 1993. Making silent stones speak. Simon & Schuster, 351 pp.

Schick, T., 1988. Cordage, basketry, and fabrics. *In* Nahal Hemar Cave, 31-43.

Schild, R., 1976. Flint mining and trade in Polish prehistory as seen from the perspective of the chocolate flint of central Poland. A second approach. Archaeologica Carpathica, tome XVI, 147-177.

Schild, R., 1983. The exploitation of chocolate flint in central Poland. The human use of flint and chert: Proceedings of the 4th International Flint Symposium, Brighton Polytechnic, 137-150.

Schild, R., and Sulgostowska, Z., 1997., eds., Man and Flint. Proceedings of the VIIth International Flint Symposium, Institute of Archaeology and Ethnology, Polish Academy of Sciences, 361 pp.

Schlingloff, D., 1974. Cotton manufacture in ancient India. Journal of the Economic and Social History of the Orient 17(1), 81-90.

Schmandt-Besserat, D., 1977. The earliest uses of clay in Syria. Expedition 19, 28-42.

Schmandt-Besserat, D., 1978. The earliest precursor of writing. Scientific American, June 1978, 50-59.

Schmandt-Besserat, D., 1980. Ochre in prehistory: 300,000 years of the use of iron ores as pigments. In The coming of the Iron Age. Yale University Press, 127-150.

Schmandt-Besserat, D., 1998. Ain Ghazal 'Monumental' Figures. Bulletin of the American Schools of Oriental Research, 310, 1-17.

Schmandt-Besserat, D., 1998b. A Neolithic female revealing her breasts. In Written on Clay and Stone. Ancient Near Eastern Studies, Presented to Krystyna Szarzyńska on the Occasion of her 80th Birthday, Agade Publishing, Warsaw, 79-83.

Schmidt, E., 1972. A Mousterian silex mine and dwelling place in the Swiss Jura. In The origin of Homo sapiens (Ecology and Conservation 3) UNESCO, 129-132.

Schmidt, K., 1995. Investigations in the Upper Mesopotamian early Neolithic: Göbekli Tepe and Gürcütepe. Neolithics; a newsletter of Southwest Asian lithics research 2, 9-10.

Schmidt, K., 1998. Frühneolithische Tempel: Ein Forschungsbericht zum präkeramischen Neolithikum Obermesopotamiens. Mitteilungen der Deustschen Orient-Gesellschaft zu Berlin 130, 17-49.

Schmidt, K., 2001. Göbekli Tepe, Southeastern Turkey: a preliminary report on the 1995-1999 excavations. Paléorient 26(1), 2645-2654.

Schnitter, N., 1979. Antike Talsperren in Anatolien. Mitteilungen Leichtweiss-Institut für Wasserbau der Technischen Universität Braunschweig 64, 3 pp. + 2 plates.

Schoenemann, P.T., 2006. Evolution of the size and functional areas of the human brain. Annual Review of Anthropology 35, 179-406.

Schultz, A.H., 1967. Notes on diseases and healed fractures of wild apes. In Diseases in Antiquity, D. Brothwell and A.T. Sandison, eds., Charles C. Thomas, Springfield, IL, 47-55.

Scobie, A., 1986. Slums, sanitation and mortality in the Roman World. Klio 68, 399-433.

Scott, K., 1980. Two hunting episodes of Middle Palaeolithic age at La Cotte de Saint Brelade, Jersey (Channel Islands). World Archaeology 12, 137-152.

Scott, R.S., et al., 2005. Dental microwear texture analysis shows within-species diet variability in fossil hominins. Nature Letters 436, 3 pp.

Scrimshaw, S.C.M., 1984. Infanticide in Human Populations: Societal and Individual Concerns. In G. Hausfater and S.B. Hardy, eds. Infanticide: Comparative and Evolutionary Perspectives, Aldine, New York, 439-462.

Searcey, N., et al., 2013. Parasitism of the Zweeloo Woman: Dicrocoeliasis evidenced in a Roman period bog mummy. International Journal of Paleopathology 3(3), 224-228.

Segal, I., Speirs, M.S., Meirav, O., Sherter, U., Feldman, H., Goring-Moris, N., 1995. Remedy for an 8500 Year-old plastered human skull from Kfar Hahoresh, Israel. Journal of Archaeological Science 22, 779-788.

Sellen, D.W., 2007. Evolution of infant and young child feeding: implications for contemporary public health. Annual Review of Nutrition 27, 123-148.

Sellers, W.I., et al., 2005. Stride lengths, speed and energy costs in walking of *Australpithecus afarensis*: using evolutionary robotics to predict locomotion of early human ancestors. Journal of the Royal Society Interface 2, 431-441.

Semaw, S., et al., 2003. 2.6-Million-year-old stone tools and associated bones from OGS-6 and OGS-7, Gona, Afar, Ethiopia. Journal of Human Evolution 45, 169-177.

Semaw, S., Renne, P., Harris, J.W.K., Feibel, C.S., Bernor, R.L., Fesseha, N., and Mowbray, K., 1997. 2.5-million-year-old stone tools from Gona, Ethiopia. Nature 385, 333-336.

Shahack-Gross, R., et al., 2014. Evidence for the repeated use of a central hearth at Middle Pleistocene (300 ky ago) Qesem Cave, Israel. Journal of Archaeological Science 44, 12-21.

Shapiro, B., et al., 2006. No proof that typhoid caused the Plague of Athens (a reply to Papagrigorakis et al.). International Journal of Infectious Diseases 10(4):334-335; author reply 335-336.

Sharma, G.R., 1985. From hunting and food gathering to domestication of plants and animals in the Belan and Ganga Valleys. *In* Recent advances in Indo-Pacific prehistory: Proceedings of the International Symposium held at Poona, December 19-21, 1978.

Sharma, G.R., et al., 1980. Beginnings of Agriculture. Abinash Prakashan, 237 pp.

Sharpless. F.F., 1908. Mercury mines at Koniah, Asia Minor. The Engineering and Mining Journal (E & MJ) 86, 601-603.

Shaw, I., 2006. 'Master of the roads': quarrying and communications networks in Egypt and Nubia. *In* L'apport de l'Égypte à l'histoire des techniques: méthodes, chronologie et comparaisons, 253-266.

Shea, J.J., 2006. The origins of lithic projectile point technology: evidence from Africa, the Levant, and Europe. Journal of Archaeological Science 33, 823-846.

Shea, J.J., 2009. The impact of projectile weaponry on Late Pleistocene Hominin evolution. *In* The Evolution of Hominin Diets: Integrating Approaches to the Study of Palaeolithic Subsistence, Springer Science+Business Media B.V., 189-199.

Sheng, A., 2010. Textiles from the Silk Road. Expedition 52, no. 3.

Shennann, S., 2008. Population processes and their consequences in Early Neolithic Central Europe. *In* The Neolithic Demographic Transition and its consequences, Springer Science+Business Media B.V., 315-329.

Shepherd, R., 1980. Prehistoric mining and allied industries. Academic Press, 265 pp.

Sherratt, A., 1977. Resources, technology and trade: an essay in early European metallurgy. *In* Problems in economic and social archaeology, 557-582.

Sherratt, A., 1981. Plough and pastoralism: aspects of the secondary products revolution. *In* Pattern of the past: studies in honour of David Clarke, Cambridge University Press, 261-305.

Sherratt, A., 1990. The genesis of megaliths: monumentality, ethnicity and social complexity in Neolithic north-west Europe. World Archaeology 22, 147-167.

Sherratt, A., 1997. Economy and Society in Prehistoric Europe. Princeton University Press, 561 pp.

Sherratt, A., 1999. Cash-crops before cash: organic consumables and trade. *In* The prehistory of food: appetite for change, Routledge, 13-34.

Sherratt, A., and Sherratt, S., 1991. Bronze Age trade in the Mediterranean. Conference publication, Paul Åströms Förlag, Jonsered, Göteborg, Studies in Mediterranean Archaeology, vol. XC, 351-386.

Sherratt, A.G., 1986. Two new finds of wooden wheels from later Neolithic and Early Bronze Age Europe. Oxford Journal of Archaeology 5, 243-248.

Sherratt, A.G., 1986b. Wool, wheels and plough marks: local developments of outside introduction in Neolithic Europe. Bulletin of the Institute of Archaeology, University of London, 1-15.

Sherratt, A.G., 1991. Sacred and profane substances: the ritual use of narcotics in Later Neolithic Europe. *In* Sacred and profane: proceedings of a conference on archaeology, ritual and religion (Oxford 1989) 50-64.

Sianto, L., et al., 2005. The Finding of Echinostoma (Trematoda: Digenea) and Hookworm Eggs in Coprolites Collected From a Brazilian Mummified Body Dated 600–1,200 Years Before Present. Journal of Parasitology 91(4), 972-975.

Sieveking, G. de G., and Newcomer, M.H., eds., 1987. The Human Uses of Flint and Chert. Cambridge University Press, Cambridge, U.K., 263 pp.

Simmons, A.H., 1998. Of tiny hippos, large cows and early colonists in Cyprus. Journal of Mediterranean Archaeology 11, 232-241.

Simmons, A.H., 1999. Faunal extinction in an Island Society: pygmy hippopotamus hunters of Cyprus. Kluwer Academic/Plenum Publishers, 381 pp.

Simoons, F.J., 1971. The antiquity of dairying in Asia and Africa. Geographical Review 61, 431-439.

Simoons, F.J., 1979. Dairying, milk use and lactose malabsorption in Eurasia: a problem in culture history. Anthropos 74, 61-80.

Simpson, G.G., 1953. The Major Features of Evolution. Columbia University Press, New York, 434 pp.

Singer, R., and Wymer, J., 1982. The Middle Stone Age at Klasies River Mouth in South Africa. University of Chicago Press, 234 pp.

Sinitsyn, A.A., 2003. A Palaeolithic 'Pompeii' at Kostenki, Russia. Antiquity 77, 9-14.

Sistiaga, A., et al., 2014. The Neanderthal meal: a new perspective using faecal biomarkers. PLoS ONE 9(6), 6 pp.

Sjøvold, T., et al., 1995. Verteilung und Grösse der Tätowierungen am Eismann von Hauslabjoch. *In* Der Mann im Eis: Neue Funde und Ergebnisse, Springer-Verlag, 279-286.

Skinner, M.M., et al, 2015. Human-like hand use in *Australopithecus africanus*. Science 347, 395-399.

Skinsnes, O.K., and Chang, P.H. 1985. Understanding of leprosy in ancient China. International Journal of Leprosy and *Other* Mycobacterial Diseases 53(2), 289-307.

Smirnov, Y., 1989. Intentional human burial: Middle Paleolithic (Last Glaciation) beginnings. Journal of World Prehistory 3, 199-233.

Smith, C.S., 1983. A Search for Structure. MIT Press, 424 pp.

Smith, G., and Williams, A., 2012. Pentrwyn copper smelting site excavation, 2011. Great Orme, Llandudno, Conway. GAT Project G2178, Preliminary Report.

Smith, N., 1971. A history of dams. Peter Davies, 279 pp.

Snir, A., et al., 2015. The origin of cultivation and proto-weeds, long before Neolithic farming. PLoS ONE 10(7), 12 pp.

Snodgrass, A.M., 1991. Bronze Age exchange: a minimalist position. *In* Bronze Age trade in the Mediterranean, Conference publication, Paul Åströms Förlag, Jonsered, Göteborg, Studies in Mediterranean Archaeology, vol. XC, 15-20.

Snodgrass, J.J., Leonard, W.R., and Robertson, M.L., 2009. The energetics of encephalization in early hominids. *In* The evolution of hominin diets: integrating approaches to the study of Palaeolithic subsistence. Springer Science+Business Media B.V., 15-29.

Snodgrass. A.M., 1983. Heavy freight in Archaic Greece. *In* Trade in the Ancient Economy, University of California Press, 16-26.

Soffer, O., 1985. The Upper Paleolithic of the Central Russian Plain. Academic Press, 539 pp.

Soffer, O., 1989. Storage, sedentism and the Eurasian Palaeolithic record. Antiquity 63, 719-732.

Soffer, O., Adovasio, J.M., and Hyland, D.C., 2000. The 'Venus' figurines. Current Anthropology 41, 511-537.

Soffer, O., Adovasio, J.M., Hyland, D.C., Klíma, B., and Svoboda, J., 1998. Perishable technologies and the genesis of the Eastern Gravettian. Anthropologie XXXVI, 43-68.

Soffer, O., Vandiver, P., Klima, B., and Svoboda, J., 1992. The pyrotechnology of performance art: Moravian Venuses and wolverines. *In* H. Knecht et al., eds., Before Lascaux: The Complex record of the Early Upper Paleolithic, CRC Press, Ch. 16, 259-275.

Solecki, R.S., 1969. A copper mineral pendant from Northern Iraq. Antiquity 43, 311-314.

Solecki, R.S., 1971. Shanidar, the Humanity of Neanderthal Man. Allen Lane, Penguin Press, 222 pp.

Solecki, R.S., 1975. Shanidar IV, A Neandertal flower burial in northern Iraq. Science 190, 880-881.

Sondaar, P.Y., et al., 1994. Middle Pleistocene faunal turnover and colonization of Flores (Indonesia) by *Homo erectus*. Comptes Rendus de l'Académie des sciences, tome 319, series II, 1255-1262.

Soodyall, H., 2006. The Prehistory of Africa. Jonathan Ball Publishers, Johannesburg & Cape Town, 189 pp.

Soressi, M., and d'Errico, F., 2007. Pigments, gravures, parures: les comportements symboliques Controversés des Néandertaliens. Les Néandertaliens: Biologie et Cultures, Éditions du CTHS, Documents préhistoriques 23, 297-309.

Soressi, M., et al., 2013. Neandertals made the first specialized bone tools in Europe. National Academy of Sciences (USA), Proceedings, Early Edition, 5 pp.

Speth, J.D., 2006. Housekeeping, Neandertal-Style: hearth placement and midden formation in Kebara Cave (Israel). *In* Transitions before the transition: evolution and stability in the Middle Paleolithic and Middle Stone Age, Springer, 171-188.

Speth, J.D., and Tchernov, E., 1998. The role of hunting and scavenging in Neandertal procurement strategies. *In* Neandertals and Modern Humans, Plenum Press, New York, 223-239.

Spoor, F., et al., 2015. Reconstructed *Homo habilis* type OH7 suggests deep-rooted species diversity in early *Homo*. Nature 519, 83-86.

Spoor, F., Wood, B., and Zonneveld, F., 1994. Implications of early hominid labyrinthine morphology for evolution of human bipedal locomotion. Nature 369, 645-648.

Spriggs, M., 1997. The Island Melanesians. Blackwell Publishers, 326 pp.

Stahl, A.B., ed., 2005. African Archaeology, a Critical Introduction. Blackwell Publishing, 490 pp.

Stapert, D., and Johansen, L., 1999. Making fire in the stone age: flint and pyrite. Geologie en Mijnbouw 78, 147-164.

Stapert, D., and Johansen, L., 1999b. Flint and pyrite: making fire in the Stone Age. Antiquity 73, 765- 777.

Stech, T., 1990. Neolithic copper metallurgy in Southwest Asia. Archeomaterials 4, 55-61.

Steele, J., 1999. Stone legacy of skilled hands. Nature 399, 24-25.

Steensberg, A., 1973. A 6000 year old ploughing instrument from Satrup Moor. Tools and Tillage 2, 105-118.

Stepanchuk, V.N., 1993. Prolom II, a Middle Palaeolithic cave site in the Eastern Crimea with non-utilitarian bone artefacts. Proceedings of the Prehistoric Society 59, 17-37.

Steudel-Numbers, K.L., and Wall-Scheffler, C.M., 2009. Optimal running speed and the evolution of hominin hunting strategies. Journal of Human Evolution 56, 355-360.

Steuer, H., and Zimmermann, U., 1993. Montanarchäologie in Europa. Jan Thorbecke Verlag Sigmaringen, 562 pp.

Stevenson, R.D., and Allaire, J.H., 1991. The development of normal feeding and swallowing. Pediatric Clinics of North America 38, 1439-1453.

Stiles, D.N., 1998. Raw material as evidence for human behaviour in the Lower Pleistocene: the Olduvai case. In Early human behavior in global context: the rise and diversity of the Lower Palaeolithic record, 133-150.

Stiles, D.N., et al., 1974. The MNK Chert Factory Site, Olduvai Gorge, Tanzania. World Archaeology 5, 285-308.

Stiner, M.C., 1999. Palaeolithic mollusk exploitation at Riparo Mochi (Balzi Rossi, Italy): food and ornaments from the Aurignacian through Epigravettian. Antiquity 73, 735-754.

Stiner, M.C., Munro, N.D., and Surovell, T.A., 2000. The tortoise and the hare. Current Anthropology 41, 39-79.

Stirland, A., 1991. Paget's disease (osteitis deformans): A classic case? International Journal of Osteoarchaeology 1(3-4), 173-177.

Stoddart, D.M. 2015. Adam's nose, and the making of humankind. Imperial College Press, 258 pp.

Stoddart, D.M., 1990. The scented ape. Cambridge University Press, 286 pp.

Stoddart, D.M., 1998. The human axillary organ: an evolutionary puzzle. Human Evolution 13, 73-89.

Stöllner, T., Cierny, J., Eibner, C., Boehnke, N., Herd, R., Maass, A., Röttger, K., Sormaz, T., Steffens, G., and Thomas, P., 2006. Der Bronzezeitliche Bergbau im Südrevier des Mitterberggebietes Bericht zu den Forschungen der Jahre 2002 bis 2006. Archaeologica Austriaca 90, 87-137.

Stöllner, T., et al., 2003. The economy of Dürrnberg-bei-Hallein: an Iron Age salt-mining centre in the Austrian Alps. Antiquaries Journal 83, 123-194.

Storck, J., and Teague, W.D., 1952. Flour for Man's Bread. University of Minnesota Press, 382 pp.

Stordeur-Yedid, D., 1979. Les Aiguilles À Chas Au Paléolithique. 'Gallia Préhistoire', XIII, suppl., 214 pp.

Stordeur, D., et al., 1996. Jerf el-Ahmar: a new Mureybetian site (PPNA) on the Middle Euphrates. Neo-lithics 2, pp. 1-2.

Stordeur, D., et al., 1997. Jerf el Ahmar un nouveau site de l'Horizon PPNA sur le moyen Euphrate Syrien. Bulletin de la Société Préhistorique Française, tome 94, 282-285.

Stos-Gale, Z.A., 1989. Cycladic copper metallurgy. In World Archaeometallurgy, 279-291.

Stout, D., et al., 2005. Raw material selectivity of the earliest stone toolmakers at Gona, Afar, Ethiopia. Journal of Human Evolution 48, 365-380.

Strait, D.S., et al., 2009. The feeding biomechanics and dietary ecology of *Australopithecus africanus.* National Academy of Sciences (USA), Proceedings 106(7), 2124-2129.

Strasser, T.F., Panapopoulou, E., Runnels, C.N., Murray, P.M., Thompson, N., Karkanas, P., McCoy, F.W., and Wegmann, K.W., 2010. Stone age seafaring in the Mediterranean. Hesperia 79, 145-190.

Straus, L.G., 1993. Upper Paleolithic hunting tactics and weapons in Western Europe. In Hunting and Animal Exploitation in the later Palaeolithic and Mesolithic of Eurasia, Archaeological Papers of the American Anthropological Association 3, 83-93.

Stringer, C., 2002. Modern human origins: progress and prospects. Philosophical Transactions of the Royal Society, London, B, 357, 563-579.

Strouhal, E., 1973. Five plastered skulls from the Pre-Pottery Neolithic B Jericho: Anthropological study. Paléorient 1, 231-247.

Strouhal, E., 1987. La tuberculose vertébrale en Égypte et Nubie anciennes. Bulletins et Mémoires de la Société d'Anthropologie de Paris 14, 261-270.

Strouhal, E., 1998. Paleopathological evidence of jaw tumors. In Dental Anthropology: Fundamentals, Limits, and Prospects, K.W. Alt, F.W. Rosing, and M. Teschler-Nicola, eds., Springer, New York, 277-292.

Stutz, L.N., and Tarlow, S., 2013. The Oxford handbook of the archaeology of death and burial. Oxford University Press, 817 pp.

Surridge, A.K., Osorio, D., and Mundy, N.I., 2003. Evolution and selection of trichromatic vision in primates. Trends in Ecology and Evolution 18, 198-205.

Susman, R.L., 1991. Who made the Oldowan tools? Fossil evidence for tool behavior in Plio-Pleistocene hominids. Journal of Anthropological Research 47, 129-151.

Susman, R.L., 1994. Fossil evidence for early hominid tool use. Science 265, 1570-1575.

Swindler, D.R., Ryan, D.P., and Rothschild, B.M., 1995. Dental remains from the Valley of the Kings, Luxor, Egypt, New Kingdom (1550 to 1070 BC). In Aspects of Dental Biology, Palaeontology, Anthropology and Evolution, J. Moggi-Cecci, ed., International Institute for the Study of Man, Florence, Italy, 365-372.

Tach, D., 2013. 5,000-year-old gaming pieces found in Turkey. Discovery News, 14 August 2013.

Talalay, L.E., 1993. Deities, dolls, and devices: Neolithic figurines from Franchthi Cave, Greece. Indiana University Press, 198 pp.

Tallet, P., 2012. Ayn Sukhna and Wadi el-Jarf: two newly discovered pharaonic harbours on the Suez Gulf. British Museum Studies in Ancient Egypt and Sudan 18, 147-168.

Tanke, D.H., and Rothschild, B.M., 1997. Paleopathology. In Encyclopedia of Dinosaurs, P.J. Currie and K. Padian, eds., Academic Press, San Diego, CA, 525-530.

Tankersley, S.L., Nagy, F., Rupert, L. Tankersley, K.O., and Tankersley, K.B., 1994. Detection of Giardia lamblia in Early Woodland Indian paleofecal specimens from Salts Cave, Mammoth Cave National Park, Kentucky. In Abstracts of the 69th Annual Meeting of the American Society of Parasitologists, August 9-13, Fort Collins, CO, 1994, Abstract 90 (Suppl. to Journal of Parasitology).

Tarantola, A., 2017. Four Thousand Years of Concepts Relating to Rabies in Animals and Humans, Its Prevention and Its Cure. Tropical Medicine and Infectious Disease Review 2, 5.

Tasić, N.N., 2000. The Neolithic Settlement pattern and salt. In Proceedings of the 8th World Salt Symposium, 2, Amsterdam, 1139-1143.

Tayles, N. 1996. Anemia, genetic diseases, and malaria in prehistoric mainland Southeast Asia. American Journal of Physical Anthropology 101(1), 11-27.

Taylor, G.M., Murphy, E., Hopkins, R., Rutland, P., and Chistov, Y., 2007. First report of *Mycobacterium bovis* DNA in human remains from the Iron Age. Microbiology 153, 1243-1249.

Tchernov, E., 1994. Biological evidence for human sedentism in Southwest Asia during the Natufian. *In* The Natufian Culture in the Levant, Archeological Series, International Monographs in Prehistory, 315-340.

Tchernov, E., 1997. Two new dogs, and other Natufian dogs from the Southern Levant. Journal of Archaeological Science 24, 65-95.

Tchernov, E., 1998. The faunal sequence of the Southwest Asian Middle Paleolithic in relation to hominid dispersal events. *In* Neandertals and Modern Humans in Western Asia, Plenum Press, New York, 77-90.

Tegel, W., et al., 2012. Early Neolithic water wells reveal the World's oldest wood architecture. PLoS ONE, vol. 7, issue 12, 8 pp.

Telford, S.R., Jr., 1984. Haemoparasites of reptiles. *In* Diseases of Amphibians and Reptiles, G.L. Hoff, F.L. Frye, and E.R. Henderson, eds., Plenum Press, New York, 385-517.

Teschler-Nicola, M., Gerold, F., Bujatti-Narbeshuber, M., Prohaska, T., Latkoczy, C., Stingeder, G., and Watkins, M., 1999. Evidence of genocide 7000 BP – Neolithic paradigm and geo-climatic reality. Collegium Antropologicum 28, 437-450.

Thackeray, J.F., 1988. Molluscan fauna from Klasies River, South Africa. South African Archaeological Bulletin 43, 27-32.

Théry-Parisot, I., and Meignen, L., 2000. Économie des combustibles (bois et lignite) dans l'Abri Moustérien des Canalettes. Gallia Préhistoire 42, 45-55.

Théry, I., et al., 1996. Coal used for fuel at two prehistoric sites in southern France: Les Canalettes (Mousterian) and Les Usclades (Mesolithic). Journal of Archaeological Science 23, 509-512.

Thieme, H., 1997. Lower Palaeolithic hunting spears from Germany. Nature 385, 807-810.

Thieme, H., 2005. The Lower Palaeolithic art of hunting. *In* The hominid individual in context: archaeological investigations of Lower and Middle Paleolithic landscapes, locales and artefacts, Routledge, 115-132.

Thieme, H., and Veil, S., 1985. Neue Untersuchungen zum eemzeitlichen Elefanten-Jagdplatz Lehringen, Ldkr. Verden. Die Kunde, N.F., 36, 11-58.

Thoms, A.V., 2009. Rocks of ages: propagation of hot-rock cookery in western North America. Journal of Archaeological Science 36, 573-591.

Thomsen, C.J., 1836. Ledetraad til Nordisk Oldkyndighed (A guide to Northern Antiquities, translated by Lord Ellesmere) London, 1848, pp. 63-68.

Tilley, L., and Oxenham, M.F., 2011. Survival against the odds: Modeling the social implications of care provision to seriously disabled individuals. International Journal of Paleopathology 1, 35-42.

Tillier, A.-M., Arensburg, B., Raky, Y., and Vandermeersch, B., 1995. Middle Palaeolithic dental caries: new evidence from Kebara (Mount Carmel, Israel). Journal of Human Evolution 29, 189-192.

Tillier, A.-M., Arensburg, B., Duday, H., and Vandermeersch, B., 2001. An early case of hydrocephalus: The Middle Paleolithic Qafzeh 12 child (Israel). American Journal of Physical Anthropology 114, 166-170.

Tishkoff, S.A., et al., 2007. Convergent adaptation of human lactase persistence in Africa and Europe. Nature Genetics 39, 31-40.

Tomasello, M., 2009. Why we cooperate. The MIT Press, 204 pp.

Torrence, R., 1986. Production and exchange of stone tools. Cambridge University Press, 256 pp.

Tosi, M., 1968. Excavations at Shahr-I Sokhta preliminary report on the second campaign. East and West 19, 283-386.

Toth, N., 1985. Archaeological evidence for preferential right-handedness in the lower and middle Pleistocene, and its possible implications. Journal of Human Evolution 14, 607-614.

Toth, N., and Schick, K., 1993. Early stone industries and inferences regarding language and cognition. *In* Tools, Language and Cognition in Human Evolution, Cambridge University Press, 346-362.

Toth, N., et al., 1993. Pan the tool-maker: investigations into the stone tool-making and tool-using capabilities of a bonobo (*Pan paniscus*). Journal of Archaeological Science 20, 81-91.

Tran, T.-N., Forestier, C.L., Drancourt, M., Raoult, D., and Aboudharam, G., 2011. Co-detection of *Bartonella quintana* and *Yersinia pestis* in an 11th–15th [century] burial site in Bondy, France. American Journal of Physical Anthropology 145. 489-494.

Trinkaus, E., 1983. The Shanidar Neanderthals. Academic Press, San Diego, CA, 502 pp.

Trinkaus, E., 1995. Neanderthal mortality patterns. Journal of Archaeological Science 223, 121-142.

Trinkaus, E., and Zimmerman, M.R., 1982. Trauma among the Shanidar Neandertals. American Journal of Physical Anthropology 57, 61-76.

Tryon, C.A., and McBrearty, S., 2002. Tephrostratigraphy and the Acheulian to Middle Stone Age transition in the Kapthurin Formation, Kenya. Journal of Human Evolution 42, 211-235.

Turk, I., Dirjec, J., and Kavur, B., 1995. The oldest musical instrument in Europe discovered in Slovenia? Razprave IV. Razreda Sazu, XXXVI, 12, 287-293.

Turk, I., et al., 1997. Mousterian 'bone flute' and other finds from Divje Babe I Cave Site in Slovenia. Znanstvenoraziskovalni Center Sazu, 223 pp.

Tyldesley, J.A., and Bahn, P.G., 1983. Use of plants in the European Palaeolithic: a review of the evidence. Quaternary Science Reviews 2, 53-81.

Tylecote, R.F., 1976. A history of metallurgy. The Metals Society, 182 pp.

Tylecote, R.F., 1987. The early history of metallurgy in Europe. Longman, 391 pp

Ullrich, H., 1995. Mortuary practices in the Palaeolithic – reflections of human-environment relations. *In* Man and environment in the Palaeolithic, Études et Recherches Archéologiques de l'Université de *Liège* (*ERAUL*) 62, 363-378.

Ungar, P.S., and Sponheimer, M., 2011. The diets of early hominins. Science 334, 190-193.

Ungar, P.S., Scott, R.S., Grine, F.E., and Teaford, M.F., 2010. Molar microwear textures and the diets of *Australopithecus anamensis* and *Australopithecus afarensis*. Philosophical Transactions of the Royal Society, London 365, 3345-3354.

Ur, J., 2005. Sennacherib's Northern Assyrian canals: new insights from satellite imagery and aerial photography. Iraq, LXVII, 317-345.

Ursulescu, N., 1995. L'utilisation des sources salées dans le néolithique de la Moldavie (Roumanie). *In* Nature et Culture, Études et Recherches Archéologiques de l'Université de *Liège* (*ERAUL*) 68, 487-495.

Urteaga, O., and Pack, G.T., 1966. On the antiquity of melanoma. Cancer 19(5), 607-610.

Valamoti, S.M., et al., 2007. Grape-pressings from northern Greece: the earliest wine in the Aegean? Antiquity 81, 54-61.

Valladas, H., Cachier, H., Maurice, P., Bernaldo de Quiros, F., Clottes, J., Cabrera Valdes, V., Uzquiano, P., and Arnold, M., 1992. Direct radiocarbon dates for prehistoric paintings at the Altamira, El Castillo and Niaux caves. Nature 357, 68-70.

Valladas, N., et al., 1999. TL dates for the Neanderthal Site of the Amud Cave, Israel. Journal of Archaeological Science 26, 259-268.

Vallois, H.-V., 1971. Le crane trépané magdalénien de Rochereil. Bulletin de la Société préhistorique française, tome 68, Études et Travaux, fasc. 2, 485-495.

Van Andel, T.H., and Runnels, C.N., 1995. The earliest farmers in Europe. Antiquity 69, 431-500.

Van den Bergh, G.D., et al., 1996. Did Homo erectus reach the Island of Flores? Bulletin of the Indo-Pacific Prehistory Association Journal 14, 27-36.

Van der Heide, G.D., 1975. Scheepsarcheologie: Scheepsopgravingen in Nederland en elders in de wereld. Strengholt, 507 pp.

Van der Velden, E., et al., 1995. The decorated body of the man from Hauslabjoch. In Der Mann im Eis: Neue Funde und Ergebnisse, Springer-Verlag, 275–278.

Van Heekeren, H.R., 1972. The Stone Age of Indonesia. Martinus Nijhoff, 247 pp.

Van Peer, P., Fullagar, R., Stokes, S., Bailey, R.M., Moeyersons, J., Steenhoudty, F., Geerts, A., Vanderbeken, T., De Dapper, M., and Geus, F., 2003. The Early to Middle Stone Age transition and the emergence of modern human behaviour at site 8-B-11, Sai Island, Sudan. Journal of Human Evolution 45, 187-193.

Van Zeist, W., 1957. De Mesolithische boot van Pesse. Niewe Drentse volksalmanak 75, 4-11.

Van Zeist, W., and Casparie, W.A., 1968. Wild einkorn wheat and barley from Tell Mureybit in northern Syria. Acta Botanica Neerlandica 17, 44-53.

Vandenabeele, F., and Olivier, J.-P., 1979. Les idéogrammes archéologiques du Linéaire B. École Française d'Athènes, Études Crétoises XXIV, 360 pp.

Vandermeersche, B., 1970. Une sépulture moustérienne avec offrandes découverte dans la grotte de Qafzeh. Comptes Rendus de l'Académie des sciences, 270, Série D, 298-301.

Vandiver, P.B., Soffer, O., Klima, B., and Svoboda, J., 1989. The origins of ceramic technology at Dolni Věstonice, Czechoslovakia. Science 246, 1002-1008.

Vanhaeren, M., 2005. Speaking with beads: the evolutionary significance of personal ornaments. In From Tools to symbols: from early hominids to modern humans. Wits University Press, 525-553.

Vanhaeren, M., d'Errico, F., Stringer, C., James, S.L., Todd, J.A., and Mienis, H.K., 2006. Middle Paleolithic shell beads in Israel and Algeria. Science 312, 1785-1788.

Vanhaeren, M., et al., 2013. Thinking strings: additional evidence for personal ornament use in the Middle Stone Age at Blombos Cave, South Africa. Journal of Human Evolution 64, 500-517.

Varki, A., 2001. Loss of N-Glycolylneuraminic Acid in humans: mechanisms, consequences, and implications for hominid evolution. Yearbook of Physical Anthropology 44, 54-69.

Vat, M.S., 1940. Excavations at Harappā: Government of India Press, vol. I, text, 488 pp., vol. II, plates.

Vencl, S., 1981. On containers in the Palaeolithic and Mesolithic. In Mesolithikum in Europa, Veröffentlichungen des Museums für Ur- und Frühgeschichte Potsdam, 14-15.

Vencl, S., 1991. Interprétation des blessures causées par les armes au Mésolithique. L'Anthropologie 95, 219-228.

Vencl, S., 1999. Stone Age warfare. In Ancient Warfare, Sutton Publishing, 57-72.

Veraprasert, M., 1992. Khlong Thom: an ancient bead-manufacturing location and an ancient entrepôt. *In* Early metallurgy: trade and urban center in Thailand and Southeast Asia, White Lotus, 149-161.

Verhoeven, M., 2002. Transformation of society: the changing role of ritual and symbolism in the PPNB and the PN in the Levant, Syria and South-East Anatolia. Paléorient 28, 5-13.

Vermeersch, P.M., and Paulissen, E., 1997. Extensive Middle Palaeolithic chert extraction in the Qena Area (Egypt). *In* Man and Flint: Proceedings of the VIIth International Flint Symposium, Warszawa-Ostrowiec, September 1995, 133-142.

Vermeersch, P.M., et al. 1998. A Middle Palaeolithic burial of a modern human at Taramsa Hill, Egypt. Antiquity 72, 475-484.

Vermeersch, P.M., et al., 1984. 33,000-yr old chert mining site and related *Homo* in the Egyptian Nile Valley. Nature 309, 342-344.

Vermeersch, P.M., Paulissen, E., and Van Peer, P., 1990. Palaeolithic chert exploitation in the limestone stretch of the Egyptian Nile Valley. The African Archaeological Review 8, 77-102.

Vermeersch, P.M., Paulissen, E., and Van Peer, P., 1995. Palaeolithic chert mining in Egypt. Archaeologia Polona 33, 11-30.

Verri, G., et al., 2004. Flint mining in prehistory recorded by *in situ*-produced cosmogenic ^{10}Be. National Academy of Sciences (USA), Proceedings 101(21), 7880-7884.

Verri, G., et al., 2005. Flint procurement strategies in the Late Lower Palaeolithic recorded by in situ produced cosmogenic ^{10}Be in Tabun and Qesem Caves (Israel). Journal of Archaeological Science 32, 207-213.

Vignaud, P., Duringer, P., Mackaye, H.T., Likias, A., Blondel, C., et al., 2002. Geology and Palaeontology of the Upper Miocene Toros-Menalla hominid locality, Chad. Nature 418, 152-155.

Vigne, J.-D., 2008. Zooarchaeological aspects of the Neolithic Diet Transition in the Near East and Europe, and their putative relationships with the Neolithic Demographic Transition. *In* The Neolithic Demographic Transition and its consequences, Springer Science+Business Media B.V., 179-205.

Vigne, J.-D., and Guilaine, J., 2004. Les premiers animaux de compagnie 8500 ans avant notre ère ? ... ou comment j'ai mangé mon chat, mon chien et mon renard. Anthropozoologica 39, 240-273.

Vigne, J.-D., et al., 2004. Early taming of the cat in Cyprus. Science 304, p. 259.

Vilà, C., et al., 1997. Multiple and ancient origins of the domestic dog. Science 276, 1687-1689.

Vilà, C., et al., 2001. Widespread origins of domestic horse lineages. Science 291, 474-477.

Villa, P., Bon, F., and Castel, J.-C., 2002. Fuel, fire and fireplaces in the Palaeolithic of Western Europe. Review of Archaeology 23, 33-42.

Villiers, J., ed., 1984. The Thailand archaeometallurgy Project 1984: Survey of base-metal resource exploitation in Loei Province, Northeastern Thailand. BiSEA Newsletter 17, pp. 1-5.

Villmoare, B., et al., 2015. Early *Homo* at 2.8 Ma from Ledi-Geraru, Afar, Ethiopia. Sciencexpress, 5 March, 7 pp.

Viollet, P.L., 2005. Water engineering in Ancient history: 5,000 years of history. International Association of Hydraulic Engineering and Research, 322 pp.

Vogel, H.U., 1982. Bergbauarchäologische Forschungen in der Volksrepublik China. Bochum, Der Anschnitt 34, 138-153.

Vogt, E., 1947. Basketry and woven fabrics of the European Stone and Bronze Ages. Ciba Review 54, 1938-1970.

Voigt, E.A., 1982. The Molluscan Fauna. *In* The Middle Stone Age at Klasies River Mouth. Cambridge University Press, 155-186.

Volman, T.P., 1984. Early prehistory of southern Africa. *In* Southern African prehistory and paleoenvironments. Cornell University Intercollegiate Program in Archaeology, 169-220.

Wachsmann, S., 1998. Seagoing ships and seamanship in the Bronze Age Levant. Texas A & M University Press, 417 pp.

Wadley, L., 1997. Where have all the dead men gone? Stone Age burial practices in South Africa. *In* Our gendered past: archaeological studies of gender in Southern Africa, Witwatersrand University Press 107-134.

Wadley, L., Hodgskiss, T., and Grant, M., 2009. Implications for complex cognition from the hafting of tools with compound adhesives in the Middle Stone Age, South Africa. National Academy of Sciences (USA), Proceedings 106, no. 24, 9590-9594.

Wadley, L., Sievers, C., Bamford, M., Goldberg, P., Berna, F., and Miller, C., 2011. Middle Stone Age bedding construction and settlement patterns at Sibudu, South Africa. Science 334, 1388-1391.

Wadley, L., Williamson, B., and Lombard, M., 2004. Ochre in hafting in Middle Stone Age southern Africa: a practical role. Antiquity 78, 661-675.

Waelkens, M., et al., 1992. Ancient stones: quarrying, trade and provenance. Leuven University Press 292 pp.

Wagner, D.B., 1986. Ancient Chinese copper smelting, sixth century BC: recent excavations and simulation experiments. Historical Metallurgy 21, 1-16.

Wagner, D.B., 1993. Iron and steel in Ancient China. E.J. Brill, 573 pp.

Wagner, D.B., 2001. The state and the iron industry in Han China. Nordic Institute of Asian Studies, NIAS Report Series 44, 148 pp.

Wagner, D.M., et al., 2014. *Yersinia pestis* and the plague of Justinian 541–543 AD: a genomic analysis. Lancet Infectious Diseases 14(4), 319-326

Wagner, G.A., et al., 1980. Early Bronze Age lead-silver mining in the Aegean: the ancient workings on Siphnos. *In* Scientific Studies in Early Mining and Extractive Metallurgy, 63-80.

Wahl, W., and König, H.G., 1987. Anthropologisch-Traumatologische Untersuchung der Menschlichen Skelettrests aus dem Bandkeramishcen Massengrab bei Talheim, Kreis Heilbronn. Fundberichte aus Baden-Württemberg, Schweizerbart, 12, 65-193.

Währen, M., 1989. Brot und Gebäck von der Jungsteinzeit bis zur Römerzeit. Helvetia Archaeologica: Archäologie in der Schweize 20, 82-110.

Waldbaum, J.C., 1980. The first archaeological appearance of iron and the transition to the Iron Age. *In* The coming of the Age of Iron, Yale University Press, pp. 69-98.

Waldron, T., 1994. Counting the Dead: The Epidemiology of Skeletal Populations. John Wiley & Sons, New York, 109 pp.

Waldron, T., 2009. Palaeopathology. Cambridge Manuals in Archaeology, 279 pp.

Walker, P.L., 2001. A bioarchaeological perspective on the history of violence. Annual Review of Anthropology 30, 573-596.

Ward-Perkin, J.B., 1971. Quarrying in antiquity: technology, tradition and social change. Proceedings of the British Academy, 137-158.

Ward, C.V., et al., 2014. Early Pleistocene third metacarpal from Kenya and the evolution of modern human-like hand morphology. National Academy of Sciences (USA), Proceedings 111, 121-124.

Ward, C.V., Kimbel, W.H., and Johanson, D.C., 2011. Complete fourth metatarsal and arches in the foot of *Australopithecus afarensis*. Science 331, 750-753.

Wasylikova, K., et al., 1991. East-Central Europe. *In* Progress in Old World palaeoethnobotany. Balkema, 207-239.

Watts, I., 2002. Ocher in the Middle Stone Age of Southern Africa: ritualized display or hide preservative? South African Archaeological Bulletin, 57, 1-14.

Watts, S., 2014. The symbolism of querns and millstones. AmS-Skrifter, Stavanger, 24, 53-66.

Watts, S., 2014b. The Life and Death of Querns. Southampton Monographs in Archaeology, N.S., 3, 189 pp.

Wayne, R.K., Leonard, J.A., and Vila, C., 2006. Genetic analysis of dog domestication. *In* Documenting Domestication, University of California Press, 279-293.

Webb, M., 2000. Lacquer: Technology and Conservation. Butterworth Heinemann, 182 pp.

Weisgerber G., and Hauptmann, A., 1988. Early copper mining and smelting in Palestine. *In* The beginnings of the use of metals and alloys. MIT Press, 52-62.

Weisgerber, G., 1978. Evidence of ancient mining sites in Oman: a preliminary report. Journal of Oman Studies 4, 15-28.

Weisgerber, G., 2006. The mineral wealth of ancient Arabia and its use I: copper mining and smelting at Feinan and Timna – comparison and evaluation of techniques, production and strategies. Arabian Archaeology and Epigraphy 17, 1-30.

Weisgerber, G., and Willies, L., 2000. The use of fire in prehistoric and ancient mining: firesetting. Paléorient 26, 131-149.

Weisgerber, G., Slotta, R., and Weiner, J., eds., 1999. 5000 Jahre Feuersteinbergbau: Die Suche nach dem Stahl der Steinzeit. Deutsches Bergbau-Museum Bochum 1980, Nr. 77, 693 pp.

Weiss, E., Kislev, M.E., Simchoni, O., Nadel, D., and Tschauner, H., 2008. Plant-food preparation area on an Upper Paleolithic brush hut floor at Ohalo II, Israel. Journal of Archaeological Science 35, 2400-2414.

Weiss, R.A., 2009. Apes, lice and prehistory. Journal of Biology 8, 20.1-20.8

Weller, O., 2000. Produire du sel par le feu: techniques et enjeux socio-économiques dans le Néolithique européen. *In* Arts du feu et productions artisanales: Actes des Rencontres, 565-584.

Weller, O., 2002. The earliest rock salt exploitation in Europe: a salt mountain in the Spanish Neolithic. Antiquity 76, 317-318.

Weller, O., and Dumitroaia, G., 2005. The earliest salt production in the world: an early Neolithic exploitation in Poiana Slatinei-Lunca, Romania. Antiquity 79, no. 306, 5 pp.

Wen, Guang, and Jing, Zhichun, 1992. Chinese Neolithic jade: a preliminary geoarchaeological study. Geoarchaeology 7, 251-275.

Wen, T., Zhaoyong, X., Zhijie, G., Yehua, X., Jianghua, S., and Zhiyi, G., 1987. Observation on the ancient lice from Loulan. Investigatio et Studium Naturae (Museum Historiae Naturae, Shanghaiense) 7, 152–155.

Wendorf, F., 1968. Site 117: A Nubian Final Paleolithic graveyard near Jebel Sahaba, Sudan. *In* Prehistory of Nubia, vol. 2, 954-995.

Wendorf, F., and Schild, R., 1986. The Wadi Kubbaniya skeleton: a Late Paleolithic burial from southern Egypt. The Prehistory of Wadi Kubbaniya, vol. I, Southern Methodist University Press, 85 pp.

Wendt, W.E., 1974. 'Art mobilier' aus der Apollo 11-Grotte in Südwest-Afrika: Die ältesten datierten Kunstwerke Afrikas. Acta Praehistorica et Archaeologica 5, 1-42.

Wendt, W.E., 1976. 'Art Mobilier' from the Apollo 11 Cave, South West Africa: Africa's oldest dated works of art. South African Archaeological Bulletin 31, 5-11.

Wertime, T.A., and Muhly, J.D., eds., 1980. The coming of the Age of Iron. Yale University Press, 555 pp.

Wheeler, P.E., 1984. The evolution of bipedality and loss of functional body hair in hominids. Journal of Human Evolution 13, 91-98.

Wheeler, P.E., 1993. The influence of stature and body form on hominid energy and water budgets: a comparison of *Australopithecus* and early *Homo* physiques. Journal of Human Evolution 24, 13-28.

White, R., 1986. Dark Caves, Bright Visions, life in Ice Age Europe. American Museum of Natural History, W.W. Norton, 176 pp.

White, R., Mensan, R., Bourrillon, R., Cretin, C., et al., 2012. Context and dating of Aurignacian vulvar representations from Abri Castanet, France. National Academy of Sciences (USA), Proceedings, 109(22), 8450-8455.

White, R., 1989a. Toward a contextual understanding of the earliest body ornaments. *In* The emergence of modern human biocultural adaptations in the later Pleistocene, Cambridge University Press, 211-231.

White, R., 1989b. Production, complexity and standardization in early Aurignacian bead and pendant manufacture: evolutionary implications. *In* The human revolution, behavioural and biological perspectives, Princeton University Press, 368-390.

White, R., 1993 [1992]. Technological and social dimensions of 'Aurignacian-Age' body ornaments across Europe. *In* Before Lascaux: the complex record of the Early Upper Paleolithic, CRC Press, 277-299.

White, T.D., 1980. Evolutionary implications of Pliocene hominid footprints. Science 208, 175-176.

Wickler, S., and Spriggs, M., 1988. Pleistocene human occupation of the Solomon Islands, Melanesia. Antiquity 62, 703-706.

Wikander, Ö., ed., 2000. Handbook of Ancient water technology. Brill, 741pp.

Wikipedia, 2015. Toilets in Japan, 12 pp.

Wilke, P.J., and Hall, H.J., 1975. Analysis of Ancient Feces: A Discussion and Annotated Bibliography. Archaeological Research Facility, Department of Anthropology, University of California, 47 pp.

Wilkins, J., Schoville, B.J., Brown, K.S., and Chazan, M., 2012. Evidence for early hafted hunting technology. Science 338, 942-946.

Williams-Thorpe, O., 1995. Obsidian in the Mediterranean and the Near East: a provenancing success story. Archaeometry 37, 217-248.

Williams-Thorpe, O., and Thorpe, R.S., 1993. Geochemistry and trade of Eastern Mediterranean millstones from the Neolithic to Roman Periods. Journal of Archaeological Science 20, 263-320.

Willies, L., et al., 1984. Ancient lead and zinc mining in Rajasthan, India. World Archaeology 16, 222-233.

Willis, C., 1996. Yellow Fever, Black Goddess. Addison-Wesley, New York, 324 pp.

Willis, N.J., 1997. Edward Jenner and the eradication of smallpox. Scottish Medical Journal 42, 118-121.

Wilson, M.I., and Daly, M., 1996. Male sexual proprietariness and violence against women. Current Directions in Psychological Science 5(1), 2-7.

Winiger, J., 1995. Die Bekleidung des Eismannes und die Anfänge der Weberei. *In* Der Mann im Eis: neue Fund und Ergebnisse, Springer-Verlag, 119-187.

Witenberg, G, 1961. Human parasites in archaeological findings. Bulletin of the Israel Exploration Society 25, 86.

Wollstonecroft, M.M., 2011. Investigating the role of food processing in human evolution: a niche construction approach. Archaeological and Anthropological Sciences 3, 141-150.

Wood, J.W., et al., 1992. The Osteological Paradox: Problems of Inferring Prehistoric Health from Skeletal Samples [and Comments and Reply]. Current Anthropology 33(4), 343-370.

Wooley, C.L., 1934. Ur Excavations: The Royal Cemetery, Vol II, text. Publications of the Joint Expedition of the British Museum and of the Museum of the University of Pennsylvania to Mesopotamia, 604 pp.

Wrangham, R., 2007. The cooking enigma. *In* The evolution of the human diet: the known, the unknown, and the unknowable, Oxford University Press, 308-323.

Wrangham, R., 2009. Catching Fire, how cooking made us human. Basic Books, 309 pp.

Wrangham, R., and Peterson, D., 1996. Demonic Males: Apes and the origins of human violence. Houghton Mifflin Company, 350 pp.

Wrangham, R.W., and Glowacki, L., 2012. Intergroup aggression in chimpanzees and war in nomadic hunter gatherers: evaluating the chimpanzee model. Human Nature 23, 5-29.

Wrangham, R.W., et al., 1999. The raw and the stolen. Current Anthropology 40, 567-594.

Wreschner, E.E., 1975. Ochre in prehistoric contexts, remarks on its implications to the understanding of human behaviour. Mitekufat haeven 13, 5-10.

Wreschner, E.E., 1980. Red ochre and human evolution: A case for discussion. Current Anthropology 21, 631-644.

Wright, E., 1990. The Ferriby Boats: Seacraft of the Bronze Age. Routledge, 206 pp.

Wright, K., and Garrard, A., 2003. Social identities and the expansion of stone bead-making in Neolithic Western Asia: new evidence from Jordan. Antiquity 77, 267-284.

Wright, L., 1963. Clean and Decent, the fascinating history of the bathroom and the water closet. Viking Press, 282 pp.

Wright, P.W.M., 1982. The bow-drill and the drilling of beads. Kabul, 1981. Afghan Studies 3-4, 95-101.

Wu, Xiaohong, Zhang, Chi, Goldberg, P., Cohen, D., Pan, Y., Arpin, T., and Bar-Yosef, O., 2012. Early pottery at 20,000 years ago in Xianrendong Cave, China. Science 336, 1696-1700.

Wyse Jackson, P.N., and Connolly, M., 2002. Fossils as Neolithic funereal adornments in County Kerry, south-west Ireland. Geology Today 18, 139-143.

Yalçin, U., and Pernicka, E., 1999. Frühneolithische Metallurgie von Aşiki Höyük. The Beginnings of Metallurgy, Der Anschnitt, Beiheft 9, 45-54.

Yamei, H., et al., 2000. Mid-Pleistocene Acheulian-like stone technology of the Bose Basin, South China. Science 287, 1622-1626.

Yan, Y., 1992. Origins of agriculture and animal husbandry in China. *In* C.M. Aikens and N.R. Song, eds., Pacific Northeast Asia in Prehistory: hunter-fisher-gatherers, farmers and socio-political elites, Washington State University Press, 113-123.

Yang Hu, Liu Guoxiang, and Tang Chung, 2007. The origin of jades in East Asia: jades of the Xinglong Wa Culture. The Chinese University of Hong Kong, 323 pp.

Yang, Yimin, et al., 2014. Proteomics evidence for kefir dairy in Early Bronze Age China. Journal of Archaeological Science 45, 178-186.

Yellen, J.E., 1998. Barbed bone points: tradition and continuity in Saharan and Sub-Saharan Africa. The African Archaeological Review 15, 173-198.

Yellen, J.E., Brooks, A.S., Cornellisser, E., Mehlman, M.J., and Stewart, K., 1995. A Middle Stone Age worked bone industry from Katanda, upper Semliki Valley, Zaire. Science 268, 553-556.

Yener, K.A., 1994. Bronze Age source of tin discovered. University of Chicago Chronicle 13(9), 45-57.

Yener, K.A., 1995. Early Bronze Age tin processing at Göltepe and Kestel, Turkey. In Civilizations of the ancient Near East, vol. 3, 1519-1521.

Yener, K.A., 2000. The domestication of metals. Brill, 210 pp.

Yener, K.A., 2008. Revisiting Kestel Mine and Göltepe: the dynamics of local provisioning of tin during the Early Bronze Age. In Ancient Mining in Turkey and the eastern Mediterranean, Turkey Historical Research Applications and Research Center Publications 2, 57-64.

Yener, K.A., and Vandiver, P.B., 1993. Reply to J.D. Muhly, 'Early Bronze Age Tin and the Taurus (with Appendix by L. Willies, Early Bronze Age Tin Working at Kestel). American Journal of Archaeology 97, 255-264.

Yener, K.A., et al., 1989. Kestel: an Early Bronze Age bronze source of tin ore in the Taurus Mountains, Turkey. Science 244, 200-203.

Yerkes, R.W., and Barkai, R., 2004. Microwear analysis of Chalcolithic bifacial tools. In Giv'at ha-Oranim: A Late Chalcolithic site in central Israel. Salvage Excavation Reports, no. 1, Publications in Archaeology, Tel Aviv, 110-124.

Yerkes, R.W., Barkai, R., Gopher, A., and Bar Yosef, O., 2003. Microwear analysis of early Neolithic (PPNA) axes and bifacial tools from Netiv Hagdud in the Jordan Valley, Israel. Journal of Archaeological Science 30, 1051-1066.

Young, R.W., 2003. Evolution of the human hand: the role of throwing and clubbing. Journal of Anatomy 202, 165-174.

Young, S.L., 2011. Craving Earth. Columbia University Press, 228 pp.`

Zarin, J., 1976. The domestication of Equidae in third millennium BC Mesopotamia. Doctoral Dissertation, Volume One, University of Chicago, 660 pp.

Zeder, M.A., 2006. A critical assessment of markers of initial domestication of goats (Capra hircus). In Documenting Domestication, University of California Press, 181-208.

Zeder, M.A., and Hesse, B., 2000. The initial domestication of goats (Capra hircus) in the Zagros Mountains 10,000 years ago. Science 287, 2254-2257.

Zeng, X.-N., et al., 1991. Analysis of characteristic odors from human male axillae. Journal of Chemical Ecology 17, 1469-1492.

Zeng, X.N., et al., 1996. Analysis of characteristic human female axillary odors: qualitative comparison to males. Journal of Chemical Ecology 22, 237-257.

Zhang, J., and Kuen, L.Y., 2005. The Magic Flutes. Natural History 114, 42-47.

Zhang, J., Harbottle, G., Wang, C., and Kong, Z., 1999. Oldest playable musical instruments found at Jiahu early Neolithic site in China. Nature 401, 66-368.

Zhao Feng, 2002. Recent excavations of textiles in China. ISAT/Costume Squad Ltd. (Hong Kong), 205 pp.

Zhou Bao Zhong, 1988. The protection of ancient Chinese lacquerware. In Urushi, Proceedings of the Urushi Study Group, Tokyo, p. 71.

Zhou Baoquan et al., 1988. Ancient copper mining and smelting at Tonglushan Daye. In The beginnings of the use of metals and alloys, MIT Press, pp. 125-129.

Zhu, R.X., et al., 2008. Early evidence of the genus *Homo* in East Asia. Journal of Human Evolution 55, 1075-1085.

Zias, J., 1995. Cannabis sativa (hashish) as an effective medicine in antiquity: the anthropological evidence. *In* The archaeology of death in the ancient Near East, Oxbow Monograph 51, 232-234.

Zias, J., and Mumcuoglu, K.Y., 1991. Pre-Pottery Neolithic B head lice from Nahal Hemar Cave, Atiqot, English Series 20, 167-168.

Zias, J.E., Tabor, J.D., and Harter-Lailheugue, S., 2006. Toilets at Qumran, the Essenes, and the scrolls: new anthropological data and old theories. Revue de Qumran 22, 631–640.

Zilhão, J., et al., 2010. Symbolic use of marine shells and mineral pigments by Iberian Neandertals. National Academy of Sciences (USA), Proceedings 107, no. 3, 1023-1028.

Zink, A.R., et al., 2006. Leishmaniasis in ancient Egypt and Upper Nubia. Emerging Infectious Diseases 12(10), 1616-1617.

Zohary, D., 1989. Domestication of the Southwest Asian Neolithic crop assemblage of cereals, pulses, and flax: the evidence from the living plants. *In* Foraging and Farming: The evolution of plant exploitation. Unwin Hyman, 358-373.

Zohary, D., 1996. The mode of domestication of the founder crops of Southwest Asian agriculture. *In* The Origins and Spread of Agriculture and Pastoralism in Eurasia, Smithsonian Institution Press, 42-158.

Zohary, D., and Hopf, M., 1993. Domestication of plants in the Old World. Clarendon Press, 279 pp.

Zohary, D., and Spiegel-Roy, P., 1975. Beginnings of fruit growing in the Old World. Science 187, 319-327.

Zollikofer, C.P.E., Ponce de León, M.S., Vandermeersch, B., and Lévêque, F., 2002. Evidence for interpersonal violence in the St. Césaire Neanderthal. National Academy of Sciences (USA), Proceedings 99(9), 6444-6448.

Zutterman, C., 2003. The bow in the ancient Near East, A re-evaluation of archery from the Late 2nd Millennium to the end of the Achaemenid Empire. Iranica Antiqua XXXVIII, 119-165.

Zvelebil, M., 1996. The agricultural frontier and the transition to farming in the circum-Baltic region. *In* The origins and spread of agriculture and pastoralism in Eurasia, Routledge, 323-345.

Zvelebil, M., ed., 1986. Hunters in Transition. Cambridge University Press, 194 pp.